Space-Time Coding

Space-Time Coding

Branka Vucetic
University of Sydney, Australia

Jinhong Yuan
University of New South Wales, Australia

WILEY

Other Wiley Editorial Offices

John Wiley & Sons Inc., 111 River Street, Hoboken, NJ 07030, USA

Jossey-Bass, 989 Market Street, San Francisco, CA 94103-1741, USA

Wiley-VCH Verlag GmbH, Boschstr. 12, D-69469 Weinheim, Germany

John Wiley & Sons Australia Ltd, 33 Park Road, Milton, Queensland 4064, Australia

John Wiley & Sons (Asia) Pte Ltd, 2 Clementi Loop #02-01, Jin Xing Distripark, Singapore 129809

John Wiley & Sons Canada Ltd, 22 Worcester Road, Etobicoke, Ontario, Canada M9W 1L1

Wiley also publishes its books in a variety of electronic formats. Some content that appears
in print may not be available in electronic books.

Library of Congress Cataloging-in-Publication Data

Vucetic, Branka.
    Space-time Coding / Branka Vucetic, Jinhong Yuan.
        p. cm.
    Includes bibliographical references and index.
    ISBN 0-470-84757-3 (alk. paper)
        1. Signal processing—Mathematics. 2. Coding theory. 3. Iterative methods
(Mathematics) 4. Wireless communication systems. I. Yuan, Jinhong, 1969– II. Title.

    TK5102.92.V82 2003
    621.382′2—dc21

                                                                        2003043054

British Library Cataloguing in Publication Data

A catalogue record for this book is available from the British Library

ISBN 0-470-84757-3

Typeset in 10/12pt Times from LaTeX files supplied by the author, processed by Laserwords Private Limited,
Chennai, India
Printed and bound in Great Britain by TJ International Ltd, Padstow, Cornwall
This book is printed on acid-free paper responsibly manufactured from sustainable forestry
in which at least two trees are planted for each one used for paper production.

Contents

List of Acronyms

3GPP	3rd Generation Partnership Project
APP	a posteriori probability
AWGN	additive white Gaussian noise
BER	bit error rate
BPSK	binary phase shift keying
CCSDS	Consultative Committee for Space Data Systems
ccdf	complementary cumulative distribution function
cdf	cumulative distribution function
CDMA	code division multiple access
CRC	cyclic redundancy check
CSI	channel state information
DAB	digital audio broadcasting
DFT	discrete Fourier transform
DLST	diagonal layered space-time
DLSTC	diagonal layered space-time code
DOA	direction of arrival
DPSK	differential phase-shift keying
DS-CDMA	direct-sequence code division multiple access
DSC	decision statistics combining
DSSS	direct-sequence spread spectrum
DVB	digital video broadcasting
EGC	equal gain combining
EIR	extrinsic information ratio
EXIT	extrinsic information transfer chart
FDMA	frequency division multiple access
FER	frame error rate
FFT	fast Fourier transform
GCD	greatest common divisor
GSM	global system for mobile
HLST	horizontal layered space-time
HLSTC	horizontal layered space-time code
ISI	intersymbol interference
LDPC	low density parity check
LLR	log-likelihood ratio
LMMSE	linear minimum mean square error

LOS	line-of-sight
LST	layered space-time
LSTC	layered space-time code
M-PSK	M-ary phase-shift keying
MAI	multiple access interference
MAP	maximum a posteriori
MGF	moment generating function
MF	matched filter
MIMO	multiple-input multiple-output
ML	maximum likelihood
MLSE	maximum likelihood sequence estimation
MMSE	minimum mean square error
MRC	maximum ratio combining
OFDM	orthogonal frequency division multiplexing
OTD	orthogonal transmit diversity
pdf	probability density function
PIC	parallel interference canceler
PN	pseudorandom number
PSK	phase shift keying
QAM	quadrature amplitude modulation
QPSK	quadrature phase-shift keying
rms	root mean square
RSC	recursive systematic convolutional
SER	symbol error rate
SISO	soft-input soft-output
SNR	signal-to-noise ratio
SOVA	soft-output Viterbi algorithm
STC	space-time code
STBC	space-time block code
STTC	space-time trellis code
STS	space-time spreading
SVD	singular value decomposition
TCM	trellis coded modulation
TDMA	time division multiple access
TLST	threaded layered space-time
TLSTC	threaded layered space-time code
TS-OTD	time-switched orthogonal transmit diversity
TS-STC	time-switched space-time code
UMTS	universal mobile telecommunication systems
VA	Viterbi algorithm
VBLAST	vertical Bell Laboratories layered space-time
VLST	vertical layered space-time
VLSTC	vertical layered space-time code
WCDMA	wideband code division multiple access
WLAN	wireless local area network
ZF	zero forcing

List of Figures

List of Tables

Preface

This book is intended to provide an introductory coverage of the subject of space-time coding. It has arisen from our research, short continuing education courses, lecture series and consulting for industry. Its purpose is to provide a working knowledge of space-time coding and its application to wireless communication systems.

With the integration of Internet and multimedia applications in next generation wireless communications, the demand for wide-band high data rate communication services is growing. As the available radio spectrum is limited, higher data rates can be achieved only by designing more efficient signaling techniques. Recent research in information theory has shown that large gains in capacity of communication over wireless channels are feasible in multiple-input multiple output (MIMO) systems [1][2]. The MIMO channel is constructed with multiple element array antennas at both ends of the wireless link. Space-time coding is a set of practical signal design techniques aimed at approaching the information theoretic capacity limit of MIMO channels. The fundamentals of space-time coding have been established by Tarokh, Seshadri and Calderbank in 1998 [3]. Space-time coding and related MIMO signal processing soon evolved into a most vibrant research area in wireless communications.

Space-time coding is based on introducing joint correlation in transmitted signals in both the space and time domains. Through this approach, simultaneous diversity and coding gains can be obtained, as well as high spectral efficiency. The initial research focused on design of joint space-time dependencies in transmitted signals with the aim of optimizing the coding and diversity gains. Lately, the emphasis has shifted towards independent multi-antenna transmissions with time domain coding only, where the major research challenge is interference suppression and cancellations in the receiver.

The book is intended for postgraduate students, practicing engineers and researchers. It is assumed that the reader has some familiarity with basic digital communications, matrix analysis and probability theory.

The book attempts to provide an overview of design principles and major space-time coding techniques starting from MIMO system information theory capacity bounds and channel models, while endeavoring to pave the way towards complex areas such as applications of space-time codes and their performance evaluation in wide-band wireless channels. Abundant use is made of illustrative examples with answers and performance evaluation results throughout the book. The examples and performance results are selected to appeal to students and practitioners with various interests. The second half of the book is targeted at a more advanced reader, providing a research oriented outlook. In organizing the material, we have tried to follow the presentation of theoretical material by appropriate applications in wireless communication systems, such as code division multiple access (CDMA) and orthogonal frequency division multiple access(OFDMA).

A consistent set of notations is used throughout the book. Proofs are included only when it is felt that they contribute sufficient insight into the problem being addressed.

Much of our unpublished work is included in the book. Examples of some new material are the performance analysis and code design principles for space-time codes that are more general and applicable to a wider range of system parameters than the known ones. The system structure, performance analysis and results of layered and space-time trellis codes in CDMA and OFDMA systems have not been published before.

Chapters 1 and 6 were written by Branka Vucetic and Chapters 3 and 7 by Jinhong Yuan. Chapter 8 was written by Jinhong Yuan, except that the content in the last two sections was joint work of Branka Vucetic and Jinhong Yuan. Most of the content in Chapters 2, 4 and 5 was joint work by Branka Vucetic and Jinhong Yuan, while the final writing was done by Jinhong Yuan for Chapters 2 and 4 and Branka Vucetic for Chapter 5.

Acknowledgements

The authors would like to express their appreciation for the assistance received during the preparation of the book. The comments and suggestions from anonymous reviewers have provided essential guidance in the early stages of the manuscript evolution. Mr Siavash Alamouti, Dr Hayoung Yang, Dr Jinho Choi, A/Prof Tadeusz Wysocki, Dr Reza Nakhai, Mr Michael Dohler, Dr Graeme Woodward and Mr Francis Chan proofread various parts of the manuscript and improved the book by many comments and suggestions. We would also like to thank Prof Vahid Tarokh, Dr Akihisa Ushirokawa, Dr Arie Reichman and Prof. Ruifeng Zhang for constructive discussions. We thank the many students, whose suggestions and questions have helped us to refine and improve the presentation. Special thanks go to Jose Manuel Dominguez Roldan for the many graphs and simulation results he provided for Chapter 1. Contributions of Zhuo Chen, Welly Firmanto, Yang Tang, Ka Leong Lo, Slavica Marinkovic, Yi Hong, Xun Shao and Michael Kuang in getting performance evaluation results are gratefully acknowledged. Branka Vucetic would like to express her gratitude to Prof Hamid Aghvami, staff and postgraduate students at King's College London, for the creative atmosphere during her study leave. She benefited from a close collaboration with Dr Reza Nakhai and Michael Dohler. The authors warmly thank Maree Belleli for her assistance in producing some of the figures.

We owe special thanks to the Australian Research Council, NEC, Optus and other companies whose support enabled graduate students and staff at Sydney University and the University of New South Wales to pursue continuing research in this field.

Mark Hammond, senior publishing editor from John Wiley, assisted and motivated us in all phases of the book preparation.

We would like to thank our families for their support and understanding during the time we devoted to writing this book.

Bibliography

[1] E. Telatar, "Capacity of multi-antenna Gaussian channels", *European Transactions on Telecommunications*, vol. 10, no. 6, Nov./Dec. 1999, pp. 585–595.

[2] G. J. Foschini and M. J. Gans, "On limits of wireless communications in a fading environment when using multiple antennas", *Wireless Personal Communications*, vol. 6, 1998, pp. 311–335.

[3] V. Tarokh, N. Seshadri and A. R. Calderbank, "Space-time codes for high data rate wireless communication: performance criterion and code construction", *IEEE Trans. Inform. Theory*, vol. 44, no. 2, Mar. 1998, pp. 744–765.

1

Performance Limits of Multiple-Input Multiple-Output Wireless Communication Systems

1.1 Introduction

Demands for capacity in wireless communications, driven by cellular mobile, Internet and multimedia services have been rapidly increasing worldwide. On the other hand, the available radio spectrum is limited and the communication capacity needs cannot be met without a significant increase in communication spectral efficiency. Advances in coding, such as turbo [5] and low density parity check codes [6][7] made it feasible to approach the Shannon capacity limit [4] in systems with a single antenna link. Significant further advances in spectral efficiency are available through increasing the number of antennas at both the transmitter and the receiver [1][2].

In this chapter we derive and discuss fundamental capacity limits for transmission over multiple-input multiple-output (MIMO) channels. They are mainly based on the theoretical work developed by Telatar [2] and Foschini [1]. These capacity limits highlight the potential spectral efficiency of MIMO channels, which grows approximately linearly with the number of antennas, assuming ideal propagation. The capacity is expressed by the maximum achievable data rate for an arbitrarily low probability of error, providing that the signal may be encoded by an arbitrarily long space-time code. In later chapters we consider some practical coding techniques which potentially approach the derived capacity limits. It has been demonstrated that the Bell Laboratories Layered Space-Time (BLAST) coding technique [3] can attain the spectral efficiencies up to 42 bits/sec/Hz. This represents a spectacular increase compared to currently achievable spectral efficiencies of 2-3 bits/sec/Hz, in cellular mobile and wireless LAN systems.

A MIMO channel can be realized with multielement array antennas. Of particular interest are propagation scenarios in which individual channels between given pairs of transmit and receive antennas are modelled by an independent flat Rayleigh fading process. In this chapter, we limit the analysis to the case of narrowband channels, so that they can be

Space-Time Coding Branka Vucetic and Jinhong Yuan
© 2003 John Wiley & Sons, Ltd ISBN: 0-470-84757-3

described by frequency flat models. The results are generalised in Chapter 8 to wide-band channels, simply by considering a wide-band channel as a set of orthogonal narrow-band channels. Rayleigh models are realistic for environments with a large number of scatterers. In channels with independent Rayleigh fading, a signal transmitted from every individual transmit antenna appears uncorrelated at each of the receive antennas. As a result, the signal corresponding to every transmit antenna has a distinct spatial signature at a receive antenna. The *independent* Rayleigh fading model can be approximated in MIMO channels where antenna element spacing is considerably larger than the carrier wavelength or the incoming wave incidence angle spread is relatively large (larger than 30°). An example of such a channel is the down link in cellular radio. In base stations placed high above the ground, the antenna signals get correlated due to a small angular spread of incoming waves and much higher antenna separations are needed in order to obtain independent signals between adjacent antenna elements than if the incoming wave incidence angle spread is large.

There have been many measurements and experiment results indicating that if two receive antennas are used to provide diversity at the base station receiver, they must be on the order of ten wavelengths apart to provide sufficient decorrelation. Similarly, measurements show that to get the same diversity improvements at remote handsets, it is sufficient to separate the antennas by about three wavelengths.

1.2   MIMO System Model

Let us consider a single point-to-point MIMO system with arrays of n_T transmit and n_R receive antennas. We focus on a complex baseband linear system model described in discrete time. The system block diagram is shown in Fig. 1.1. The transmitted signals in each symbol period are represented by an $n_T \times 1$ column matrix $\mathbf{x}$, where the ith component x_i, refers to the transmitted signal from antenna i. We consider a Gaussian channel, for which, according to information theory [4], the optimum distribution of transmitted signals is also Gaussian. Thus, the elements of $\mathbf{x}$ are considered to be zero mean independent identically distributed (i.i.d.) Gaussian variables. The covariance matrix of the transmitted signal is given by

$$\mathbf{R}_{xx} = E\{\mathbf{x}\mathbf{x}^H\} \tag{1.1}$$

where $E\{\cdot\}$ denotes the expectation and the operator $\mathbf{A}^H$ denotes the Hermitian of matrix $\mathbf{A}$, which means the transpose and component-wise complex conjugate of $\mathbf{A}$. The total

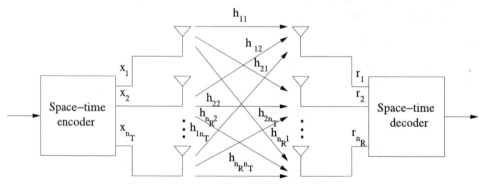

Figure 1.1   Block diagram of a MIMO system

transmitted power is constrained to P, regardless of the number of transmit antennas n_T. It can be represented as

$$P = \text{tr}(\mathbf{R}_{xx}) \tag{1.2}$$

where $\text{tr}(\mathbf{A})$ denotes the trace of matrix $\mathbf{A}$, obtained as the sum of the diagonal elements of $\mathbf{A}$. If the channel is unknown at the transmitter, we assume that the signals transmitted from individual antenna elements have equal powers of P/n_T. The covariance matrix of the transmitted signal is given by

$$\mathbf{R}_{xx} = \frac{P}{n_T}\mathbf{I}_{n_T} \tag{1.3}$$

where $\mathbf{I}_{n_T}$ is the $n_T \times n_T$ identity matrix. The transmitted signal bandwidth is narrow enough, so its frequency response can be considered as flat. In other words, we assume that the channel is memoryless.

The channel is described by an $n_R \times n_T$ complex matrix, denoted by $\mathbf{H}$. The ij-th component of the matrix $\mathbf{H}$, denoted by h_{ij}, represents the channel fading coefficient from the jth transmit to the ith receive antenna. For normalization purposes we assume that the received power for each of n_R receive branches is equal to the total transmitted power. Physically, it means that we ignore signal attenuations and amplifications in the propagation process, including shadowing, antenna gains etc. Thus we obtain the normalization constraint for the elements of $\mathbf{H}$, on a channel with fixed coefficients, as

$$\sum_{j=1}^{n_T} |h_{ij}|^2 = n_T, \quad i = 1, 2, \ldots, n_R \tag{1.4}$$

When the channel matrix elements are random variables, the normalization will apply to the expected value of the above expression.

We assume that the channel matrix is known to the receiver, but not always at the transmitter. The channel matrix can be estimated at the receiver by transmitting a training sequence. The estimated channel state information (CSI) can be communicated to the transmitter via a reliable feedback channel.

The elements of the channel matrix $\mathbf{H}$ can be either deterministic or random. We will focus on examples relevant to wireless communications, which involve the Rayleigh and Rician distributions of the channel matrix elements. In most situations we consider the Rayleigh distribution, as it is most representative for non-line-of-sight (NLOS) radio propagation.

The noise at the receiver is described by an $n_R \times 1$ column matrix, denoted by $\mathbf{n}$. Its components are statistically independent complex zero-mean Gaussian variables, with independent and equal variance real and imaginary parts. The covariance matrix of the receiver noise is given by

$$\mathbf{R}_{nn} = E\{\mathbf{nn}^H\} \tag{1.5}$$

If there is no correlation between components of $\mathbf{n}$, the covariance matrix is obtained as

$$\mathbf{R}_{nn} = \sigma^2 \mathbf{I}_{n_R} \tag{1.6}$$

Each of n_R receive branches has identical noise power of σ^2.

The receiver is based on a maximum likelihood principle operating jointly over n_R receive antennas. The received signals are represented by an $n_R \times 1$ column matrix, denoted by $\mathbf{r}$, where each complex component refers to a receive antenna. We denote the average power at the output of each receive antenna by P_r. The average signal-to-noise ratio (SNR) at each receive antenna is defined as

$$\gamma = \frac{P_r}{\sigma^2} \tag{1.7}$$

As we assumed that the total received power per antenna is equal to the total transmitted power, the SNR is equal to the ratio of the total transmitted power and the noise power per receive antenna and it is independent of n_T. Thus it can be written as

$$\gamma = \frac{P}{\sigma^2} \tag{1.8}$$

By using the linear model the received vector can be represented as

$$\mathbf{r} = \mathbf{Hx} + \mathbf{n} \tag{1.9}$$

The received signal covariance matrix, defined as $E\{\mathbf{rr}^H\}$, by using (1.9), is given by

$$\mathbf{R}_{rr} = \mathbf{HR}_{xx}\mathbf{H}^H, \tag{1.10}$$

while the total received signal power can be expressed as tr($\mathbf{R}_{rr}$).

1.3   MIMO System Capacity Derivation

The system capacity is defined as the maximum possible transmission rate such that the probability of error is arbitrarily small.

Initially, we assume that the channel matrix is not known at the transmitter, while it is perfectly known at the receiver.

By the singular value decomposition (SVD) theorem [11] any $n_R \times n_T$ matrix $\mathbf{H}$ can be written as

$$\mathbf{H} = \mathbf{UDV}^H \tag{1.11}$$

where $\mathbf{D}$ is an $n_R \times n_T$ non-negative and diagonal matrix, $\mathbf{U}$ and $\mathbf{V}$ are $n_R \times n_R$ and $n_T \times n_T$ unitary matrices, respectively. That is, $\mathbf{UU}^H = \mathbf{I}_{n_R}$ and $\mathbf{VV}^H = \mathbf{I}_{n_T}$, where $\mathbf{I}_{n_R}$ and $\mathbf{I}_{n_T}$ are $n_R \times n_R$ and $n_T \times n_T$ identity matrices, respectively. The diagonal entries of $\mathbf{D}$ are the non-negative square roots of the eigenvalues of matrix $\mathbf{HH}^H$. The eigenvalues of $\mathbf{HH}^H$, denoted by λ, are defined as

$$\mathbf{HH}^H\mathbf{y} = \lambda\mathbf{y}, \quad \mathbf{y} \neq 0 \tag{1.12}$$

where $\mathbf{y}$ is an $n_R \times 1$ vector associated with λ, called an eigenvector.

The non-negative square roots of the eigenvalues are also referred to as the singular values of $\mathbf{H}$. Furthermore, the columns of $\mathbf{U}$ are the eigenvectors of $\mathbf{HH}^H$ and the columns of $\mathbf{V}$ are the eigenvectors of $\mathbf{H}^H\mathbf{H}$. By substituting (1.11) into (1.9) we can write for the received vector $\mathbf{r}$

$$\mathbf{r} = \mathbf{UDV}^H\mathbf{x} + \mathbf{n} \tag{1.13}$$

Let us introduce the following transformations

$$\mathbf{r}' = \mathbf{U}^H \mathbf{r}$$

$$\mathbf{x}' = \mathbf{V}^H \mathbf{x} \tag{1.14}$$

$$\mathbf{n}' = \mathbf{U}^H \mathbf{n}$$

as $\mathbf{U}$ and $\mathbf{V}$ are invertible. Clearly, multiplication of vectors $\mathbf{r}$, $\mathbf{x}$ and $\mathbf{n}$ by the corresponding matrices as defined in (1.14) has only a scaling effect. Vector $\mathbf{n}'$ is a zero mean Gaussian random variable with i.i.d real and imaginary parts. Thus, the original channel is equivalent to the channel represented as

$$\mathbf{r}' = \mathbf{D}\mathbf{x}' + \mathbf{n}' \tag{1.15}$$

The number of nonzero eigenvalues of matrix $\mathbf{H}\mathbf{H}^H$ is equal to the rank of matrix $\mathbf{H}$, denoted by r. For the $n_R \times n_T$ matrix $\mathbf{H}$, the rank is at most $m = \min(n_R, n_T)$, which means that at most m of its singular values are nonzero. Let us denote the singular values of $\mathbf{H}$ by $\sqrt{\lambda_i}$, $i = 1, 2, \ldots, r$. By substituting the entries $\sqrt{\lambda_i}$ in (1.15), we get for the received signal components

$$r_i' = \sqrt{\lambda_i} x_i' + n_i', \quad i = 1, 2, \ldots, r$$

$$r_i' = n_i', \quad i = r+1, r+2, \ldots, n_R \tag{1.16}$$

As (1.16) indicates, received components, r_i', $i = r+1, r+2, \ldots, n_R$, do not depend on the transmitted signal, i.e. the channel gain is zero. On the other hand, received components r_i', for $i = 1, 2, \ldots, r$ depend only on the transmitted component x_i'. Thus the equivalent MIMO channel from (1.15) can be considered as consisting of r uncoupled parallel sub-channels. Each sub-channel is assigned to a singular value of matrix $\mathbf{H}$, which corresponds to the amplitude channel gain. The channel power gain is thus equal to the eigenvalue of matrix $\mathbf{H}\mathbf{H}^H$. For example, if $n_T > n_R$, as the rank of $\mathbf{H}$ cannot be higher than n_R, Eq. (1.16) shows that there will be at most n_R nonzero gain sub-channels in the equivalent MIMO channel, as shown in Fig. 1.2.

On the other hand if $n_R > n_T$, there will be at most n_T nonzero gain sub-channels in the equivalent MIMO channel, as shown in Fig. 1.3. The eigenvalue spectrum is a MIMO channel representation, which is suitable for evaluation of the best transmission paths.

The covariance matrices and their traces for signals $\mathbf{r}'$, $\mathbf{x}'$ and $\mathbf{n}'$ can be derived from (1.14) as

$$\mathbf{R}_{r'r'} = \mathbf{U}^H \mathbf{R}_{rr} \mathbf{U}$$

$$\mathbf{R}_{x'x'} = \mathbf{V}^H \mathbf{R}_{xx} \mathbf{V} \tag{1.17}$$

$$\mathbf{R}_{n'n'} = \mathbf{U}^H \mathbf{R}_{nn} \mathbf{U}$$

$$\mathrm{tr}(\mathbf{R}_{r'r'}) = \mathrm{tr}(\mathbf{R}_{rr})$$

$$\mathrm{tr}(\mathbf{R}_{x'x'}) = \mathrm{tr}(\mathbf{R}_{xx}) \tag{1.18}$$

$$\mathrm{tr}(\mathbf{R}_{n'n'}) = \mathrm{tr}(\mathbf{R}_{nn})$$

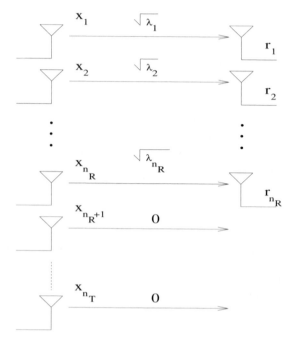

Figure 1.2   Block diagram of an equivalent MIMO channel if $n_T > n_R$

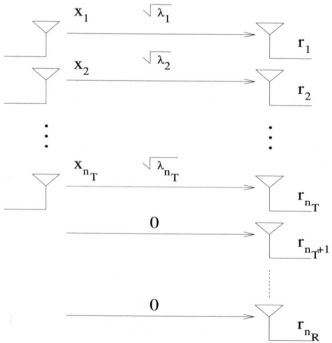

Figure 1.3   Block diagram of an equivalent MIMO channel if $n_R > n_T$

The above relationships show that the covariance matrices of $\mathbf{r'}$, $\mathbf{x'}$ and $\mathbf{n'}$, have the same sum of the diagonal elements, and thus the same powers, as for the original signals, $\mathbf{r}$, $\mathbf{x}$ and $\mathbf{n}$, respectively.

Note that in the equivalent MIMO channel model described by (1.16), the sub-channels are uncoupled and thus their capacities add up. Assuming that the transmit power from each antenna in the equivalent MIMO channel model is P/n_T, we can estimate the overall channel capacity, denoted by C, by using the Shannon capacity formula

$$C = W \sum_{i=1}^{r} \log_2 \left(1 + \frac{P_{ri}}{\sigma^2} \right) \tag{1.19}$$

where W is the bandwidth of each sub-channel and P_{ri} is the received signal power in the ith sub-channel. It is given by

$$P_{ri} = \frac{\lambda_i P}{n_T} \tag{1.20}$$

where $\sqrt{\lambda_i}$ is the singular value of channel matrix $\mathbf{H}$. Thus the channel capacity can be written as

$$
\begin{aligned}
C &= W \sum_{i=1}^{r} \log_2 \left(1 + \frac{\lambda_i P}{n_T \sigma^2} \right) \\
&= W \log_2 \prod_{i=1}^{r} \left(1 + \frac{\lambda_i P}{n_T \sigma^2} \right)
\end{aligned}
\tag{1.21}
$$

Now we will show how the channel capacity is related to the channel matrix $\mathbf{H}$. Assuming that $m = \min(n_R, n_T)$, Eq. (1.12), defining the eigenvalue-eigenvector relationship, can be rewritten as

$$(\lambda \mathbf{I}_m - \mathbf{Q})\mathbf{y} = 0, \quad \mathbf{y} \neq 0 \tag{1.22}$$

where $\mathbf{Q}$ is the Wishart matrix defined as

$$\mathbf{Q} = \begin{cases} \mathbf{H}\mathbf{H}^H, & n_R < n_T \\ \mathbf{H}^H\mathbf{H}, & n_R \geq n_T \end{cases} \tag{1.23}$$

That is, λ is an eigenvalue of $\mathbf{Q}$, if and only if $\lambda \mathbf{I}_m - \mathbf{Q}$ is a singular matrix. Thus the determinant of $\lambda \mathbf{I}_m - \mathbf{Q}$ must be zero

$$\det(\lambda \mathbf{I}_m - \mathbf{Q}) = 0 \tag{1.24}$$

The singular values λ of the channel matrix can be calculated by finding the roots of Eq. (1.24).

We consider the characteristic polynomial $p(\lambda)$ from the left-hand side in Eq. (1.24)

$$p(\lambda) = \det(\lambda \mathbf{I}_m - \mathbf{Q}) \tag{1.25}$$

It has degree equal to m, as each row of $\lambda \mathbf{I}_m - \mathbf{Q}$ contributes one and only one power of λ in the Laplace expansion of $\det(\lambda \mathbf{I}_m - \mathbf{Q})$ by minors. As a polynomial of degree m

with complex coefficients has exactly m zeros, counting multiplicities, we can write for the characteristic polynomial

$$p(\lambda) = \Pi_{i=1}^{m}(\lambda - \lambda_i) \tag{1.26}$$

where λ_i are the roots of the characteristic polynomial $p(\lambda)$, equal to the channel matrix singular values. We can now write Eq. (1.24) as

$$\Pi_{i=1}^{m}(\lambda - \lambda_i) = 0 \tag{1.27}$$

Further we can equate the left-hand sides of (1.24) and (1.27)

$$\Pi_{i=1}^{m}(\lambda - \lambda_i) = \det(\lambda \mathbf{I}_m - \mathbf{Q}) \tag{1.28}$$

Substituting $-\frac{n_T \sigma^2}{P}$ for λ in (1.28) we get

$$\Pi_{i=1}^{m}\left(1 + \frac{\lambda_i P}{n_T \sigma^2}\right) = \det\left(\mathbf{I}_m + \frac{P}{n_T \sigma^2}\mathbf{Q}\right) \tag{1.29}$$

Now the capacity formula from (1.21) can be written as

$$C = W \log_2 \det\left(\mathbf{I}_m + \frac{P}{n_T \sigma^2}\mathbf{Q}\right) \tag{1.30}$$

As the nonzero eigenvalues of $\mathbf{HH}^H$ and $\mathbf{H}^H\mathbf{H}$ are the same, the capacities of the channels with matrices $\mathbf{H}$ and $\mathbf{H}^H$ are the same. Note that if the channel coefficients are random variables, formulas (1.21) and (1.30), represent instantaneous capacities or mutual information. The mean channel capacity can be obtained by averaging over all realizations of the channel coefficients.

1.4 MIMO Channel Capacity Derivation for Adaptive Transmit Power Allocation

[1]When the channel parameters are known at the transmitter, the capacity given by (1.30) can be increased by assigning the transmitted power to various antennas according to the "water-filling" rule [2]. It allocates more power when the channel is in good condition and less when the channel state gets worse. The power allocated to channel i is given by (Appendix 1.1)

$$P_i = \left(\mu - \frac{\sigma^2}{\lambda_i}\right)^+, \quad i = 1, 2, \ldots, r \tag{1.31}$$

where a^+ denotes $\max(a, 0)$ and μ is determined so that

$$\sum_{i=1}^{r} P_i = P \tag{1.32}$$

---

[1]In practice, transmit power is constrained by regulations and hardware costs.

We consider the singular value decomposition of channel matrix $\mathbf{H}$, as in (1.11). Then, the received power at sub-channel i in the equivalent MIMO channel model is given by

$$P_{ri} = (\lambda_i \mu - \sigma^2)^+ \tag{1.33}$$

The MIMO channel capacity is then

$$C = W \sum_{i=1}^{r} \log_2 \left(1 + \frac{P_{ri}}{\sigma^2}\right) \tag{1.34}$$

Substituting the received signal power from (1.33) into (1.34) we get

$$C = W \sum_{i=1}^{r} \log_2 \left[1 + \frac{1}{\sigma^2}(\lambda_i \mu - \sigma^2)^+\right] \tag{1.35}$$

The covariance matrix of the transmitted signal is given by

$$\mathbf{R}_{xx} = \mathbf{V} \, \text{diag}(P_1, P_2, \ldots, P_{n_T})\mathbf{V}^H \tag{1.36}$$

1.5 MIMO Capacity Examples for Channels with Fixed Coefficients

In this section we examine the maximum possible transmission rates in a number of various channel settings. First we focus on examples of channels with constant matrix elements. In most examples the channel is known only at the receiver, but not at the transmitter. All other system and channel assumptions are as specified in Section 1.2.

Example 1.1: Single Antenna Channel

Let us consider a channel with $n_T = n_R = 1$ and $\mathbf{H} = h = 1$. The Shannon formula gives the capacity of this channel

$$C = W \log_2 \left(1 + \frac{P}{\sigma^2}\right) \tag{1.37}$$

The same expression can be obtained by applying formula (1.30). Note that for high SNRs, the capacity grows logarithmically with the SNR. Also in this region, a 3 dB increase in SNR gives a normalized capacity C/W increase of 1 bit/sec/Hz. Assuming that the channel coefficient is normalized so that $|h|^2 = 1$, and for the SNR (P/σ^2) of 20 dB, the capacity of a single antenna link is 6.658 bits/s/Hz.

Example 1.2: A MIMO Channel with Unity Channel Matrix Entries

For this channel the matrix elements h_{ij} are

$$h_{ij} = 1, \quad i = 1, 2, \ldots, n_R, \quad j = 1, 2, \ldots, n_T \tag{1.38}$$

Coherent Combining

In this channel, with the channel matrix given by (1.38), the same signal is transmitted simultaneously from n_T antennas. The received signal at antenna i is given by

$$r_i = n_T x \tag{1.39}$$

and the received signal power at antenna i is given by

$$P_{ri} = n_T^2 \frac{P}{n_T} = n_T P \tag{1.40}$$

where P/n_T is the power transmitted from one antenna. Note that though the power per transmit antenna is P/n_T, the total received power per receive antenna is $n_T P$. The power gain of n_T in the total received power comes due to coherent combining of the transmitted signals.

The rank of channel matrix $\mathbf{H}$ is 1, so there is only one received signal in the equivalent channel model with the power

$$P_r = n_R n_T P \tag{1.41}$$

Thus applying formula (1.19) we get for the channel capacity

$$C = W \log_2 \left(1 + n_R n_T \frac{P}{\sigma^2} \right) \tag{1.42}$$

In this example, the multiple antenna system reduces to a single effective channel that only benefits from higher power achieved by transmit and receive diversity. This system achieves a diversity gain of $n_R n_T$ relative to a single antenna link. The cost of this gain is the system complexity required to implement coordinated transmissions and coherent maximum ratio combining. However, the capacity grows logarithmically with the total number of antennas $n_T n_R$. For example, if $n_T = n_R = 8$ and $10 \log_{10} P/\sigma^2 = 20$ dB, the normalized capacity C/W is 12.65 bits/sec/Hz.

Noncoherent Combining

If the signals transmitted from various antennas are different and all channel entries are equal to 1, there is only one received signal in the equivalent channel model with the power of $n_R P$. Thus the capacity is given by

$$C = W \log_2 \left(1 + n_R \frac{P}{\sigma^2} \right) \tag{1.43}$$

For an SNR of 20 dB and $n_R = n_T = 8$, the capacity is 9.646 bits/sec/Hz.

Example 1.3: A MIMO Channel with Orthogonal Transmissions

In this example we consider a channel with the same number of transmit and receive antennas, $n_T = n_R = n$, and that they are connected by orthogonal parallel sub-channels, so there is no interference between individual sub-channels. This could be achieved for example, by linking each transmitter with the corresponding receiver by a separate waveguide, or by spreading transmitted signals from various antennas by orthogonal spreading sequences. The channel matrix is given by

$$\mathbf{H} = \sqrt{n}\,\mathbf{I}_n$$

The scaling by $\sqrt{n}$ is introduced to satisfy the power constraint in (1.4).

Since

$$\mathbf{H}\mathbf{H}^H = n\mathbf{I}_n$$

by applying formula (1.30) we get for the channel capacity

$$C = W \log_2 \det \left(\mathbf{I}_n + \frac{nP}{n\sigma^2}\mathbf{I}_n \right)$$

$$= W \log_2 \det \left[\mathrm{diag}\left(1 + \frac{P}{\sigma^2} \right) \right]$$

$$= W \log_2 \left(1 + \frac{P}{\sigma^2} \right)^n$$

$$= nW \log_2 \left(1 + \frac{P}{\sigma^2} \right)$$

For the same numerical values $n_T = n_R = n = 8$ and SNR of 20 dB, as in Example 1.2, the normalized capacity C/W is 53.264 bits/sec/Hz. Clearly, the capacity is much higher than in Example 1.2, as the sub-channels are uncoupled giving a multiplexing gain of n.

Example 1.4: Receive Diversity

Let us assume that there is only one transmit and n_R receive antennas. The channel matrix can be represented by the vector

$$\mathbf{H} = (h_1, h_2, \ldots, h_{n_R})^T$$

where the operator $(\cdot)^T$ denotes the matrix transpose. As $n_R > n_T$, formula (1.30) should be written as

$$C = W \log_2 \left[\det \left(\mathbf{I}_{\mathbf{n_T}} + \frac{P}{n_T\sigma^2}\mathbf{H}^H\mathbf{H} \right) \right] \tag{1.44}$$

As $\mathbf{H}^H\mathbf{H} = \sum_{i=1}^{n_R} |h_i|^2$, by applying formula (1.30) we get for the capacity

$$C = W \log_2 \left(1 + \sum_{i=1}^{n_R} |h_i|^2 \frac{P}{\sigma^2} \right) \tag{1.45}$$

This capacity corresponds to linear maximum combining at the receiver. In the case when the channel matrix elements are equal and normalized as follows

$$|h_1|^2 = |h_2|^2 = \cdots |h_{n_R}|^2 = 1$$

the capacity in (1.45) becomes

$$C = W \log_2 \left(1 + n_R \frac{P}{\sigma^2} \right) \tag{1.46}$$

This system achieves the diversity gain of n_R relative to a single antenna channel. For $n_R = 8$ and SNR of 20 dB, the receive diversity capacity is 9.646 bits/s/Hz.

Selection diversity is obtained if the best of the n_R channels is chosen. The capacity of this system is given by

$$
\begin{aligned}
C &= \max_i \left\{ W \log_2 \left(1 + \frac{P}{\sigma^2} |h_i|^2 \right) \right\} \\
&= W \log_2 \left(1 + \frac{P}{\sigma^2} \max_i \{ |h_i|^2 \} \right)
\end{aligned} \tag{1.47}
$$

where the maximization is performed over i, $i = 1, 2, \dots, n_R$.

Example 1.5: Transmit Diversity

In this system there are n_T transmit and only one receive antenna. The channel is represented by the vector

$$\mathbf{H} = (h_1, h_2, \dots, h_{n_T})$$

As $\mathbf{H}\mathbf{H}^H = \sum_{j=1}^{n_T} |h_j|^2$, by applying formula (1.30) we get for the capacity

$$C = W \log_2 \left(1 + \sum_{j=1}^{n_T} |h_j|^2 \frac{P}{n_T \sigma^2} \right) \tag{1.48}$$

If the channel coefficients are equal and normalized as in (1.4), the transmit diversity capacity becomes

$$C = W \log_2 \left(1 + \frac{P}{\sigma^2} \right) \tag{1.49}$$

The capacity does not increase with the number of transmit antennas. This expression applies to the case when the transmitter does not know the channel. For coordinated transmissions, when the transmitter knows the channel, we can apply the capacity formula from (1.35). As the rank of the channel matrix is one, there is only one term in the sum in (1.35) and only one nonzero eigenvalue given by

$$\lambda = \sum_{j=1}^{n_T} |h_j|^2$$

The value for μ from the normalization condition is given by

$$\mu = P + \frac{\sigma^2}{\lambda}$$

So we get for the capacity

$$C = W \log_2 \left(1 + \sum_{j=1}^{n_T} |h_j|^2 \frac{P}{\sigma^2} \right) \tag{1.50}$$

If the channel coefficients are equal and normalized as in (1.4), the capacity becomes

$$C = W \log_2 \left(1 + n_T \frac{P}{\sigma^2} \right) \tag{1.51}$$

For $n_T = 8$ and SNR of 20 dB, the transmit diversity with the channel knowledge at the transmitter is 9.646 bits/s/Hz.

1.6 Capacity of MIMO Systems with Random Channel Coefficients

Now we turn to a more realistic case when the channel matrix entries are random variables. Initially, we assume that the channel coefficients are perfectly estimated at the receiver but unknown at the transmitter. Furthermore, we assume that the entries of the channel matrix are zero mean Gaussian complex random variables. Its real and imaginary parts are independent zero mean Gaussian i.i.d. random variables, each with variance of 1/2. Each entry of the channel matrix thus has a Rayleigh distributed magnitude, uniform phase and expected magnitude square equal to unity, $E[|h_{ij}|^2] = 1$.

The probability density function (pdf) for a Rayleigh distributed random variable $z = \sqrt{z_1^2 + z_2^2}$, where z_1 and z_2 are zero mean statistically independent orthogonal Gaussian random variables each having a variance σ_r^2, is given by

$$p(z) = \frac{z}{\sigma_r^2} e^{\frac{-z^2}{2\sigma_r^2}} \qquad z \geq 0 \tag{1.52}$$

In this analysis σ_r^2 is normalized to 1/2. The antenna spacing is large enough to ensure uncorrelated channel matrix entries. According to frequency of channel coefficient changes, we will distinguish three scenarios.

1. Matrix **H** is random. Its entries change randomly at the beginning of each symbol interval T and are constant during one symbol interval. This channel model is referred to as *fast fading* channel.
2. Matrix **H** is random. Its entries are random and are constant during a fixed number of symbol intervals, which is much shorter than the total transmission duration. We refer to this channel model as *block fading* .
3. Matrix **H** is random but is selected at the start of transmission and kept constant all the time. This channel model is referred to as *slow* or *quasi-static fading* model.

In this section we will estimate the maximum transmission rate in various propagation scenarios and give relevant examples.

1.6.1   Capacity of MIMO Fast and Block Rayleigh Fading Channels

In the derivation of the expression for the MIMO channel capacity on fast Rayleigh fading channels, we will start from the simple single antenna link. The coefficient $|h|^2$ in the capacity expression for a single antenna link (1.37), is a chi-squared distributed random variable, with two degrees of freedom, denoted by χ_2^2. This random variable can be expressed as $y = \chi_2^2 = z_1^2 + z_2^2$, where z_1 and z_2 are zero mean statistically independent orthogonal Gaussian variables, each having a variance σ_r^2, which is in this analysis normalized to $1/2$. Its pdf is given by

$$p(y) = \frac{1}{2\sigma_r^2} e^{-\frac{y}{2\sigma_r^2}}, \quad y \geq 0 \tag{1.53}$$

The capacity for a fast fading channel can then be obtained by estimating the mean value of the capacity given by formula (1.37)

$$C = E\left[W \log_2\left(1 + \chi_2^2 \frac{P}{\sigma^2}\right)\right] \tag{1.54}$$

where $E[\cdot]$ denotes the expectation with respect to the random variable χ_2^2.

By using the singular value decomposition approach, the MIMO fast fading channel, with the channel matrix $\mathbf{H}$, can be represented by an equivalent channel consisting of $r \leq \min(n_T, n_R)$ decoupled parallel sub-channels, where r is the rank of $\mathbf{H}$. Thus the capacities of these sub-channels add up, giving for the overall capacity

$$C = E\left[W \sum_{i=1}^{r} \log_2\left(1 + \lambda_i \frac{P}{n_T \sigma^2}\right)\right] \tag{1.55}$$

where $\sqrt{\lambda_i}$ are the singular values of the channel matrix. Alternatively, by using the same approach as in the capacity derivation in Section 1.3, we can write for the mean MIMO capacity on fast fading channels

$$C = E\left\{W \log_2 \det\left[\left(\mathbf{I}_r + \frac{P}{\sigma^2 n_T}\mathbf{Q}\right)\right]\right\} \tag{1.56}$$

where $\mathbf{Q}$ is defined as

$$\mathbf{Q} = \begin{cases} \mathbf{HH}^H, & n_R < n_T \\ \mathbf{H}^H\mathbf{H}, & n_R \geq n_T \end{cases} \tag{1.57}$$

For block fading channels, as long as the expected value with respect to the channel matrix in formulas (1.55) and (1.56) can be observed, i.e. the channel is ergodic, we can calculate the channel capacity by using the same expressions as in (1.55) and (1.56).

While the capacity can be easily evaluated for $n_T = n_R = 1$, the expectation in formulas (1.55) or (1.56) gets quite complex for larger values of n_T and n_R. They can be evaluated with the aid of Laguerre polynomials [2][13] as follows

$$C = W \int_0^\infty \log_2\left(1 + \frac{P}{n_T \sigma^2}\lambda\right) \sum_{k=0}^{m-1} \frac{k!}{(k+n+m)!}[L_k^{n-m}(\lambda)]^2 \lambda^{n-m} e^{-\lambda} d\lambda$$

where

$$m = \min(n_T, n_R) \tag{1.58}$$

$$n = \max(n_T, n_R) \tag{1.59}$$

and $L_k^{n-m}(x)$ is the associate Laguerre polynomial of order k, defined as [13]

$$L_k^{n-m}(x) = \frac{1}{k!} e^x x^{m-n} \frac{d^k}{dx^k}(e^{-x} x^{n-m+k}) \tag{1.60}$$

Let us define

$$\tau = \frac{n}{m}$$

By increasing m and n and keeping their ratio τ constant, the capacity, normalized by m, approaches

$$\lim_{n \to \infty} \frac{C}{m} = \frac{W}{2\pi} \int_{v_1}^{v_2} \log_2\left(1 + \frac{Pm}{n_T \sigma^2} v\right) \sqrt{\left(\frac{v_2}{v} - 1\right)\left(1 - \frac{v_1}{v}\right)} dv \tag{1.61}$$

where

$$v_2 = (\sqrt{\tau} + 1)^2$$

and

$$v_1 = (\sqrt{\tau} - 1)^2$$

Example 1.6: A Fast and Block Fading Channel with Receive Diversity

For a receive diversity system with one transmit and n_R receive antennas on a fast Rayleigh fading channel, specified by the channel matrix

$$\mathbf{H} = (h_1, h_2, \dots, h_{n_R})^T$$

Formula (1.56) gives the capacity expression for maximum ratio combining at the receiver

$$C = E\left[W \log_2\left(1 + \frac{P}{\sigma^2} \chi_{2n_R}^2\right)\right] \tag{1.62}$$

where

$$\chi_{2n_R}^2 = \sum_{i=1}^{n_R} |h_i|^2$$

is a chi-squared random variable with $2n_R$ degrees of freedom. It can be represented as

$$y = \chi_{2n_R}^2 = \sum_{i=1}^{2n_R} z_i^2 \tag{1.63}$$

where z_i, $i = 1, 2, \ldots, 2n_R$, are statistically independent, identically distributed zero mean Gaussian random variables, each having a variance σ_r^2, which is in this analysis normalized to $1/2$. Its pdf is given by

$$p(y) = \frac{1}{\sigma_r^{2n_R} 2^{n_R} \Gamma(n_R)} y^{n_R - 1} e^{-\frac{y}{2\sigma_r^2}}, \quad y \geq 0 \tag{1.64}$$

where $\Gamma(p)$ is the gamma function, defined as

$$\Gamma(p) = \int_0^\infty t^{p-1} e^{-t} dt, \quad p > 0 \tag{1.65}$$

$$\Gamma(p) = (p - 1)!, \quad p \text{ is an integer}, \ p > 0 \tag{1.66}$$

$$\Gamma\left(\frac{1}{2}\right) = \sqrt{\pi} \tag{1.67}$$

$$\Gamma\left(\frac{1}{3}\right) = \frac{\sqrt{\pi}}{2} \tag{1.68}$$

If a selection diversity receiver is used, the capacity is given by

$$C = E\left\{W \log_2\left[1 + \frac{P}{\sigma^2} \max_i(|h_i|^2)\right]\right\} \tag{1.69}$$

The channel capacity curves for receive diversity with maximum ratio combining are shown in Fig. 1.4 and with selection combining in Fig. 1.5.

Example 1.7: A Fast and Block Fading Channel with Transmit Diversity

For a transmit diversity system with n_T transmit and one receive antenna on a fast Rayleigh fading channel, specified by the channel matrix

$$\mathbf{H} = (h_1, h_2, \ldots, h_{n_T}),$$

formula (1.56) gives the capacity expression for uncoordinated transmission

$$C = E\left[W \log_2\left(1 + \frac{P}{n_T \sigma^2} \chi_{2n_T}^2\right)\right] \tag{1.70}$$

where

$$\chi_{2n_T}^2 = \sum_{j=1}^{n_T} |h_j|^2$$

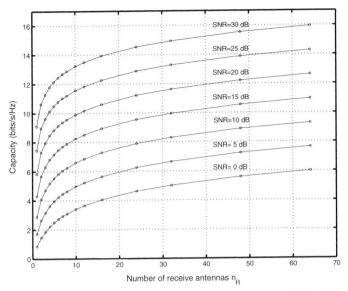

Figure 1.4 Channel capacity curves for receive diversity on a fast and block Rayleigh fading channel with maximum ratio diversity combining

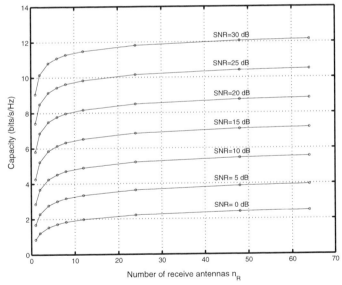

Figure 1.5 Channel capacity curves for receive diversity on a fast and block Rayleigh fading channel with selection diversity combining

is a chi-squared random variable with $2n_T$ degrees of freedom. As the number of transmit antennas increases, the capacity approaches the asymptotic value

$$\lim_{n_T \to \infty} C = W \log_2 \left(1 + \frac{P}{\sigma^2} \right) \qquad (1.71)$$

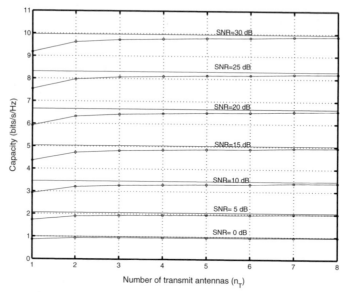

Figure 1.6   Channel capacity curves for uncoordinated transmit diversity on a fast and block Rayleigh fading channel

That is, the system behaves as if the total power is transmitted over a single unfaded channel. In other words, the transmit diversity is able to remove the effect of fading for a large number of antennas.

The channel capacity curves for transmit diversity with uncoordinated transmissions are shown in Fig. 1.6. The capacity is plotted against the number of transmit antennas n_T. The curves are shown for various values of the signal-to-noise ratio, in the range of 0 to 30 dB. The capacity of transmit diversity saturates for $n_T \geq 2$. That is, the capacity asymptotic value from (1.71) is achieved for the number of transmit antennas of 2 and there is no point in increasing it further.

In coordinated transmissions, when all transmitted signals are the same and synchronous, the capacity is given by

$$C = E\left[W \log_2\left(1 + \frac{P}{\sigma^2}\chi^2_{2n_T}\right)\right] \tag{1.72}$$

Example 1.8: A MIMO Fast and Block Fading Channel with Transmit-Receive Diversity

We consider a MIMO system with n transmit and n receive antennas, over a fast Rayleigh fading channel, assuming that the channel parameters are known at the receiver but not at the transmitter. In this case

$$m = n = n_R = n_T$$

so that the asymptotic capacity, from (1.61), is given by

$$\lim_{n \to \infty} \frac{C}{Wn} = \frac{1}{\pi} \int_0^4 \log_2\left(1 + \frac{P}{\sigma^2}v\right)\sqrt{\frac{1}{v} - \frac{1}{4}}\,dv \tag{1.73}$$

or in a closed form [2]

$$\lim_{n \to \infty} \frac{C}{Wn} = \log_2 \frac{P}{\sigma^2} - 1 + \frac{\sqrt{1 + \frac{4P}{\sigma^2}} - 1}{2\frac{P}{\sigma^2}} + 2 \tanh^{-1} \frac{1}{\sqrt{1 + \frac{4P}{\sigma^2}}} \qquad (1.74)$$

Expression (1.73) can be bounded by observing that $\log(1 + x) \geq \log x$, as

$$\lim_{n \to \infty} \frac{C}{Wn} \geq \frac{1}{\pi} \int_0^4 \log_2 \left(\frac{P}{\sigma^2} v \right) \sqrt{\frac{1}{v} - \frac{1}{4}} \, dv \qquad (1.75)$$

This bound can be expressed in a closed form as

$$\lim_{n \to \infty} \frac{C}{Wn} \geq \log_2 \frac{P}{\sigma^2} - 1 \qquad (1.76)$$

The bound in (1.76) shows that the capacity increases linearly with the number of antennas and logarithmically with the SNR. In this example there is a multiplexing gain of n, as there are n independent sub-channels which can be identified by their coefficients, perfectly estimated at the receiver.

The capacity curves obtained by using the bound in (1.76), are shown in Fig. 1.7, for the signal-to-noise ratio as a parameter, varying between 0 and 30 dB.

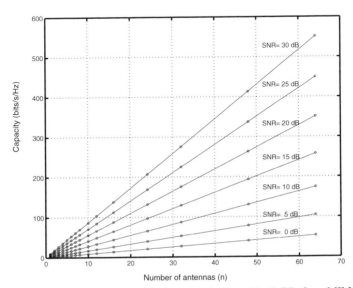

Figure 1.7 Channel capacity curves obtained by using the bound in (1.76), for a MIMO system with transmit/receive diversity on a fast and block Rayleigh fading channel

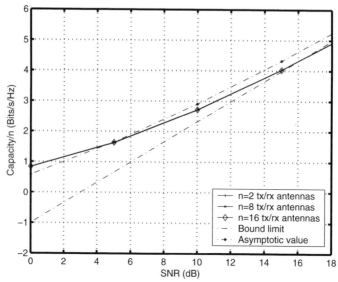

Figure 1.8 Normalized capacity bound curves for a MIMO system on a fast and block Rayleigh fading channel

The normalized capacity bound C/n from (1.76), the asymptotic capacity from (1.74) and the simulated average capacity by using (1.56), versus the SNR and with the number of antennas as a parameter, are shown in Fig. 1.8. Note that in the figure the curves for $n = 2$, 8, and 16 antennas coincide. As this figure indicates, the simulation curves are very close to the bound. This confirms that the bound in (1.76) is tight and can be used for channel capacity estimation on fast fading channels with a large n.

Example 1.9: A MIMO Fast and Block Fading Channel with Transmit-Receive Diversity and Adaptive Transmit Power Allocation

The instantaneous MIMO channel capacity for adaptive transmit power allocation is given by formula (1.35). The average capacity for an ergodic channel can be obtained by averaging over all realizations of the channel coefficients. Figs. 1.9 and 1.10 show the capacities estimated by simulation of an adaptive and a nonadaptive system, for a number of receive antennas as a parameter and a variable number of transmit antennas over a Rayleigh MIMO channel, at an SNR of 25 dB. In the adaptive system the transmit powers were allocated according to the water-filling principle and in the nonadaptive system the transmit powers from all antennas were the same. As the figures shows, when the number of the transmit antennas is the same or lower than the number of receive antennas, there is almost no gain in adaptive power allocation. However, when the numbers of transmit antennas is larger than the number of receive antennas, there is a significant potential gain to be achieved by water-filling power distribution. For four transmit and two receive antennas, the gain is about 2 bits/s/Hz and for fourteen transmit and two receive antennas it is about 5.6 bits/s/Hz. The benefit obtained by adaptive power distribution is higher for a lower SNR and diminishes at high SNRs, as demonstrated in Fig. 1.11.

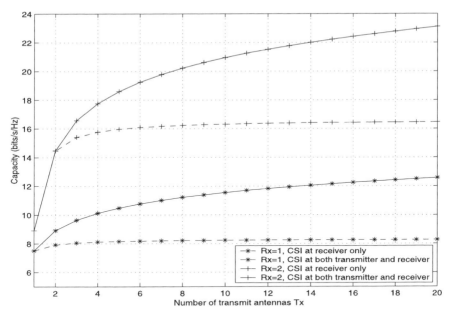

Figure 1.9 Achievable capacities for adaptive and nonadaptive transmit power allocations over a fast MIMO Rayleigh channel, for SNR of 25 dB, the number of receive antennas $n_R = 1$ and $n_R = 2$ and a variable number of transmit antennas

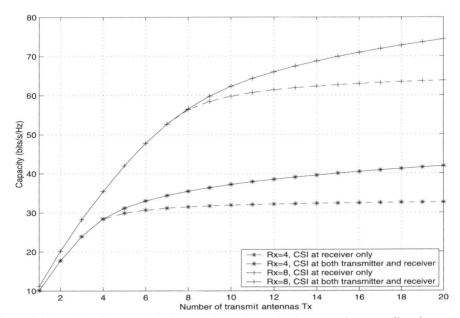

Figure 1.10 Achievable capacities for adaptive and nonadaptive transmit power allocations over a fast MIMO Rayleigh channel, for SNR of 25 dB, the number of receive antennas $n_R = 4$ and $n_R = 8$ and a variable number of transmit antennas

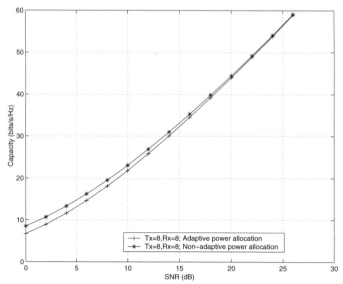

Figure 1.11   Capacity curves for a MIMO slow Rayleigh fading channel with eight transmit and eight receive antennas with and without transmit power adaptation and a variable SNR

1.6.2   Capacity of MIMO Slow Rayleigh Fading Channels

Now we consider a MIMO channel for which **H** is chosen randomly, according to a Rayleigh distribution, at the beginning of transmission and held constant for a transmission block. An example of such a system is wireless LANs with high data rates and low fade rates, so that a fade might last over more than a million symbols. As before, we consider that the channel is perfectly estimated at the receiver and unknown at the transmitter.

In this system, the capacity, estimated by (1.30), is a random variable. It may even be zero, as there is a nonzero probability that a particular realization of **H** is incapable of supporting arbitrarily low error rates, no matter what codes we choose. In this case we estimate the capacity complementary cumulative distribution function (ccdf). The ccdf defines the probability that a specified capacity level is provided. We denote it by P_c. The outage capacity probability, denoted by P_{out}, specifies the probability of not achieving a certain level of capacity. It is equal to the capacity cumulative distribution function (cdf) or $1 - P_c$.

1.6.3   Capacity Examples for MIMO Slow Rayleigh Fading Channels

Example 1.10: Single Antenna Link

In this system $n_T = n_R = 1$. The capacity is given by

$$C = W \log_2 \left(1 + \frac{P}{\sigma^2} \chi_2^2 \right) \tag{1.77}$$

where χ_2^2 is a chi-squared random variable with two degrees of freedom.

Example 1.11: Receive Diversity

In this system there is one transmit and n_R receive antennas. The capacity for receivers with maximum ratio combining is given by

$$C = W \log_2 \left(1 + \frac{P}{\sigma^2} \chi_{2n_R}^2 \right) \tag{1.78}$$

where $\chi_{2n_R}^2$ is a chi-squared random variable with $2n_R$ degrees of freedom.

Example 1.12: Transmit Diversity

In this system there is n_T transmit and one receive antenna. The capacity for receivers with uncoordinated transmissions is given by

$$C = W \log_2 \left(1 + \frac{P}{n_T \sigma^2} \chi_{2n_T}^2 \right) \tag{1.79}$$

where $\chi_{2n_T}^2$ is a chi-squared random variable with $2n_T$ degrees of freedom.

Example 1.13: Combined Transmit-Receive Diversity

In this system there is n_T transmit and n_R receive antennas. Assuming that $n_T \geq n_R$ we can write for the lower bound of the capacity as

$$C > W \sum_{i=n_T-(n_R-1)}^{n_T} \log_2 \left(1 + \frac{P}{n_T \sigma^2} (\chi_2^2)_i \right) \tag{1.80}$$

where $(\chi_2^2)_i$ is a chi-squared random variable with 2 degrees of freedom. The upper bound is

$$C < W \sum_{i=1}^{n_T} \log_2 \left(1 + \frac{P}{n_T \sigma^2} (\chi_{2n_R}^2)_i \right) \tag{1.81}$$

where $(\chi_{2n_R}^2)_i$ is a chi-squared random variable with $2n_R$ degrees of freedom. This case corresponds to a system of uncoupled parallel transmissions, where each of n_T transmit antennas is received by a separate set of n_R receive antennas, so that there is no interference.

For $n = n_R = n_T$ and n very large, the capacity is lower bounded as [1]

$$\frac{C}{Wn} > \left(1 + \frac{\sigma^2}{P} \right) \log_2 \left(1 + \frac{P}{\sigma^2} \right) - \log_2 e + \varepsilon_n \tag{1.82}$$

where the random variable ε_n has a Gaussian distribution with mean

$$E\{\varepsilon_n\} = \frac{1}{n} \log_2 \left(1 + \frac{P}{\sigma^2} \right)^{-1/2} \tag{1.83}$$

and variance

$$\text{Var}\{\varepsilon_n\} = \left(\frac{1}{n \ln 2} \right)^2 \cdot \left[\ln \left(1 + \frac{P}{\sigma^2} \right) - \frac{P/\sigma^2}{1 + P/\sigma^2} \right] \tag{1.84}$$

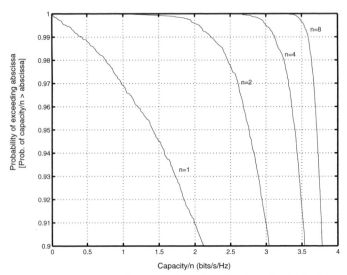

Figure 1.12 Capacity per antenna ccdf curves for a MIMO slow Rayleigh fading channel with constant SNR of 15 dB and a variable number of antennas

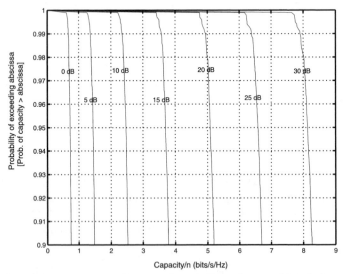

Figure 1.13 Capacity per antenna ccdf curves for a MIMO slow Rayleigh fading channel with a constant number of antennas $n_T = n_R = 8$ and a variable SNR

Figs. 1.12 and 1.13 show the ccdf capacity per antenna curves on a slow Rayleigh fading channel obtained by simulation from (1.19).

Figure 1.12 is plotted for various numbers of antennas and a constant SNR. It demonstrates that the probability that the capacity achieves a given level improves markedly when the number of antennas increases. Figure 1.13 shows the ccdf curves for a constant number of antennas and variable SNRs. For large values of antennas, the ccdf curves, shown in

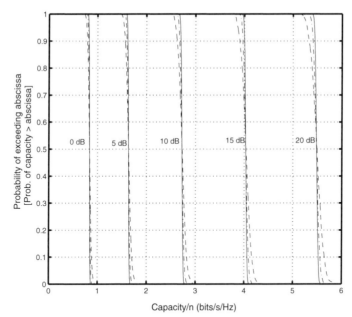

Figure 1.14 Capacity per antenna ccdf curves for a MIMO slow Rayleigh fading channel with a large number of antennas $n_R = n_T = n = 64$ (solid line), 32 (next to the solid line) and 16 (second to the solid line) and a variable SNR of 0, 5, 10, 15 and 20 dB

Fig. 1.14, exhibit asymptotic behavior which is more pronounced for lower SNRs. That is, the capacities per antenna remain approximately constant for increasing numbers of antennas. In other words, in MIMO channels with a large number of antennas, the capacity grows linearly with the number of antennas. That agrees with the analytical bound in (1.82). The analytical bounds on C/Wn in (1.82) are plotted in Figs. 1.15 and 1.16 for similar system configurations as in Figs. 1.12 and 1.13, respectively. Comparing the simulated and analytical ccdf curves it is clear that the simulated capacities are slightly higher for a given probability level. Therefore, the bound in (1.82) can be used for reasonably accurate capacity estimations. Figure 1.17 depicts the capacity that can be achieved on slow Rayleigh fading channels 99% of the time, i.e. 1% outage. It shows that even with a relatively small number of antennas, large capacities are available. For example, at an SNR of 20 dB and with 8 antennas at each site, about 37 bits/sec/Hz could be achieved. This compares to the achievable spectral efficiencies between 1-2 bits/sec/Hz in the second generation cellular mobile systems.

1.7   Effect of System Parameters and Antenna Correlation on the Capacity of MIMO Channels

The capacity gain of MIMO channels, derived under the idealistic assumption that the channel matrix entries are independent complex Gaussian variables, might be reduced on real channels. The high spectral efficiency predicted by (1.30) is diminished if the signals arriving at the receivers are correlated. The effect of spatial fading correlation on the MIMO

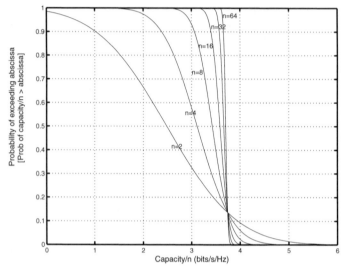

Figure 1.15   Analytical capacity per antenna ccdf bound curves for a MIMO slow Rayleigh fading channel with a fixed SNR of 15 dB and a variable number of transmit/receive antennas

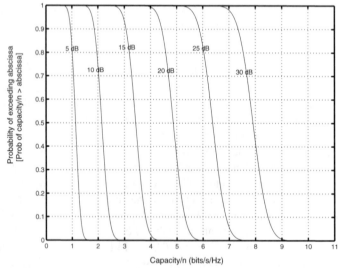

Figure 1.16   Analytical capacity per antenna ccdf bound curves for a MIMO slow Rayleigh fading channel with 8 transmit/receive antennas and variable SNRs

channel capacity has been addressed in [8][16]. Correlation between antenna elements can be reduced by separating antennas spatially [8][16][17]. However, low correlation between antenna elements does not guarantee high spectral efficiency. In real channels, degenerate propagation conditions, known as "keyholes", have been observed, leading to a considerable decrease in MIMO channel capacity. In indoor environments such propagation effect can occur in long hallways. A similar condition exists in tunnels or in systems with large separations between the transmit and the receive antennas in outdoor environments [10][15][16].

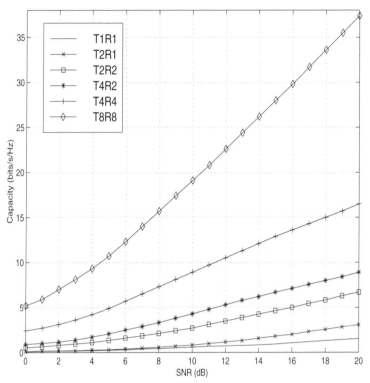

Figure 1.17 Achievable capacity for a MIMO slow Rayleigh fading channel for 1% outage, versus SNR for a variable number of transmit/receive antennas

The "keyholes" reduce the rank of the channel matrix and thus lower the capacity.

In this section, we first define the correlation coefficients and introduce the correlation matrix models in MIMO systems. Then we proceed with the estimation of the correlation coefficients and the effect of correlation between antennas on the capacity in the MIMO channels with a line of sight (LOS) path and in the absence of scattering. It is followed by a correlation model for a Rician fading channel and a channel with no LOS propagation. Subsequently, we will demonstrate the "keyhole" effect and its influence on the channel matrix and capacity. Then we will derive a channel model for an outdoor channel with scattering, described by the system parameters such as the angular spread, transmit and receive scattering radii and the distance between the transmitter and receiver. By using this channel model we will discuss under what conditions degenerate propagation occurs and its effects on channel capacity. The discussion is supported by capacity curves for various system parameters.

A MIMO channel with n_T transmit and n_R receive antennas can be described by an $n_R \times n_T$ channel matrix **H**. It can be represented in this form

$$\mathbf{H} = [\mathbf{h}_1, \mathbf{h}_2, \ldots, \mathbf{h}_i, \ldots, \mathbf{h}_{n_R}]^T \qquad (1.85)$$

where $\mathbf{h}_i$, $i = 1, 2, \ldots, n_R$, is given by

$$\mathbf{h}_i = [h_{i,1}, h_{i,2}, \ldots, h_{i,n_T}]$$

For the purpose of the calculation of the antenna correlation coefficients, we arrange vectors $\mathbf{h}_i$ in a vector $\mathbf{h}$ with $n_R n_T$ elements, as follows

$$\mathbf{h} = [\mathbf{h}_1, \mathbf{h}_2, \ldots, \mathbf{h}_i, \ldots, \mathbf{h}_{n_R}] \tag{1.86}$$

We define an $n_R n_T \times n_R n_T$ correlation matrix Θ as follows

$$\Theta = E[\mathbf{h}^H \mathbf{h}] \tag{1.87}$$

where $\mathbf{h}^H$ denotes the Hermitian of $\mathbf{h}$.

If the entries of the channel matrix $\mathbf{H}$ are independent identically distributed (iid) variables, Θ is an identity matrix which produces a maximum capacity.

In order to simplify the analysis, we assume that the correlation between the receive antenna elements does not depend on the transmit antennas and vice versa. This assumption can be justified by the fact that only immediate antenna surroundings cause the correlation between array elements and have no impact on correlation observed between the elements of the array at the other end of the link [8][18]. In such a case we can define an $n_R \times n_R$ correlation coefficient matrix, denoted by $\Theta_\mathbf{R}$, for the receive antennas and an $n_T \times n_T$ correlation matrix, denoted by Θ_T, for the transmit antennas.

Assuming that we model the correlation of the receive and the transmit array elements independently, their respective correlation matrices can be represented as

$$\Theta_R = \mathbf{K}_R \mathbf{K}_R^H \tag{1.88}$$

where $\mathbf{K}_R$ is an $n_R \times n_R$ matrix and

$$\Theta_T = \mathbf{K}_T \mathbf{K}_T^H \tag{1.89}$$

where $\mathbf{K}_T$ is an $n_T \times n_T$ matrix. Matrices $\mathbf{K}_R$ and $\mathbf{K}_T$ are $n_R \times n_R$ and $n_T \times n_T$ lower triangular matrices, respectively, with positive diagonal elements. They can be obtained from their respective correlation matrices Θ_R and Θ_T, by Cholesky decomposition [11] (Appendix 1.2). A correlated MIMO channel matrix, denoted by $\mathbf{H}_c$, can be represented as

$$\mathbf{H}_c = \mathbf{K}_R \mathbf{H} \mathbf{K}_T \tag{1.90}$$

where $\mathbf{H}$ is the channel matrix with uncorrelated complex Gaussian entries.

1.7.1   Correlation Model for LOS MIMO Channels

Let us consider a MIMO channel with a linear array of n_T transmit and n_R receive antennas, with the respective antenna element separation of d_t and d_r, as shown in Fig. 1.18. We assume that the separation between the transmitter and the receiver, denoted by R, is much larger than d_r or d_t.

We first consider a system with line of sight (LOS) propagation without scattering. The channel matrix entries, denoted by h_{ki}, are given by

$$h_{ki} = e^{-j2\pi \frac{R_{ki}}{\lambda}} \quad k = 1, 2, \ldots, n_R, \quad i = 1, 2, \ldots, n_T, \tag{1.91}$$

as their amplitudes are normalized and R_{ki} is the distance between receive antenna k and transmit antenna i.

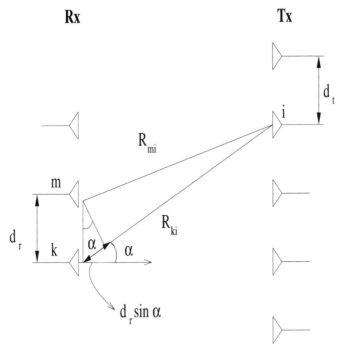

Rx   **Tx**

Figure 1.18 Propagation model for a LOS nonfading system

We assume that the receive and the transmit antenna element correlation coefficients are independent. That is, we calculate the receive antenna element correlation coefficients, for a fixed transmit antenna element, for example the first one, $i = 1$. Clearly this does not limit the generality of the analysis. In this case, the receive antenna correlation coefficients, for the LOS system model, shown in Fig. 1.18, are given by

$$\theta_{mk} = E[h_{m1}^* h_{k1}] \quad m = 1, 2, \ldots, n_R, \quad k = 1, 2, \ldots, n_R \tag{1.92}$$

The correlation coefficients can be obtained by substituting the channel entries from (1.91) into the expression (1.92)

$$\theta_{mk} = E\left[e^{-j2\pi \frac{R_{k1} - R_{m1}}{\lambda}}\right] \tag{1.93}$$

For the broadside array considered here with the angle of orientation of $\pi/2$ and large distances R_{k1} and R_{m1}, the correlation coefficient in (1.93) can be approximated as

$$\theta_{mk} = \begin{cases} e^{-j2\pi \frac{d_{mk}\sin\alpha}{\lambda}} & m \neq k \\ 1 & m = k \end{cases} \tag{1.94}$$

where d_{mk} is the distance between the receive antenna elements m and k, α is the plane-wave direction of arrival (DOA).

As the expression (1.94) shows, the correlation coefficient is governed by the antenna element separation and will be largest between the adjacent antenna elements with the separation of d_r. If the antenna separation is small compared to the wavelength λ, all correlation coefficients will be the same and equal to one. According to the expression in (1.94) the correlation coefficients are the same and equal to one, if the direction of arrival α is small, i.e., $\alpha \simeq 0$.

In both cases, for small antenna element separations and small directions of arrival, the channel matrix, denoted by $\mathbf{H}_1$, is of rank one.

Assuming that $n_T = n_R = n$, matrix $\mathbf{H}_1\mathbf{H}_1^H$ has only one eigenvalue equal to n^2 from (1.24). The capacity of this channel (1.21) is given by

$$C = W \log_2 \left(1 + n \frac{P}{\sigma^2} \right) \tag{1.95}$$

In the case that the antenna elements are more widely separated, the channel matrix entries h_{ij} will have different values. If they are chosen in such a way that the channel matrix, denoted by $\mathbf{H}_n$, is of rank n and that the matrix $\mathbf{H}_n\mathbf{H}_n^H$ is given by

$$\mathbf{H}_n\mathbf{H}_n^H = n\mathbf{I}_n \tag{1.96}$$

the capacity can be expressed as

$$C = W \log_2 \left(1 + \frac{P}{\sigma^2} \right)^n$$
$$= nW \log_2 \left(1 + \frac{P}{\sigma^2} \right) \tag{1.97}$$

This can be achieved, for example, if the values of the channel matrix entries are given by

$$h_{ik} = e^{j\gamma_{ik}}, \quad i = 1, 2, \ldots, n, \quad k = 1, 2, \ldots, n \tag{1.98}$$

where

$$\gamma_{ik} = \frac{\pi}{n}[(i - i_0) - (k - k_0)]^2 \tag{1.99}$$

where i_0 and k_0 are integers. If $n = 2$, then $i_0 = k_0 = 0$, giving for the channel matrix

$$\mathbf{H} = \left[\begin{array}{cc} 1 & j \\ j & 1 \end{array} \right] \tag{1.100}$$

This corresponds to two linear arrays broadside to each other.

1.7.2   Correlation Model for a Rayleigh MIMO Fading Channel

We consider a linear array of n_R omnidirectional receive antennas, spaced at a distance d_r, surrounded by clutter, as shown in Fig. 1.19. There are n_T transmit antenna radiating signals which are reflected by the scatterers surrounding the receiver. The plane-wave directions of arrival of signals coming from the scatterers towards the receive antennas is α.

For a linear array with regularly spaced antennas and the orientation angle of $\pi/2$, the correlation coefficient of the signals received by antennas m and k, separated by distance d_{mk}, can be obtained as

$$\theta_{mk} = E \left[e^{-j2\pi \frac{R_{k1} - R_{m1}}{\lambda}} \right] \tag{1.101}$$

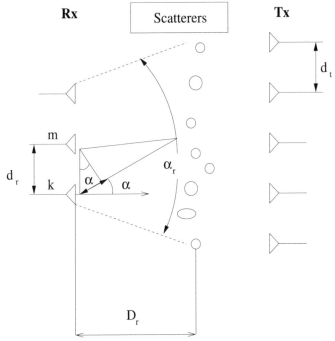

Figure 1.19   Propagation model for a MIMO fading channel

where $(R_{k1} - R_{m1})$ is approximated as

$$R_{k1} - R_{m1} \simeq d_{mk} \sin \alpha \qquad (1.102)$$

The correlation coefficient is given by

$$\theta_{mk} = \left\{ \begin{array}{ll} \int_{-\alpha_r/2}^{\alpha_r/2} e^{-j2\pi \frac{d_{mk}}{\lambda} \sin(\alpha)} p(\alpha) d\alpha, & m \neq k \\ 1, & m = k \end{array} \right. \qquad (1.103)$$

where $p(\alpha)$ is the probability distribution of the direction of arrival or the angular spectrum, and α_r is the receive antenna angular spread. For a uniformly distributed angular spectrum between $-\pi$ and π

$$p(\alpha) = \frac{1}{2\pi}, \qquad (1.104)$$

the correlation coefficient θ_{mk} is given by [14]

$$\theta_{mk} = J_0\left(2\pi \frac{d_{mk}}{\lambda}\right), \quad m \neq k \qquad (1.105)$$

where $J_0(\cdot)$ is the zeroth order Bessel function. To achieve a zero correlation coefficient in this case, the antenna elements should be spaced by $\lambda/2$, as $J_0(\pi) \simeq 0$. Base stations are typically positioned high above the ground and have a narrow angular spread, which makes the correlation coefficient high even for large antenna separations. It has been shown

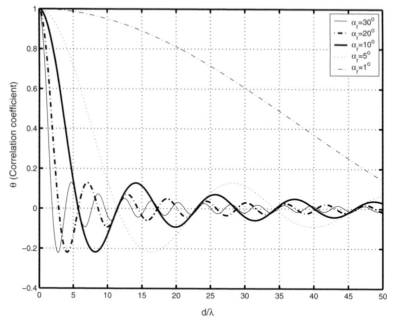

Figure 1.20 Correlation coefficient in a fading MIMO channel with a uniformly distributed direction of arrival α

that the angular spread at base stations with cell radii of 1km is about 2°. The correlation coefficients for a uniform direction of arrival distribution and various angle spreads are shown in Fig. 1.20. For small values of the angle spread very large antenna separations are needed to obtain low correlation. On the other hand, if the angle spread is reasonably large, for example 30°, low correlation (<0.2) can be obtained for antenna spacing not higher than two wavelengths. For low element separation (<$\lambda/2$), the correlation coefficient is high (>0.5) even for large angle spreads.

If the angular spectrum is Gaussian, the correlation coefficient goes monotonically down with antenna separation [8]. The correlation coefficient for a Gaussian distribution of the direction of arrival is shown in Fig. 1.21, for the same angle spreads as for the uniform distribution shown in Fig. 1.20. The pdf for the zero mean Gaussian distributed direction of arrival, denoted by $p(\alpha)$, is given by

$$p(\alpha) = \begin{cases} \frac{1}{\sqrt{2\pi}\sigma}e^{-\frac{\alpha^2}{2\sigma^2}} & -\frac{\alpha_r}{2} \leq \alpha \leq \frac{\alpha_r}{2} \\ 0 & |\alpha| > \frac{\alpha_r}{2} \end{cases} \qquad (1.106)$$

The standard deviation for the Gaussian distributed direction of arrival is calculated in such a way as to obtain the same rms values for the uniform and Gaussian distributions for a given angle spread α_r and is given by

$$\sigma = \alpha_r k \qquad (1.107)$$

where $k = 1/2\sqrt{3}$.

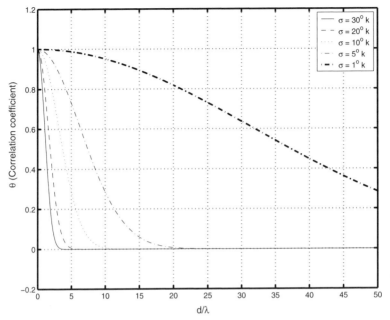

Figure 1.21 Correlation coefficient in a fading MIMO channel with a Gaussian distributed direction of arrival and the standard deviation $\sigma = \alpha_r k$, where $k = 1/2\sqrt{3}$

Assuming that $n_R = n_T = n$, the capacity of a correlated MIMO fading channel can be expressed as

$$C = \log_2 \left[\det \left(\mathbf{I}_n + \frac{P}{n\sigma^2} \mathbf{K}_R \mathbf{H} \mathbf{K}_T \mathbf{K}_T^H \mathbf{H}^H \mathbf{K}_R^H \right) \right] \qquad (1.108)$$

By using the identity

$$\det(\mathbf{I} + \mathbf{AB}) = \det(\mathbf{I} + \mathbf{BA}) \qquad (1.109)$$

we get for the capacity

$$C = \log_2 \left[\det \left(\mathbf{I}_n + \frac{P}{n\sigma^2} \Theta_R^H \mathbf{H} \Theta_T \mathbf{H}^H \right) \right] \qquad (1.110)$$

The MIMO channel capacity for a system with $n_R = n_T = 4$ in a fast Rayleigh fading channel, with uniform distribution of the direction of arrival, obtained by averaging the expression in (1.110) and variable receive antenna angle spreads and antenna element separations, are shown in Fig. 1.22.

In order to consider the effect of the receive antenna elements correlation in a slow Rayleigh fading channel, a system with four receive and four transmit antennas is simulated. The receive antenna correlation coefficients are calculated by expression (1.105) assuming a uniform distribution of the direction of arrival. The remote antenna array elements are assumed to be uncorrelated, which is realistic for the case of an array immersed in clutter with a separation of a half wavelength between elements. Using (1.90) to impose correlation on a random uncorrelated matrix $\mathbf{H}$ we calculate the ccdf of the capacity for angle spreads

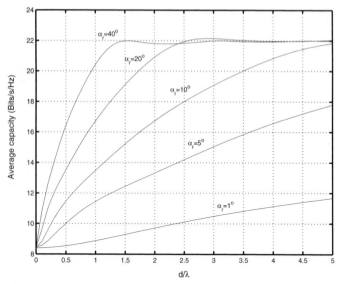

Figure 1.22 Average capacity in a fast MIMO fading channel for variable antenna separations and receive antenna angle spread with constant SNR of 20 dB and $n_T = n_R = 4$ antennas

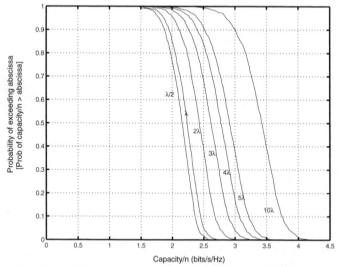

Figure 1.23 Capacity ccdf curves for a correlated slow fading channel, receive antenna angle spread of $1°$ and variable antenna element separations

of $1°$, $5°$, $40°$ and SNR $= 20$ dB and $n_R = n_T = 4$. The ccdf capacity curves are shown in Figs. 1.23–1.25 for variable antenna element separations. They show that for the angle spread of $1°$ the capacity of 3 bits/sec/Hz is exceeded with the probability of 90% for the antenna separation of 10λ, while for the angle spreads of $5°$ and $40°$ the same capacity is exceeded with the same probability for the antenna separations of 2λ and $<\lambda/2$ wavelengths, respectively.

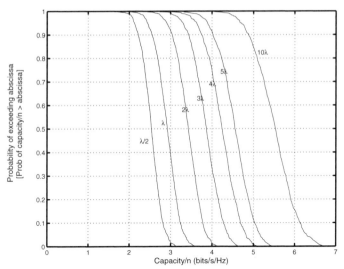

Figure 1.24 Capacity ccdf curves for a correlated slow fading channel, receive antenna angle spread of $5°$ and variable antenna element separations

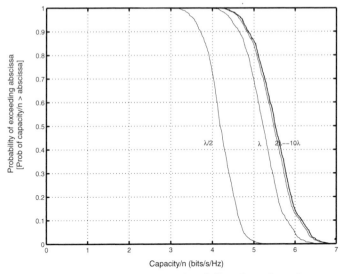

Figure 1.25 Capacity ccdf curves for a correlated slow fading channel, receive antenna angle spread of $40°$ and variable antenna element separations

1.7.3 Correlation Model for a Rician MIMO Channel

A Rician model is obtained in a system with LOS propagation and scattering. The model is characterised by the Rician factor, denoted by K and defined as the ratio of the line of sight and the scatter power components. The pdf for a Rician random variable x is given by

$$p(x) = 2x(1 + K)e^{-K-(1+K)x^2} I_0 \left(2x\sqrt{K(K + 1)}\right) \quad x \geq 0 \qquad (1.111)$$

where

$$K = \frac{D^2}{2\sigma_r^2} \tag{1.112}$$

and D^2 and $2\sigma_r^2$ are the powers of the LOS and scattered components, respectively. The powers are normalized such that

$$D^2 + 2\sigma_r^2 = 1 \tag{1.113}$$

The channel matrix for a Rician MIMO model can be decomposed as [1]

$$\mathbf{H} = D\mathbf{H}_{LOS} + \sqrt{2\sigma_r}\mathbf{H}_{Rayl} \tag{1.114}$$

where $\mathbf{H}_{LOS}$ is the channel matrix for the LOS propagation with no scattering and $\mathbf{H}_{Rayl}$ is the channel matrix for the case with scattering only.

In one extreme case in LOS propagation, when the receive antenna elements are fully correlated, its LOS channel matrix, denoted by $\mathbf{H}_1$, has all entries equal to one, as in (1.38), and its rank is one. The capacity curves for this case are shown in Fig. 1.26 for $n_T = n_R = 3$. For the Rician factor of zero (K in dB $\rightarrow -\infty$), which defines a Rayleigh channel, the capacity is equal to the capacity of the fully correlated Rayleigh fading channel. As the Rician factor increases (K $\rightarrow +\infty$), the capacity reaches the logarithmic expression in (1.95).

For the other LOS extreme case, when the receive antenna elements are uncorrelated, the LOS channel matrix, denoted by $\mathbf{H}_n$, is of rank n, and with the entries given by (1.98). The capacity curves for this case with $n_T = n_R = 3$ are shown in Fig. 1.27. For the Rician factor of zero (K $\rightarrow -\infty$), the capacity is equal to the capacity of an uncorrelated Rayleigh fading channel. As the Rician factor increases (K $\rightarrow +\infty$), the capacity approaches the linear expression in (1.97).

1.7.4 Keyhole Effect

Let us consider a system with two transmit and two receive uncorrelated antennas surrounded by clutter. This system would under normal propagation conditions produce a matrix with independent complex Gaussian variables, giving a high capacity. However, if these two sets of antennas are separated by a screen with a small hole in it, as shown in Fig. 1.28, we get a propagation situation known as "keyhole". The only way for the transmitted signals to propagate is to pass through the keyhole. If the transmitted signals are arranged in a vector

$$\mathbf{x} = (x_1, x_2)^T \tag{1.115}$$

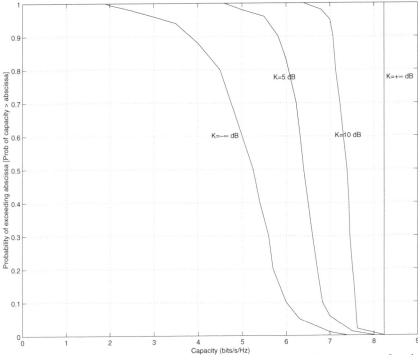

Figure 1.26 Ccdf capacity per antenna curves on a Rician channel with $n_R = n_T = 3$ and $SNR = 20$ dB, with a variable Rician factor and fully correlated receive antenna elements

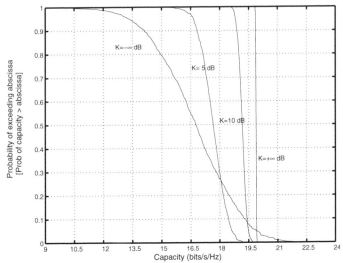

Figure 1.27 Ccdf capacity per antenna curves on a Rician channel with $n_R = n_T = 3$ and $SNR = 20$ dB, with a variable Rician factor and independent receive antenna elements

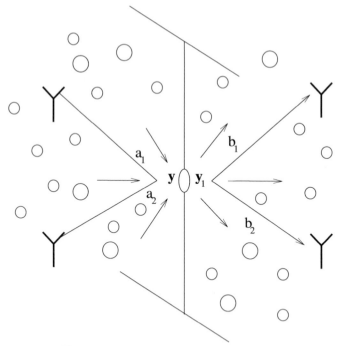

Figure 1.28 A keyhole propagation scenario

where x_1 and x_2 are signals transmitted from the first and second antenna, respectively, the signal incident at the keyhole, denoted by y, is given by

$$y = \mathbf{H}_1 \mathbf{x} \tag{1.116}$$

where

$$\mathbf{H}_1 = (a_1, a_2) \tag{1.117}$$

and a_1 and a_2 are the channel coefficients corresponding to transmitted signals x_1 and x_2, respectively. They can be described by independent complex Gaussian variables. The signal at the other side of the keyhole, denoted by y_1, is given by

$$y_1 = gy \tag{1.118}$$

where g is the keyhole attenuation.

The signal vector at the receive antennas on the other side of the keyhole, denoted by $\mathbf{r}$, is given by

$$\mathbf{r} = \mathbf{H}_2 y_1 \tag{1.119}$$

where $\mathbf{H}_2$ is the channel matrix describing the propagation on the right hand side of the keyhole. It can be represented as

$$\mathbf{H}_2 = \begin{bmatrix} b_1 \\ b_2 \end{bmatrix} \tag{1.120}$$

where b_1 and b_2 are the channel coefficients corresponding to the first and second receive antennas, respectively. Thus the received signal vector at the right hand side of the keyhole can be written as

$$\mathbf{r} = g\mathbf{H}_2\mathbf{H}_1\mathbf{x} \tag{1.121}$$

In (1.121) we can identify the equivalent channel matrix, denoted by $\mathbf{H}$, as $g\mathbf{H}_2\mathbf{H}_1$. It is given by

$$\mathbf{H} = g \begin{bmatrix} a_1b_1 & a_2b_1 \\ a_1b_2 & a_2b_2 \end{bmatrix} \tag{1.122}$$

The rank of this channel matrix is one and thus there is no multiplexing gain in this channel. The capacity is given by

$$C = \log_2 \left(1 + \lambda \frac{P}{2\sigma^2} \right) \tag{1.123}$$

where λ is the singular value of the channel matrix $\mathbf{H}$ and is given by

$$\lambda = g^2(a_1^2 + a_2^2)(b_1^2 + b_2^2) \tag{1.124}$$

1.7.5 MIMO Correlation Fading Channel Model with Transmit and Receive Scatterers

Now we focus on a MIMO fading channel model with no LOS path. The propagation model is illustrated in Fig. 1.29. We consider a linear array of n_R receive omnidirectional antennas and a linear array of n_T omnidirectional transmit antennas. Both the receive and transmit antennas are surrounded by clutter and large objects obstructing the LOS path. The scattering radius at the receiver side is denoted by D_r and at the transmitted side by D_t. The distance between the receiver and the transmitter is R. It is assumed to be much larger than the scattering radii D_r and D_t. The receive and transmit scatterers are placed at the distance R_r and R_t from their respective antennas. These distances are assumed large enough from the antennas for the plane-wave assumption to hold. The angle spreads at the receiver, denoted by α_r, and at the transmitter, denoted by α_t, are given by

$$\alpha_r = 2 \tan^{-1} \frac{D_r}{R_r} \tag{1.125}$$

$$\alpha_t = 2 \tan^{-1} \frac{D_t}{R_t} \tag{1.126}$$

Let us assume that there are S scatterers surrounding both the transmitter and the receiver. The receive scatterers are subject to an angle spread of

$$\alpha_S = 2 \tan^{-1} \frac{D_t}{R} \tag{1.127}$$

The elements of the correlation matrix of the received scatterers, denoted by Θ_S, depend on the value of the respective angle spread α_S.

The signals radiated from the transmit antennas are arranged into an n_T dimensional vector

$$\mathbf{x} = (x_1, x_2, \ldots, x_i, \ldots, x_{n_T}) \tag{1.128}$$

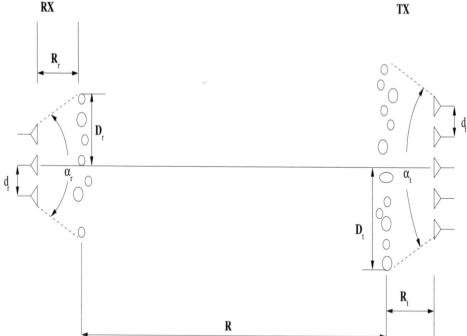

Figure 1.29 Propagation model for a MIMO correlated fading channel with receive and transmit scatterers

The S transmit scatterers capture and re-radiate the captured signal from the transmitted antennas. The S receive scatterers capture the signals transmitted from the S transmit scatterers.

We denote by $\mathbf{y}_i$ an S-dimensional vector of signals originating from antenna i and captured by the S receive scatterers

$$\mathbf{y}_i = (y_{1,i}, y_{2,i}, \ldots, y_{S,i})^T \tag{1.129}$$

It can be represented as

$$\mathbf{y}_i = \mathbf{K}_S \mathbf{g}_i x_i \tag{1.130}$$

where the scatterer correlation matrix Θ_S is defined as

$$\Theta_S = \mathbf{K}_S \mathbf{K}_S^H \tag{1.131}$$

and $\mathbf{g}_i$ is a vector column consisting of S uncorrelated complex Gaussian components. It represents the channel coefficients from the transmit antenna to the S transmit scatterers. All the signal vectors coming from n_T antennas, $\mathbf{y}_i$, for $i = 1, 2, \ldots, n_T$, captured and re-radiated by S receive scatterers, can be collected into an $S \times n_T$ matrix, denoted by $\mathbf{Y}$, given by

$$\mathbf{Y} = \mathbf{K}_S \mathbf{G}_T \mathbf{X} \tag{1.132}$$

where $\mathbf{G}_T = [\mathbf{g}_1, \mathbf{g}_2, \ldots, \mathbf{g}_{n_T}]$ is an $S \times n_T$ matrix with independent complex Gaussian random variable entries and $\mathbf{X}$ is the matrix of transmitted signals arranged as the diagonal elements of an $n_T \times n_T$ matrix, with $x_{i,i} = x_i$, $i = 1, 2, \ldots, n_T$, while $x_{i,j} = 0$, for $i \neq j$.

Taking into account correlation between the transmit antenna elements, we get for the matrix $\mathbf{Y}$

$$\mathbf{Y} = \mathbf{K}_S \mathbf{G}_T \mathbf{K}_T \mathbf{X} \qquad (1.133)$$

where the transmit correlation matrix Θ_T is defined as

$$\Theta_T = \mathbf{K}_T \mathbf{K}_T^H \qquad (1.134)$$

The receive scatterers also re-radiate the captured signals. The vector of n_R received signals, coming from antenna i, denoted by $\mathbf{r}_i$, can be represented as

$$\mathbf{r}_i = (r_{i,1}, r_{i,2}, \ldots, r_{i,n_R})^T \quad i = 1, 2, \ldots, n_T \qquad (1.135)$$

It is given by

$$\mathbf{r}_i = \mathbf{K}_R \mathbf{G}_R \mathbf{y}_i \qquad (1.136)$$

where $\mathbf{G}_R$ is an $n_R \times S$ matrix with independent complex Gaussian random variables. The receive correlation matrix Θ_R is defined as

$$\Theta_R = \mathbf{K}_R \mathbf{K}_R^H \qquad (1.137)$$

The received signal vectors $\mathbf{r}_i$, $i = 1, 2, \ldots, n_T$, can be arranged into an $n_R \times n_T$ matrix $\mathbf{R} = [\mathbf{r}_1, \mathbf{r}_2, \ldots, \mathbf{r}_i, \ldots, \mathbf{r}_{n_T}]$, given by

$$\mathbf{R} = \mathbf{K}_R \mathbf{G}_R \mathbf{Y} \qquad (1.138)$$

Substituting $\mathbf{Y}$ from (1.133) into (1.138) we get

$$\mathbf{R} = \frac{1}{\sqrt{S}} \mathbf{K}_R \mathbf{G}_R \mathbf{K}_S \mathbf{G}_T \mathbf{K}_T \mathbf{X} \qquad (1.139)$$

where the received signal vector is divided by a factor $\sqrt{S}$ for the normalization purposes. As the channel input-output relationship can in general be written as

$$\mathbf{R} = \mathbf{H} \mathbf{X} \qquad (1.140)$$

where $\mathbf{H}$ is the channel matrix, by comparing the relationships in (1.139) and (1.140), we can identify the overall channel matrix as

$$\mathbf{H} = \frac{1}{\sqrt{S}} \mathbf{K}_R \mathbf{G}_R \mathbf{K}_S \mathbf{G}_T \mathbf{K}_T \qquad (1.141)$$

A similar analysis can be performed when there are only transmit scatterers, or both transmit and receive scatterers.

1.7.6 The Effect of System Parameters on the Keyhole Propagation

As the expression for the channel matrix in (1.141) indicates, the behavior of the MIMO fading channel is controlled by the three matrices $\mathbf{K}_R$, $\mathbf{K}_S$ and $\mathbf{K}_T$. Matrices $\mathbf{K}_R$ and $\mathbf{K}_T$

are directly related to the respective antenna correlation matrices and govern the receive and transmit antenna correlation properties.

The rank of the overall channel matrix depends on the ranks of all three matrices $\mathbf{K}_R$, $\mathbf{K}_S$ and $\mathbf{K}_T$ and a low rank of any of them can cause a low channel matrix rank. The scatterer matrix $\mathbf{K}_S$ will have a low rank if the receive scatterers angle spread is low, which will happen if the ratio D_t/R is low. That is, if the distance between the transmitter and the receiver R is high, the elements of K_S are likely to be the same, so the rank of $\mathbf{K}_S$ and thus the rank of $\mathbf{H}$ will be low. In the extreme case when the rank is one, there is only one thin radio pipe between the transmitter and the receiver and this situation is equivalent to the keyhole effect. Note that if there is no scattering at the transmitter side, the parameter relevant for the low rank is the transmit antenna radius, instead of D_t.

The rank of the channel matrix can also be one when either the transmit or receive array antenna elements are fully correlated, which happens if either the corresponding antenna elements separations or angle spreads are low.

The fading statistics is determined by the distribution of the entries of the matrix obtained as the product of $\mathbf{G}_R\mathbf{K}_S\mathbf{G}_T$ in (1.141). To determine the fading statistics of the correlated fading MIMO channel in (1.141) we consider the two extreme cases, when the channel matrix is of full rank and of rank one. In the first case, matrix $\mathbf{K}_S$ becomes an identity matrix and the fading statistics is determined by the product of the two $n_R \times S$ and $S \times n_T$ complex Gaussian matrices $\mathbf{G}_R$ and $\mathbf{G}_T$. Each entry in the resulting matrix $\mathbf{H}$, being a sum of S independent random variables, according to the central limit theorem, is also a complex Gaussian matrix, if S is large. Thus the signal amplitudes undergo a Rayleigh fading distribution.

In the other extreme case, when the matrix $\mathbf{K}_S$ has a rank of one, the MIMO channel matrix entries are products of two independent complex Gaussian variables. Thus their amplitude distribution is the product of two independent Rayleigh distributions, each with the power of $2\sigma_r^2$, called the double Rayleigh distribution. The pdf for the double Rayleigh distribution is given by

$$f(z) = \int_0^\infty \frac{z}{w\sigma_r^4} e^{-\frac{w^4+z^2}{2w^2\sigma_r^2}}\, dw, \quad z \geq 0, \tag{1.142}$$

For the channel matrix ranks between one and the full rank, the fading distribution will range smoothly between Rayleigh and double Rayleigh distributions.

The probability density functions for single and double Rayleigh distributions are shown in Fig. 1.30.

The channel matrices, given by (1.141), are simulated in slow fading channels for various system parameters and the capacity is estimated by using (1.30). It is assumed in all simulations that the scattering radii are the same and equal to the distances between the antenna and the scatterers on both sides in order to maintain high local angle spreads and thus low antenna element correlations. It is assumed that the number of scatterers is high (32 in simulations). The capacity increases as the number of scatterers increases, but above a certain value its influence on capacity is negligible. Now we focus on examining the effect of the scattering radii and the distance between antennas on the keyhole effect. The capacity

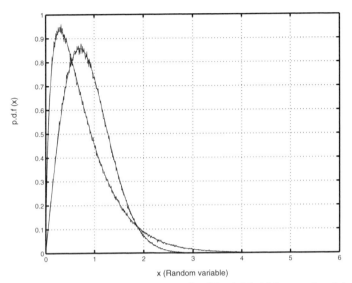

Figure 1.30 Probability density functions for normalized Rayleigh (right curve) and double Rayleigh distributions (left curve)

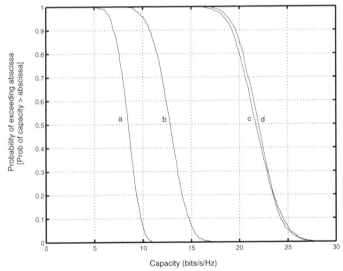

Figure 1.31 Capacity ccdf obtained for a MIMO slow fading channel with receive and transmit scatterers and SNR $= 20$ dB (a) $D_r = D_t = 50$ m, $R = 1000$ km, (b) $D_r = D_t = 50$ m, $R = 50$ km, (c) $D_r = D_t = 100$ m, $R = 5$ km, SNR $= 20$ dB; (d) Capacity ccdf curve obtained from (1.30) (without correlation or keyholes considered)

curves for various combination of system parameters in a MIMO channel with $n_R = n_T = 4$ are shown in Fig. 1.31. The first left curve corresponds to a low rank matrix, obtained for a low ratio of D_t/R, while the rightmost curve corresponds to a high rank channel matrix, in a system with a high D_t/R ratio.

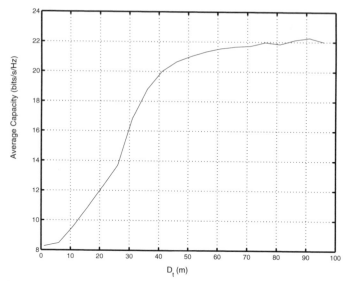

Figure 1.32 Average capacity on a fast MIMO fading channel for a fixed range of $R = 10$ km between scatterers, the distance between the receive antenna elements 3λ, the distance between the antennas and the scatterers $R_t = R_r = 50$ m, SNR $= 20$ dB and a variable scattering radius $D_t = D_r$

The average capacity increase in a fast fading channel, as the scattering radius D_t increases, while keeping the distance R constant, is shown in Fig. 1.32. For a distance of 10 km, 80% of the capacity is attained if the scattering radius increases to 35m.

Appendix 1.1 Water-filling Principle

Let us consider a MIMO channel where the channel parameters are known at the transmitter. The allocation of power to various transmitter antennas can be obtained by a "water-filling" principle. The "water-filling principle" can be derived by maximizing the MIMO channel capacity under the power constraint [20]

$$\sum_{i=1}^{n_T} P_i = P \quad i = 1, 2, \ldots, n_T \tag{1.143}$$

where P_i is the power allocated to antenna i and P is the total power, which is kept constant. The normalized capacity of the MIMO channel is determined as

$$C/W = \sum_{i=1}^{n_T} \log_2 \left[1 + \frac{P_i \lambda_i}{\sigma^2} \right] \tag{1.144}$$

Following the method of Lagrange multipliers, we introduce the function

$$Z = \sum_{i=1}^{n_T} \log_2 \left[1 + \frac{P_i \lambda_i}{\sigma^2} \right] + L \left(P - \sum_{i=1}^{n_T} P_i \right) \tag{1.145}$$

where L is the Lagrange multiplier, λ_i is the ith channel matrix singular value and σ^2 is the noise variance. The unknown transmit powers P_i are determined by setting the partial derivatives of Z to zero

$$\frac{\delta Z}{\delta P_i} = 0 \tag{1.146}$$

$$\frac{\delta Z}{\delta P_i} = \frac{1}{ln2} \frac{\lambda_i/\sigma^2}{1 + P_i\lambda_i/\sigma^2} - L = 0 \tag{1.147}$$

Thus we obtain for P_i

$$P_i = \mu - \frac{\sigma^2}{\lambda_i} \tag{1.148}$$

where μ is a constant, given by $1/Lln2$. It can be determined from the power constraint (1.143).

Appendix 1.2: Cholesky Decomposition

A symmetric and positive definite matrix can be decomposed into a lower and upper triangular matrix $A = LL^T$, where L (which can be seen as a square root of A) is a lower triangular matrix with positive diagonal elements. To solve $Ax = b$ one solves first $Ly = b$ and then $L^Tx = y$ for x.

$A = LL^T$

$$
\begin{bmatrix}
a_{11} & a_{12} & a_{1n} \\
a_{21} & \cdots & a_{2n} \\
\vdots & & \\
a_{n1} & a_{n2} & a_{nn}
\end{bmatrix}
=
\begin{bmatrix}
l_{11} & 0 & \cdots & 0 \\
l_{21} & l_{22} & \cdots & 0 \\
l_{31} & l_{32} & \vdots & 0 \\
\vdots & \vdots & \cdots & \vdots \\
l_{n1} & l_{n2} & \cdots & l_{nn}
\end{bmatrix}
\begin{bmatrix}
l_{11} & l_{21} & \cdots & l_{n1} \\
0 & l_{22} & \cdots & l_{n2} \\
0 & 0 & \vdots & l_{n3} \\
\vdots & \vdots & \cdots & \vdots \\
0 & 0 & \cdots & l_{nn}
\end{bmatrix}
$$

where a_{ij}, and l_{ij} are the entries of A and L, respectively.

$$a_{11} = l_{11}^2 \rightarrow l_{11} = \sqrt{a_{11}}$$

$$a_{21} = l_{21}l_{11} \rightarrow l_{21} = a_{21}/l_{11}, \ldots l_{n1} = a_{n1}/l_{11}$$

$$a_{22} = l_{21}^2 + l_{22}^2 \rightarrow l_{22} = \sqrt{(a_{22} - l_{21}^2)}$$

$$a_{32} = l_{31}l_{21} + l_{32}l_{22} \rightarrow l_{32} = (a_{32} - l_{31}l_{21})/l_{22}$$

In general, for $i = 1, 2, \ldots n, \ j = i + 1, \ldots n$

$$l_{ii} = \sqrt{a_{ii} - \sum_{k=1}^{i-1} l_{ik}^2}$$

$$l_{ji} = \left(a_{ji} - \sum_{k=1}^{i-1} l_{jk}l_{ik} \right) \Big/ l_{ii}$$

Because **A** is symmetric and positive, the expression under the square root is always positive.

Bibliography

[1] G.J. Foschini and M.J. Gans, "On limits of wireless communications in a fading environment when using multiple antennas", *Wireless Personal Communications*, vol. 6, 1998, pp. 311–335.

[2] E. Telatar, "Capacity of multi-antenna Gaussian channels", *European Transactions on Telecommunications*, vol. 10, no. 6, Nov./Dec. 1999, pp. 585–595.

[3] G.J. Foschini, "Layered space-time architecture for wireless communications in a fading environment when using multiple antennas", *Bell Labs. Tech. J.*, vol. 6, no. 2, pp. 41–59, 1996.

[4] C.E. Shannon, "A mathematical theory of communication", *Bell Syst. Tech. J.*, vol. 27, pp. 379–423 (Part one), pp. 623–656 (Part two), Oct. 1948, reprinted in book form, University of Illinois Press, Urbana, 1949.

[5] C. Berrou, A. Glavieux and P. Thitimajshima, "Near Shannon limit error-correcting coding and decoding: turbo codes", in *Proc. 1993 Inter. Conf Commun.*, 1993, pp. 1064–1070.

[6] R.G. Gallager, *Low Density Parity Check Codes*, MIT Press, Cambridge, Massachusets, 1963.

[7] D.C. MacKay, "Near Shannon limit performance of low density parity check codes", *Electronics Letters*, vol. 32, pp. 1645–1646, Aug. 1966.

[8] D. Chizhik, F. Rashid-Farrokhi, J. Ling and A. Lozano, "Effect of antenna separation on the capacity of BLAST in correlated channels", *IEEE Commun. Letters*, vol. 4, no. 11, Nov. 2000, pp. 337–339.

[9] A. Grant, S. Perreau, J. Choi and M. Navarro, "Improved radio access for cdma2000", *Technical Report A9.1*, July 1999.

[10] D. Chizhik, G. Foschini, M. Gans and R. Valenzuela, "Keyholes, correlations and capacities of multielement transmit and receive antennas", *Proc. Vehicular Technology Conf.*, VTC'2001, May 2001, Rhodes, Greece.

[11] R. Horn and C. Johnson, *Matrix Analysis*, Cambridge University Press, 1985.

[12] R. Galager, *Information Theory and Reliable Communication*, John Wiley and Sons, Inc., 1968.

[13] I.S. Gradshteyn and I.M. Ryzhik, *Table of Integrals, Series and Products*, New York, Academic Press, 1980.

[14] W. Jakes, *Microwave Mobile Communications*, IEEE Press, 1993.

[15] D. Gesbert, H. Boelcskei, D. Gore and A. Paulraj, "MIMO wireless channels: capacity and performance prediction", *Proc. Globecom'2000*, pp. 1083–1088, 2000.

[16] D.S. Shiu, G. Foschini, M. Gans and J. Kahn, "Fading correlation and effect on the capacity of multielement antenna systems", *IEEE Trans. Commun.*, vol. 48, no. 3, March 2000, pp. 502–512.

[17] P. Driessen and G. Foschini, "On the capacity for multiple input-multiple output wireless channels: a geometric interpretation", *IEEE Trans. Commun*, vol. 47, no. 2, Feb. 1999, pp. 173–176.

[18] A. Moustakas, H. Baranger, L. Balents, A. Sengupta and S. Simon, "Communication through a diffusive medium: coherence and capacity", *Science*, vol. 287, pp. 287–290, Jan. 2000.

[19] T.M. Cover and J.A. Thomas, *Elements of Information Theory* , New York, Wiley, 1991.

[20] N. Milosavljevic, Adaptive Space-Time Block Codes, Final Year Project, The University of Belgrade, 2002.

2

Space-Time Coding Performance Analysis and Code Design

2.1 Introduction

In Chapter 1, we showed that the information capacity of wireless communication systems can be increased considerably by employing multiple transmit and receive antennas. For a system with a large number of transmit and receive antennas and an independent flat fading channel known at the receivers, the capacity grows linearly with the minimum number of antennas.

An effective and practical way to approaching the capacity of multiple-input multiple-output (MIMO) wireless channels is to employ *space-time (ST) coding* [6]. Space-time coding is a coding technique designed for use with multiple transmit antennas. Coding is performed in both spatial and temporal domains to introduce correlation between signals transmitted from various antennas at various time periods. The spatial-temporal correlation is used to exploit the MIMO channel fading and minimize transmission errors at the receiver. Space-time coding can achieve transmit diversity and power gain over spatially uncoded systems without sacrificing the bandwidth. There are various approaches in coding structures, including space-time block codes (STBC), space-time trellis codes (STTC), space-time turbo trellis codes and layered space-time (LST) codes. A central issue in all these schemes is the exploitation of multipath effects in order to achieve high spectral efficiencies and performance gains. In this chapter, we start with a brief review of fading channel models and diversity techniques. Then, we proceed with the analysis of the performance of space-time codes on fading channels. The analytical pairwise error probability upper bounds over Rician and Rayleigh channels with independent fading are derived. They are followed by the presentation of the code design criteria on slow and fast Rayleigh fading channels.

Space-Time Coding Branka Vucetic and Jinhong Yuan
© 2003 John Wiley & Sons, Ltd ISBN: 0-470-84757-3

2.2 Fading Channel Models

2.2.1 Multipath Propagation

In a cellular mobile radio environment, the surrounding objects, such as houses, building or trees, act as reflectors of radio waves. These obstacles produce reflected waves with attenuated amplitudes and phases. If a modulated signal is transmitted, multiple reflected waves of the transmitted signal will arrive at the receiving antenna from different directions with different propagation delays. These reflected waves are called *multipath waves* [47]. Due to the different arrival angles and times, the multipath waves at the receiver site have different phases. When they are collected by the receiver antenna at any point in space, they may combine either in a constructive or a destructive way, depending on the random phases. The sum of these multipath components forms a spatially varying standing wave field. The mobile unit moving through the multipath field will receive a signal which can vary widely in amplitude and phase. When the mobile unit is stationary, the amplitude variations in the received signal are due to the movement of surrounding objects in the radio channel. The amplitude fluctuation of the received signal is called *signal fading*. It is caused by the time-variant multipath characteristics of the channel.

2.2.2 Doppler Shift

Due to the relative motion between the transmitter and the receiver, each multipath wave is subject to a shift in frequency. The frequency shift of the received signal caused by the relative motion is called the *Doppler shift*. It is proportional to the speed of the mobile unit. Consider a situation when only a single tone of frequency f_c is transmitted and a received signal consists of only one wave coming at an incident angle θ with respect to the direction of the vehicle motion. The Doppler shift of the received signal, denoted by f_d, is given by

$$f_d = \frac{vf_c}{c} \cos\theta \tag{2.1}$$

where v is the vehicle speed and c is the speed of light. The Doppler shift in a multipath propagation environment spreads the bandwidth of the multipath waves within the range of $f_c \pm f_{d_{max}}$, where $f_{d_{max}}$ is the maximum Doppler shift, given by

$$f_{d_{max}} = \frac{vf_c}{c} \tag{2.2}$$

The maximum Doppler shift is also referred as the maximum *fade rate*. As a result, a single tone transmitted gives rise to a received signal with a spectrum of nonzero width. This phenomenon is called *frequency dispersion* of the channel.

2.2.3 Statistical Models for Fading Channels

Because of the multiplicity of factors involved in propagation in a cellular mobile environment, it is convenient to apply statistical techniques to describe signal variations.

In a narrowband system, the transmitted signals usually occupy a bandwidth smaller than the channel's *coherence bandwidth*, which is defined as the frequency range over which the channel fading process is correlated. That is, all spectral components of the transmitted signal are subject to the same fading attenuation. This type of fading is referred to as *frequency nonselective* or *frequency flat*. On the other hand, if the transmitted signal bandwidth is greater than the channel coherence bandwidth, the spectral components of the transmitted signal with a frequency separation larger than the coherence bandwidth are faded independently. The received signal spectrum becomes distorted, since the relationships between various spectral components are not the same as in the transmitted signal. This phenomenon is known as *frequency selective* fading. In wideband systems, the transmitted signals usually undergo frequency selective fading.

In this section we introduce Rayleigh and Rician fading models to describe signal variations in a narrowband multipath environment. The frequency selective fading models for a wideband system are addressed in Chapter 8.

Rayleigh Fading

We consider the transmission of a single tone with a constant amplitude. In a typical land mobile radio channel, we may assume that the direct wave is obstructed and the mobile unit receives only reflected waves. When the number of reflected waves is large, according to the central limit theorem, two quadrature components of the received signal are uncorrelated Gaussian random processes with a zero mean and variance σ_s^2. As a result, the envelope of the received signal at any time instant undergoes a Rayleigh probability distribution and its phase obeys a uniform distribution between $-\pi$ and π. The probability density function (pdf) of the Rayleigh distribution is given by

$$p(a) = \begin{cases} \frac{a}{\sigma_s^2} \cdot e^{-a^2/2\sigma_s^2} & a \geq 0 \\ 0 & a < 0 \end{cases} \tag{2.3}$$

The mean value, denoted by m_a, and the variance, denoted by σ_a^2, of the Rayleigh distributed random variable are given by

$$\begin{aligned} m_a &= \sqrt{\frac{\pi}{2}} \cdot \sigma_s = 1.2533\sigma_s \\ \sigma_a^2 &= \left(2 - \frac{\pi}{2}\right)\sigma_s^2 = 0.4292\sigma_s^2 \end{aligned} \tag{2.4}$$

If the probability density function in (2.3) is normalized so that the average signal power $(E[a^2])$ is unity, then the normalized Rayleigh distribution becomes

$$p(a) = \begin{cases} 2ae^{-a^2} & a \geq 0 \\ 0 & a < 0 \end{cases} \tag{2.5}$$

The mean value and the variance are

$$\begin{aligned} m_a &= 0.8862 \\ \sigma_a^2 &= 0.2146 \end{aligned} \tag{2.6}$$

The pdf for a normalized Rayleigh distribution is shown in Fig. 2.1.

In fading channels with a maximum Doppler shift of $f_{d_{\max}}$, the received signal experiences a form of frequency spreading and is band-limited between $f_c \pm f_{d_{\max}}$. Assuming an omni-directional antenna with waves arriving in the horizontal plane, a large number of reflected waves and a uniform received power over incident angles, the power spectral density of the faded amplitude, denoted by $|P(f)|$, is given by

$$|P(f)| = \begin{cases} \dfrac{1}{2\pi\sqrt{f_{d_{\max}}^2 - f^2}} & \text{if } |f| \leq |f_{d_{\max}}| \\ 0 & \text{otherwise} \end{cases} \tag{2.7}$$

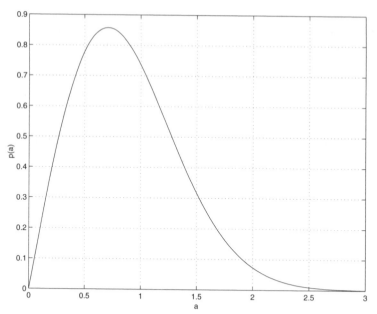

Figure 2.1 The pdf of Rayleigh distribution

where f is the frequency and $f_{d_{\max}}$ is the maximum fade rate. The value of $f_{d_{\max}} T_s$ is the maximum fade rate normalized by the symbol rate. It serves as a measure of the channel memory. For correlated fading channels this parameter is in the range $0 < f_{d_{\max}} T_s < 1$, indicating a finite channel memory. The autocorrelation function of the fading process is given by

$$R(\tau) = J_0 \left(2\pi f_{d_{\max}} \tau \right) \tag{2.8}$$

where $J_0(\cdot)$ is the zero-order Bessel function of the first kind.

Rician Fading

In some propagation scenarios, such as satellite or microcellular mobile radio channels, there are essentially no obstacles on the line-of-sight path. The received signal consists of a direct wave and a number of reflected waves. The direct wave is a stationary nonfading signal with a constant amplitude. The reflected waves are independent random signals. Their sum is called the *scattered component* of the received signal.

When the number of reflected waves is large, the quadrature components of the scattered signal can be characterized as a Gaussian random process with a zero mean and variance σ_s^2. The envelope of the scattered component has a Rayleigh probability distribution.

The sum of a constant amplitude direct signal and a Rayleigh distributed scattered signal results in a signal with a Rician envelope distribution. The pdf of the Rician distribution is given by

$$p(a) = \begin{cases} \dfrac{a}{\sigma_s^2} e^{-\frac{(a^2 + D^2)}{2\sigma_s^2}} I_0 \left(\dfrac{aD}{\sigma_s^2} \right) & a \geq 0 \\ 0 & a < 0 \end{cases} \tag{2.9}$$

where D^2 is the direct signal power and $I_0(\cdot)$ is the modified Bessel function of the first kind and zero-order.

Assuming that the total average signal power is normalized to unity, the pdf in (2.9) becomes

$$p(a) = \begin{cases} 2a(1+K)e^{-K-(1+K)a^2} I_0\left(2a\sqrt{K(K+1)}\right) & a \ge 0 \\ 0 & a < 0 \end{cases}$$

where K is the Rician factor, denoting the power ratio of the direct and the scattered signal components. The Rician factor is given by

$$K = \frac{D^2}{2\sigma_s^2} \tag{2.10}$$

The mean and the variance of the Rician distributed random variable are given by

$$\begin{aligned} m_a &= \tfrac{1}{2}\sqrt{\tfrac{\pi}{1+K}}e^{-\tfrac{K}{2}}\left[(1+K)I_0\left(\tfrac{K}{2}\right)+KI_1\left(\tfrac{K}{2}\right)\right] \\ \sigma_a^2 &= 1 - m_a^2 \end{aligned} \tag{2.11}$$

where $I_1(\cdot)$ is the first order modified Bessel function of the first kind. Small values of K indicate a severely faded channel. For $K = 0$, there is no direct signal component and the Rician pdf becomes a Rayleigh pdf. On the other hand, large values of K indicate a slightly faded channel. For K approaching infinity, there is no fading at all resulting in an AWGN channel. The Rician distributions with various K are shown in Fig. 2.2.

These two models can be applied to describe the received signal amplitude variations when the signal bandwidth is much smaller than the coherence bandwidth.

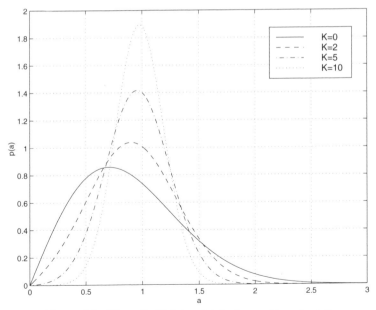

Figure 2.2   The pdf of Rician distributions with various K

2.3 Diversity

2.3.1 Diversity Techniques

In wireless mobile communications, diversity techniques are widely used to reduce the effects of multipath fading and improve the reliability of transmission without increasing the transmitted power or sacrificing the bandwidth [49] [48]. The diversity technique requires multiple replicas of the transmitted signals at the receiver, all carrying the same information but with small correlation in fading statistics. The basic idea of diversity is that, if two or more independent samples of a signal are taken, these samples will fade in an uncorrelated manner, e.g., some samples are severely faded while others are less attenuated. This means that the probability of all the samples being simultaneously below a given level is much lower than the probability of any individual sample being below that level. Thus, a proper combination of the various samples results in greatly reduced severity of fading, and correspondingly, improved reliability of transmission.

In most wireless communication systems a number of diversity methods are used in order to get the required performance. According to the domain where diversity is introduced, diversity techniques are classified into *time*, *frequency* and *space diversity*.

Time Diversity

Time diversity can be achieved by transmitting identical messages in different time slots, which results in uncorrelated fading signals at the receiver. The required time separation is at least the coherence time of the channel, or the reciprocal of the fading rate $1/f_d = c/vf_c$. The coherence time is a statistical measure of the period of time over which the channel fading process is correlated. Error control coding is regularly used in digital communication systems to provide a coding gain relative to uncoded systems. In mobile communications, error control coding is combined with interleaving to achieve time diversity. In this case, the replicas of the transmitted signals are usually provided to the receiver in the form of redundancy in the time domain introduced by error control coding [15]. The time separation between the replicas of the transmitted signals is provided by time interleaving to obtain independent fades at the input of the decoder. Since time interleaving results in decoding delays, this technique is usually effective for fast fading environments where the coherence time of the channel is small. For slow fading channels, a large interleaver can lead to a significant delay which is untolerable for delay sensitive applications such as voice transmission. This constraint rules out time diversity for some mobile radio systems. For example, when a mobile radio station is stationary, time diversity cannot help to reduce fades. One of the drawbacks of the scheme is that due to the redundancy introduced in the time domain, there is a loss in bandwidth efficiency.

Frequency Diversity

In frequency diversity, a number of different frequencies are used to transmit the same message. The frequencies need to be separated enough to ensure independent fading associated with each frequency. The frequency separation of the order of several times the channel coherence bandwidth will guarantee that the fading statistics for different frequencies are essentially uncorrelated. The coherence bandwidth is different for different propagation environments. In mobile communications, the replicas of the transmitted signals are usually provided to the receiver in the form of redundancy in the frequency domain introduced by spread spectrum such as direct sequence spread spectrum (DSSS), multicarrier modulation and frequency hopping. Spread spectrum techniques are effective when the coherence bandwidth of the channel is small. However, when the coherence bandwidth of the channel is larger than the spreading bandwidth, the multipath delay spread will be small relative to the symbol period. In this case, spread spectrum is ineffective to provide frequency diversity. Like time diversity, frequency diversity induces a loss in bandwidth efficiency due to a redundancy introduced in the frequency domain.

Space Diversity

Space diversity has been a popular technique in wireless microwave communications. Space diversity is also called *antenna diversity*. It is typically implemented using multiple antennas or antenna arrays arranged together in space for transmission and/or reception. The multiple antennas are separated physically by a proper distance so that the individual signals are uncorrelated. The separation requirements vary with antenna height, propagation environment and frequency. Typically a separation of a few wavelengths is enough to obtain uncorrelated signals. In space diversity, the replicas of the transmitted signals are usually provided to the receiver in the form of redundancy in the space domain. Unlike time and frequency diversity, space diversity does not induce any loss in bandwidth efficiency. This property is very attractive for future high data rate wireless communications.

Polarization diversity and *angle diversity* are two examples of space diversity. In polarization diversity, horizontal and vertical polarization signals are transmitted by two different polarized antennas and received by two different polarized antennas. Different polarizations ensure that the two signals are uncorrelated without having to place the two antennas far apart [15]. Angle diversity is usually applied for transmissions with carrier frequency larger than 10 GHz. In this case, as the transmitted signals are highly scattered in space, the received signals from different directions are independent to each other. Thus, two or more directional antennas can be pointed in different directions at the receiver site to provide uncorrelated replicas of the transmitted signals [52].

Depending on whether multiple antennas are used for transmission or reception, we can classify space diversity into two categories: *receive diversity* and *transmit diversity* [40]. In receive diversity, multiple antennas are used at the receiver site to pick up independent copies of the transmit signals. The replicas of the transmitted signals are properly combined to increase the overall received SNR and mitigate multipath fading. In transmit diversity, multiple antennas are deployed at the transmitter site. Messages are processed at the transmitter and then spread across multiple antennas. The details of transmit diversity is discussed in Section 2.3.3.

In practical communication systems, in order to meet the system performance requirements, two or more conventional diversity schemes are usually combined to provide *multidimensional diversity* [48]. For example, in GSM cellular systems multiple receive antennas at base stations are used in conjunction with interleaving and error control coding to simultaneously exploit both space and time diversity.

2.3.2   Diversity Combining Methods

In the previous section, diversity techniques were classified according to the domain where the diversity is introduced. The key feature of all diversity techniques is a low probability of simultaneous deep fades in various diversity subchannels. In general, the performance of communication systems with diversity techniques depends on how multiple signal replicas are combined at the receiver to increase the overall received SNR. Therefore, diversity schemes can also be classified according to the type of combining methods employed at the receiver. According to the implementation complexity and the level of channel state information required by the combining method at the receiver, there are four main types of combining techniques, including *selection combining*, *switched combining*, *equal-gain combining* (EGC) and *maximal ratio combining* (MRC) [48] [49].

Selection Combining

Selection combining is a simple diversity combining method. Consider a receive diversity system with n_R receive antennas. The block diagram of the selection combining scheme is shown in Fig. 2.3. In such a system, the signal with the largest instantaneous signal-to-noise ratio (SNR) at every symbol interval is selected as the output, so that the output SNR is equal to that of the best incoming signal. In practice, the signal with the highest sum of the signal and noise power $(S + N)$ is usually used, since it is difficult to measure the SNR.

Switched Combining

In a switched combining diversity system as shown in Fig. 2.4, the receiver scans all the diversity branches and selects a particular branch with the SNR above a certain predetermined threshold. This signal is selected as the output, until its SNR drops below the threshold. When this happens, the receiver starts scanning again and switches to another branch. This scheme is also called *scanning diversity*.

Compared to selection diversity, switched diversity is inferior since it does not continually pick up the best instantaneous signal. However, it is simpler to implement as it does not require simultaneous and continuous monitoring of all the diversity branches [50].

For both the selection and switched diversity schemes, the output signal is equal to only one of all the diversity branches. In addition, they do not require any knowledge of channel state information. Therefore, these two schemes can be used in conjunction with coherent as well as noncoherent modulations [48].

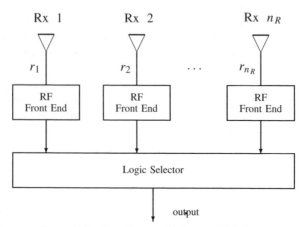

Figure 2.3   Selection combining method

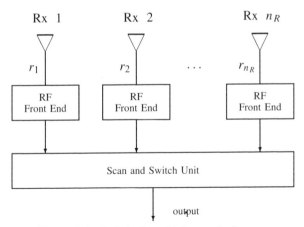

Figure 2.4   Switched combining method

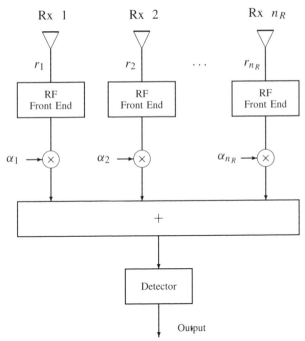

Figure 2.5   Maximum ratio combining method

Maximal Ratio Combining

Maximum ratio combining is a linear combining method. In a general linear combining process, various signal inputs are individually weighted and added together to get an output signal. The weighting factors can be chosen in several ways.

Figure 2.5 shows a block diagram of a maximum ratio combining diversity. The output signal is a linear combination of a weighted replica of all of the received signals. It is given by

$$r = \sum_{i=1}^{n_R} \alpha_i \cdot r_i \tag{2.12}$$

where r_i is the received signal at receive antenna i, and α_i is the weighting factor for receive antenna i. In maximum ratio combining, the weighting factor of each receive antenna is chosen to be in proportion to its own signal voltage to noise power ratio. Let A_i and ϕ_i be the amplitude and phase of the received signal r_i, respectively. Assuming that each receive antenna has the same average noise power, the weighting factor α_i can be represented as

$$\alpha_i = A_i e^{-j\phi_i} \tag{2.13}$$

This method is called optimum combining since it can maximize the output SNR. It is shown that the maximum output SNR is equal to the sum of the instantaneous SNRs of the individual signals [49].

In this scheme, each individual signal must be co-phased, weighted with its corresponding amplitude and then summed. This scheme requires the knowledge of channel fading amplitude and signal phases. So, it can be used in conjunction with coherent detection, but it is not practical for noncoherent detection [48].

Equal Gain Combining

Equal gain combining is a suboptimal but simple linear combining method. It does not require estimation of the fading amplitude for each individual branch. Instead, the receiver sets the amplitudes of the weighting factors to be unity.

$$\alpha_i = e^{-j\phi_i} \tag{2.14}$$

In this way all the received signals are co-phased and then added together with equal gain. The performance of equal-gain combining is only marginally inferior to maximum ratio combining. The implementation complexity for equal-gain combining is significantly less than the maximum ratio combining.

Example 2.1

In order to illustrate the effects of multipath fading on system error performance, we consider an uncoded BPSK system with and without multipath fading.

The bit error probability of BPSK signals on AWGN channels with coherent detection is given by [47]

$$P_b(e) = Q\left(\sqrt{\frac{2E_b}{N_0}}\right) \tag{2.15}$$

where $\frac{E_b}{N_0}$ is the ratio of the bit energy to the noise power spectral density.

In a fading channel, we assume that a fading coefficient is constant within each signalling interval so that coherent detection can be achieved. For a given fading attenuation a, the conditional bit error probability of coherent BPSK signals is given by

$$P_b(e|a) = Q\left(\sqrt{2\gamma_b}\right) \tag{2.16}$$

where $\gamma_b = a^2 \frac{E_b}{N_0}$ is the received SNR per bit. To obtain the average error probability when a is random, we need to average (2.16) over the probability density function of γ_b. Let us define the average SNR per bit as

$$\overline{\gamma}_b = E(a^2)\frac{E_b}{N_0} \tag{2.17}$$

where $E(\cdot)$ denotes the expectation operation. For a Rayleigh fading channel, the average bit error probability of BPSK signals is given by [47]

$$P_b(e) = \frac{1}{2}\left(1 - \sqrt{\frac{\overline{\gamma}_b}{1+\overline{\gamma}_b}}\right) \tag{2.18}$$

In order to compare the performance of the coherent BPSK signalling on channels with and without fading, we plot the bit error probabilities (2.15) and (2.18) in Fig. 2.6. From the figure, we can observe that the error rate decreases exponentially with the increasing SNR for a nonfading channel. However, for a Rayleigh fading channel, the error rate decreases inversely with the SNR. In order to achieve the same bit error rate of 10^{-4}, the required transmission power for a fading channel must increase by more than 25 dB relative to that for a nonfading channel due to the impairment of the multipath fading.

To show the effectiveness of the diversity techniques in combating the multipath fading, we consider an uncoded BPSK system with receive diversity on fading channels in the following example.

Let us assume that the receiver employs n_R receive antennas. The transmitted BPSK signals are received over n_R independent and identically distributed (i.i.d.) Rayleigh fading channels corrupted by AWGN. The n_R received signals are combined by using an MRC

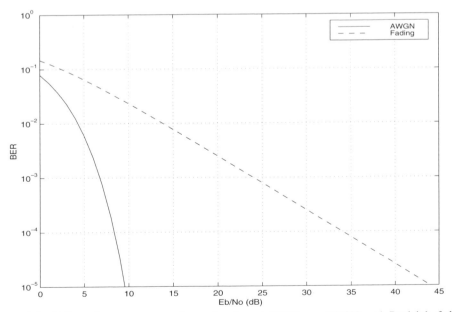

Figure 2.6 BER performance comparison of coherent BPSK on AWGN and Rayleigh fading channels

method. Let $\overline{\gamma}_k = E[\gamma_k]$ be the average value of SNR per bit on the k-th channel. For independent and identically distributed channels, $\overline{\gamma} = \overline{\gamma}_k$. The average SNR per bit after the MRC is $n_R\overline{\gamma}$. The average bit error probability of the coherent BPSK with n_R receive antennas and MRC diversity on Rayleigh i.i.d. fading channels is given by [47]

$$P_b(e) = \left[\frac{1}{2}\left(1 - \sqrt{\frac{\overline{\gamma}}{1+\overline{\gamma}}}\right)\right]^{n_R} \sum_{k=0}^{n_R-1} \binom{n_R-1+k}{k} \left[\frac{1}{2}\left(1 + \sqrt{\frac{\overline{\gamma}}{1+\overline{\gamma}}}\right)\right]^k \quad (2.19)$$

When the average SNR on each diversity channel is high, the above average bit error probability can be approximated as

$$P_b(e) \approx \left(\frac{1}{4\overline{\gamma}}\right)^{n_R} \binom{2n_R-1}{n_R} \quad (2.20)$$

The bit error rate curves for various numbers of receive antennas n_R are depicted in Fig. 2.7. The receive antenna diversity dramatically improves the error performance compared to the case without diversity ($n_R = 1$). In particular, we observe that the error probability decreases inversely with the n_R-th power of the SNR. For the same error rate of 10^{-4}, the MRC receive diversity technique reduces the transmission power by about 17 dB, 6 dB, 3 dB, 2 dB and 1.6 dB, when the number of receive antennas is increased from one to six successively.

2.3.3 Transmit Diversity

In present cellular mobile communications systems multiple receive antennas are used for the base stations with the aim to both suppress co-channel interference and minimize the

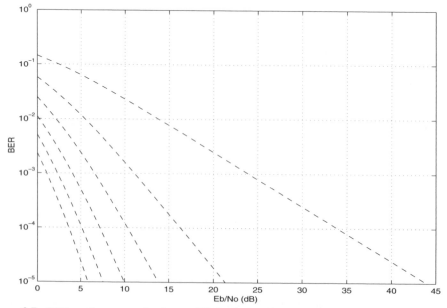

Figure 2.7 BER performance of coherent BPSK on Rayleigh fading channels with MRC receive diversity; the top curve corresponds to the performance without diversity; the other lower curves correspond to systems with 2, 3, 4, 5 and 6 receive antennas, respectively, starting from the top

fading effects. For example, in GSM and IS-136, multiple antennas are used at the base station to create uplink (from mobiles to base stations) receive diversity, compensating for the relatively low transmission power from the mobile. This improves the quality and range in the uplink. But for the downlink (from base stations to mobiles), it is difficult to utilize receive diversity at the mobile. Firstly, it is hard to place more than two antennas in a small-sized portable mobile. Secondly, multiple receive antennas imply multiple sets of RF down convertors and, as a result, more processing power, which is limited for mobile units. For the downlink, it is more practical to consider transmit diversity. It is easy to install multiple transmit antennas in the base station and provide the extra power for multiple transmissions. Transmit diversity decreases the required processing power of the receivers, resulting in a simpler system structure, lower power consumption and lower cost. Furthermore, transmit diversity can be combined with receive diversity to further improve the system performance.

In contrast to receiver diversity which is widely applied in cellular mobile systems, transmit diversity has received little attention as the behavior of transmit antenna diversity is dramatically different from that of receive antenna diversity and it is more difficult to exploit transmit diversity [15]. The difficulties mainly include that: (1) since the transmitted signals from multiple antennas are mixed spatially before they arrive at the receiver, some additional signal processing is required at both the transmitter and the receiver in order to separate the received signals and exploit diversity; and (2) unlike the receiver that can usually estimate fading channels, the transmitter does not have instantaneous information about the channel unless the information is fed back from the receiver to the transmitter [40].

Transmit diversity can increase the channel capacity considerably as has been shown in Chapter 1. A number of transmit diversity schemes have been proposed in the literature. These schemes can be divided into two categories: schemes with and without feedback. The difference between the two types of schemes is that the former relies on the channel information at the transmitter, which is obtained via feedback channels, while the latter does not require any channel information at the transmitter [6] [15] [40].

For transmit diversity systems with feedback, modulated signals are transmitted from multiple transmit antennas with different weighting factors. The weighting factors for the transmit antennas are chosen adaptively so that the received signal power or channel capacity is maximized. Switched diversity proposed in [16] is an example of such transmit diversity schemes. In practical cellular mobile systems, mobility and environment change cause fast channel variations, making channel estimation and tracking difficult. The imperfect channel estimation and mismatch between the previous channel state and current channel condition will decrease the received signal SNR and affect the system performance.

For transmit diversity schemes without feedback, messages to be transmitted are usually processed at the transmitter and then sent from multiple transmit antennas. Signal processing at the transmitter is designed appropriately to enable the receiver exploiting the embedded diversity from the received signals. At the receiver, messages are recovered by using a signal detection technique. A typical example is a delay diversity scheme [17] [18] [19]. In this scheme, copies of the same symbol are transmitted through multiple antennas in different times as shown in Fig. 2.8. At the receiver side, the delays of the second upto the n_T-th transmit antennas introduce a multipath-like distortion for the signal transmitted from the first antenna. The multipath distortion can be resolved or exploited at the receiver by using a maximum likelihood sequence estimator (MLSE) or a minimum mean square error (MMSE) equalizer to obtain a diversity gain. In some sense, the delay diversity is an optimal

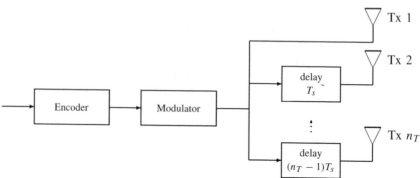

Figure 2.8 Delay transmit diversity scheme

transmit diversity scheme since it can achieve the maximum possible transmit diversity order determined by the number of transmit antennas without bandwidth expansion [20] [21].

Consider an n_T-transmit diversity system with a single receive antenna and no feedback. The average bit error probability of this scheme with BPSK modulation on Rayleigh i.i.d. fading channels is given by

$$
P_b(e) = \left[\frac{1}{2} \left(1 - \sqrt{\frac{\overline{\gamma}}{1 + \overline{\gamma}}} \right) \right]^{n_T} \sum_{k=0}^{n_T - 1} \left(\begin{array}{c} n_T - 1 + k \\ k \end{array} \right) \left[\frac{1}{2} \left(1 + \sqrt{\frac{\overline{\gamma}}{1 + \overline{\gamma}}} \right) \right]^k \quad (2.21)
$$

where

$$
\overline{\gamma} = \frac{E_b}{n_T N_0} \quad (2.22)
$$

and $\frac{E_b}{N_0}$ is the average bit energy to noise power spectral density ratio at the receive antenna.

In Fig. 2.9, we plot the bit error rate performance of the scheme against $\frac{E_b}{N_0}$ for various numbers of the transmit diversity n_T. From this figure, we can observe that at the BER of 10^{-4} the error performance is improved by about 14.5 dB, 4 dB and 2 dB, when the transmit diversity order is increased from one to two, two to three and three to four, respectively. However, the performance curves suggest that a further increase in the transmit diversity can only improve the performance by less than 1 dB. For a large number of diversity branches, the fading channel converges towards an AWGN channel, as the error performance curve for a large n_T almost approaches the one for the AWGN channel. It is important to mention that this feature plays an important role in deriving the space-time code design criteria, which are discussed later in this chapter.

In order to improve the error performance of the multiple antennas transmission, it is possible to combine error control coding with the transmit diversity design. Various schemes have been proposed to use error control coding in conjunction with multiple transmit antennas [22] [23]. Error control coding in combination with transmit diversity schemes can achieve a coding gain in addition to the diversity benefit, but suffers a loss in bandwidth due to code redundancy.

A better alternative is a joint design of error control coding, modulation and transmit diversity with no bandwidth expansion. This can be done by viewing coding, modulation and multiple transmission as one signal processing module. Coding techniques designed

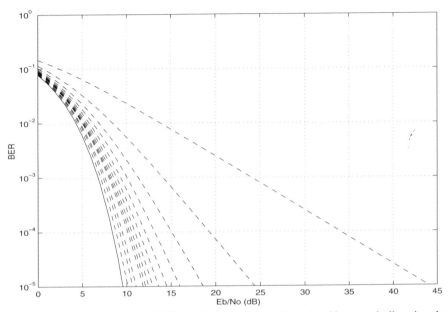

Figure 2.9 BER performance of BPSK on Rayleigh fading channels with transmit diversity; the top curve corresponds to the performance without diversity, and the bottom curve indicates the performance on AWGN channels; the curves in between correspond to systems with 2, 3, 4, 5, 6, 7, 8, 9, 10, 15, 20 and 40 transmit antennas, respectively, starting from the top

for multiple antenna transmission are called *space-time coding* [6]. In particular, coding is performed by adding properly designed redundancy in both spatial and temporal domains, which introduces correlation into the transmitted signals. Due to joint design, space-time codes can achieve transmit diversity as well as a coding gain without sacrificing bandwidth. Space-time codes can be further combined with multiple receive antennas to minimize the effects of multipath fading and to achieve the capacity of MIMO systems.

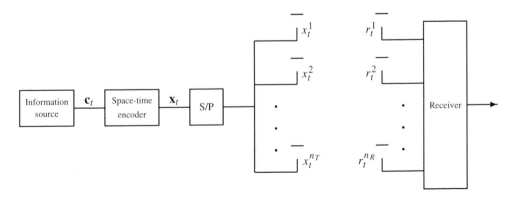

Figure 2.10 A baseband system model

2.4   Space-Time Coded Systems

We consider a baseband space-time coded communication system with n_T transmit antennas and n_R receive antennas, as shown in Fig. 2.10. The transmitted data are encoded by a space-time encoder. At each time instant t, a block of m binary information symbols, denoted by

$$\mathbf{c}_t = (c_t^1, c_t^2, \ldots, c_t^m) \tag{2.23}$$

is fed into the space-time encoder. The space-time encoder maps the block of m binary input data into n_T modulation symbols from a signal set of $M = 2^m$ points. The coded data are applied to a serial-to-parallel (S/P) converter producing a sequence of n_T parallel symbols, arranged into an $n_T \times 1$ column vector

$$\mathbf{x}_t = (x_t^1, x_t^2, \ldots, x_t^{n_T})^T, \tag{2.24}$$

where T means the transpose of a matrix. The n_T parallel outputs are simultaneously transmitted by n_T different antennas, whereby symbol x_t^i, $1 \leq i \leq n_T$, is transmitted by antenna i and all transmitted symbols have the same duration of $T sec$. The vector of coded modulation symbols from different antennas, as shown in (2.24), is called a *space-time symbol*. The spectral efficiency of the system is

$$\eta = \frac{r_b}{B} = m \quad \text{bits/sec/Hz} \tag{2.25}$$

where r_b is the data rate and B is the channel bandwidth. The spectral efficiency in (2.25) is equal to the spectral efficiency of a reference uncoded system with one transmit antenna.

The multiple antennas at both the transmitter and the receiver create a MIMO channel.

For wireless mobile communications, each link from a transmit antenna to a receive antenna can be modeled by flat fading, if we assume that the channel is memoryless. The MIMO channel with n_T transmit and n_R receive antennas can be represented by an $(n_R \times n_T)$ channel matrix $\mathbf{H}$. At time t, the channel matrix is given by

$$\mathbf{H}_t = \begin{bmatrix} h_{1,1}^t & h_{1,2}^t & \cdots & h_{1,n_T}^t \\ h_{2,1}^t & h_{2,2}^t & \cdots & h_{2,n_T}^t \\ \vdots & \vdots & \ddots & \vdots \\ h_{n_R,1}^t & h_{n_R,2}^t & \cdots & h_{n_R,n_T}^t \end{bmatrix} \tag{2.26}$$

where the ji-th element, denoted by $h_{j,i}^t$, is the fading attenuation coefficient for the path from transmit antenna i to receive antenna j.

In the analysis, we assume that the fading coefficients $h_{j,i}^t$ are independent complex Gaussian random variables with mean $\mu_h^{j,i}$ and variance $1/2$ per dimension, implying that the amplitude of the path coefficients are modeled as Rician fading. In terms of the coefficient variation speed, we consider fast and slow fading channels. For slow fading, it is assumed that the fading coefficients are constant during a frame and vary from one frame to another, which means that the symbol period is small compared to the channel coherence time. The slow fading is also referred to as *quasi-static fading* [6]. In a fast fading channel, the fading coefficients are constant within each symbol period and vary from one symbol to another.

At the receiver, the signal at each of the n_R receive antennas is a noisy superposition of the n_T transmitted signals degraded by channel fading. At time t, the received signal at antenna j, $j = 1, 2, \ldots, n_R$, denoted by r_t^j, is given by

$$r_t^j = \sum_{i=1}^{n_T} h_{j,i}^t x_t^i + n_t^j \tag{2.27}$$

where n_t^j is the noise component of receive antenna j at time t, which is an independent sample of the zero-mean complex Gaussian random variable with the one sided power spectral density of N_0.

Let us represent the received signals from n_R receive antennas at time t by an $n_R \times 1$ column vector.

$$\mathbf{r}_t = (r_t^1, r_t^2, \ldots, r_t^{n_R})^T \tag{2.28}$$

The noise at the receiver can be described by an $n_R \times 1$ column vector, denoted by $\mathbf{n}_t$

$$\mathbf{n}_t = (n_t^1, n_t^2, \ldots, n_t^{n_R})^T \tag{2.29}$$

where each component refers to a sample of the noise at a receive antenna. Thus, the received signal vector can be represented as

$$\mathbf{r}_t = \mathbf{H}_t \mathbf{x}_t + \mathbf{n}_t \tag{2.30}$$

We assume that the decoder at the receiver uses a maximum likelihood algorithm to estimate the transmitted information sequence and that the receiver has ideal channel state information (CSI) on the MIMO channel. On the other hand, the transmitter has no information about the channel. At the receiver, the decision metric is computed based on the squared Euclidean distance between the hypothesized received sequence and the actual received sequence as

$$\sum_{t} \sum_{j=1}^{n_R} \left| r_t^j - \sum_{i=1}^{n_T} h_{j,i}^t x_t^i \right|^2 \tag{2.31}$$

The decoder selects a codeword with the minimum decision metric as the decoded sequence.

2.5   Performance Analysis of Space-Time Codes

In the performance analysis we assume that the transmitted data frame length is L symbols for each antenna. We define an $n_T \times L$ space-time codeword matrix, obtained by arranging the transmitted sequence in an array, as

$$\mathbf{X} = [\mathbf{x}_1, \mathbf{x}_2, \ldots, \mathbf{x}_L] = \begin{bmatrix} x_1^1 & x_2^1 & \cdots & x_L^1 \\ x_1^2 & x_2^2 & \cdots & x_L^2 \\ \vdots & \vdots & \ddots & \vdots \\ x_1^{n_T} & x_2^{n_T} & \cdots & x_L^{n_T} \end{bmatrix} \tag{2.32}$$

where the i-th row $\mathbf{x}^i = [x_1^i, x_2^i, \ldots, x_L^i]$ is the data sequence transmitted from the i-th transmit antenna, and the t-th column $\mathbf{x}_t = [x_t^1, x_t^2, \ldots, x_t^{n_T}]^T$ is the space-time symbol at time t.

The pairwise error probability $P(\mathbf{X}, \hat{\mathbf{X}})$ is the probability that the decoder selects as its estimate an erroneous sequence $\hat{\mathbf{X}} = (\hat{\mathbf{X}}_1, \hat{\mathbf{X}}_2, \dots, \hat{\mathbf{X}}_L)$ when the transmitted sequence was in fact $\mathbf{X} = (\mathbf{x}_1, \mathbf{x}_2, \dots, \mathbf{x}_L)$. In maximum likelihood decoding, this occurs if

$$\sum_{t=1}^{L} \sum_{j=1}^{n_R} \left| r_t^j - \sum_{i=1}^{n_T} h_{j,i}^t x_t^i \right|^2 \geq \sum_{t=1}^{L} \sum_{j=1}^{n_R} \left| r_t^j - \sum_{i=1}^{n_T} h_{j,i}^t \hat{x}_t^i \right|^2 \tag{2.33}$$

The above inequality is equivalent to

$$\sum_{t=1}^{L} \sum_{j=1}^{n_R} 2Re \left\{ (n_t^j)^* \sum_{i=1}^{n_T} h_{j,i}^t (\hat{x}_t^i - x_t^i) \right\} \geq \sum_{t=1}^{L} \sum_{j=1}^{n_R} \left| \sum_{i=1}^{n_T} h_{j,i}^t (\hat{x}_t^i - x_t^i) \right|^2 \tag{2.34}$$

where $Re\{\cdot\}$ means the real part of a complex number.

Assuming that ideal CSI is available at the receiver, for a given realization of the fading variable matrix sequence $\mathbf{H} = (\mathbf{H}_1, \mathbf{H}_2, \dots, \mathbf{H}_L)$, the term on the right hand side of (2.34) is a constant equal to $d_h^2(\mathbf{X}, \hat{\mathbf{X}})$ and the term on the left hand side of (2.34) is a zero-mean Gaussian random variable. $d_h^2(\mathbf{X}, \hat{\mathbf{X}})$ is a modified Euclidean distance between the two space-time codeword matrices $\mathbf{X}$ and $\hat{\mathbf{X}}$, given by

$$d_h^2(\mathbf{X}, \hat{\mathbf{X}}) = \|\mathbf{H} \cdot (\hat{\mathbf{X}} - \mathbf{X})\|^2$$

$$= \sum_{t=1}^{L} \sum_{j=1}^{n_R} \left| \sum_{i=1}^{n_T} h_{j,i}^t (\hat{x}_t^i - x_t^i) \right|^2 \tag{2.35}$$

The pairwise error probability conditioned on $\mathbf{H}$ is given by

$$P(\mathbf{X}, \hat{\mathbf{X}}|\mathbf{H}) = Q\left(\sqrt{\frac{E_s}{2N_0} d_h^2(\mathbf{X}, \hat{\mathbf{X}})} \right) \tag{2.36}$$

where E_s is the energy per symbol at each transmit antenna and $Q(x)$ is the complementary error function defined by

$$Q(x) = \frac{1}{\sqrt{2\pi}} \int_x^\infty e^{-t^2/2} dt \tag{2.37}$$

By using the inequality

$$Q(x) \leq \frac{1}{2} e^{-x^2/2}, \quad x \geq 0 \tag{2.38}$$

the conditional pairwise error probability (2.36) can be upper bounded by

$$P(\mathbf{X}, \hat{\mathbf{X}}|\mathbf{H}) \leq \frac{1}{2} \exp\left(-d_h^2(\mathbf{X}, \hat{\mathbf{X}}) \frac{E_s}{4N_0} \right) \tag{2.39}$$

2.5.1 Error Probability on Slow Fading Channels

On slow fading channels, the fading coefficients within each frame are constant. So we can ignore the superscript of the fading coefficients

$$h_{j,i}^1 = h_{j,i}^2 = \cdots = h_{j,i}^L = h_{j,i}; \quad i = 1, 2, \dots, n_T, \quad j = 1, 2, \dots, n_R \tag{2.40}$$

Let us define a *codeword difference matrix* $\mathbf{B}(\mathbf{X}, \hat{\mathbf{X}})$ as

$$
\mathbf{B}(\mathbf{X}, \hat{\mathbf{X}}) = \mathbf{X} - \hat{\mathbf{X}}
$$

$$
= \begin{bmatrix}
x_1^1 - \hat{x}_1^1 & x_2^1 - \hat{x}_2^1 & \cdots & x_L^1 - \hat{x}_L^1 \\
x_1^2 - \hat{x}_1^2 & x_2^2 - \hat{x}_2^2 & \cdots & x_L^2 - \hat{x}_L^2 \\
\vdots & \vdots & \ddots & \vdots \\
x_1^{n_T} - \hat{x}_1^{n_T} & x_2^{n_T} - \hat{x}_2^{n_T} & \cdots & x_L^{n_T} - \hat{x}_L^{n_T}
\end{bmatrix} \tag{2.41}
$$

We can construct an $n_T \times n_T$ *codeword distance matrix* $\mathbf{A}(\mathbf{X}, \hat{\mathbf{X}})$, defined as

$$
\mathbf{A}(\mathbf{X}, \hat{\mathbf{X}}) = \mathbf{B}(\mathbf{X}, \hat{\mathbf{X}}) \cdot \mathbf{B}^H(\mathbf{X}, \hat{\mathbf{X}}) \tag{2.42}
$$

where H denotes the Hermitian (transpose conjugate) of a matrix. It is clear that $\mathbf{A}(\mathbf{X}, \hat{\mathbf{X}})$ is nonnegative definite Hermitian, as $\mathbf{A}(\mathbf{X}, \hat{\mathbf{X}}) = \mathbf{A}^H(\mathbf{X}, \hat{\mathbf{X}})$ and the eigenvalues of $\mathbf{A}(\mathbf{X}, \hat{\mathbf{X}})$ are nonnegative real numbers [53]. Therefore, there exist a unitary matrix $\mathbf{V}$ and a real diagonal matrix Δ such that

$$
\mathbf{V}\mathbf{A}(\mathbf{X}, \hat{\mathbf{X}})\mathbf{V}^H = \Delta \tag{2.43}
$$

The rows of $\mathbf{V}$, $\{\mathbf{v}_1, \mathbf{v}_2, \ldots, \mathbf{v}_{n_T}\}$, are the eigenvectors of $\mathbf{A}(\mathbf{X}, \hat{\mathbf{X}})$, forming a complete orthonormal basis of an N-dimensional vector space. The diagonal elements of Δ are the eigenvalues $\lambda_i \geq 0$, $i = 1, 2, \ldots, n_T$, of $\mathbf{A}(\mathbf{X}, \hat{\mathbf{X}})$. The diagonal matrix Δ can be represented as

$$
\Delta = \begin{bmatrix}
\lambda_1 & 0 & \cdots & 0 \\
0 & \lambda_2 & \cdots & 0 \\
\vdots & \vdots & \ddots & \vdots \\
0 & 0 & \cdots & \lambda_{n_T}
\end{bmatrix} \tag{2.44}
$$

For simplicity, we assume that $\lambda_1 \geq \lambda_2 \geq \cdots \geq \lambda_{n_T} \geq 0$.

Next, let

$$
\mathbf{h}_j = (h_{j,1}, h_{j,2}, \ldots, h_{j,n_T}) \tag{2.45}
$$

Equation (2.35) can be rewritten as

$$
d_h^2(\mathbf{X}, \hat{\mathbf{X}}) = \sum_{j=1}^{n_R} \mathbf{h}_j \mathbf{A}(\mathbf{X}, \hat{\mathbf{X}}) \mathbf{h}_j{}^H
$$

$$
= \sum_{j=1}^{n_R} \sum_{i=1}^{n_T} \lambda_i |\beta_{j,i}|^2 \tag{2.46}
$$

where

$$
\beta_{j,i} = \mathbf{h}_j \cdot \mathbf{v}_i \tag{2.47}
$$

and $\cdot$ denotes the inner product of complex vectors. Substituting (2.46) into (2.39), we get

$$P(\mathbf{X}, \hat{\mathbf{X}}|\mathbf{H}) \leq \frac{1}{2} \exp\left(-\sum_{j=1}^{n_R}\sum_{i=1}^{n_T} \lambda_i |\beta_{j,i}|^2 \frac{E_s}{4N_0}\right) \tag{2.48}$$

Inequality (2.48) is an upper bound on the conditional pairwise error probability expressed as a function of $|\beta_{j,i}|$, which is contingent upon $h_{j,i}$. Next, we determine the distribution of $|\beta_{j,i}|$ assuming the knowledge of $h_{j,i}$.

Note that $h_{j,i}$ are complex Gaussian random variables with mean $\mu_h^{j,i}$ and variance $1/2$ per dimension and $\{\mathbf{v}_1, \mathbf{v}_2, \ldots, \mathbf{v}_{n_T}\}$ is an orthonormal basis of an N-dimensional vector space. It is obvious that $\beta_{j,i}$ in (2.47) are independent complex Gaussian random variables with variance $1/2$ per dimension and mean $\mu_\beta^{j,i}$, where

$$\mu_\beta^{j,i} = E[\mathbf{h}_j] \cdot E[\mathbf{v}_i]$$

$$= [\mu_h^{j,1}, \mu_h^{j,2}, \ldots, \mu_h^{j,n_T}] \cdot \mathbf{v}_i \tag{2.49}$$

where $E[\cdot]$ denotes the expectation. Let $K^{j,i} = |\mu_\beta^{j,i}|^2$, then $|\beta_{j,i}|$ has a Rician distribution with the probability density function (pdf) [6]

$$p(|\beta_{j,i}|) = 2|\beta_{j,i}| \exp(-|\beta_{j,i}|^2 - K^{j,i}) I_0(2|\beta_{j,i}|\sqrt{K^{j,i}}) \tag{2.50}$$

$I_0(\cdot)$ is the zero-order modified Bessel function of the first kind.

In order to get an upper bound on the unconditional pairwise error probability, we need to average (2.48) with respect to the Rician random variables $|\beta_{j,i}|$. Let r denote the rank of the matrix $\mathbf{A}(\mathbf{X}, \hat{\mathbf{X}})$. In this analysis, we distinguish two cases, depending on the value of rn_R.

The Pairwise Error Probability Upper Bound for Large Values of rn_R

Since $|\beta_{j,i}|$ has a Rician distribution, $|\beta_{j,i}|^2$ has a noncentral chi-square distribution with 2 degrees of freedom and noncentrality parameter $S = |\mu_\beta^{j,i}|^2 = K^{j,i}$. The mean value and the variance of the noncentral chi-square-distributed random variables $|\beta_{j,i}|^2$ are given by

$$\mu_{|\beta_{j,i}|^2} = 1 + K^{j,i} \tag{2.51}$$

and

$$\sigma_{|\beta_{j,i}|^2}^2 = 1 + 2K^{j,i}, \tag{2.52}$$

respectively.

At the right hand side of inequality (2.48), there are rn_R independent chi-square-distributed random variables. For a large value of rn_R ($rn_R \geq 4$), which corresponds to a large number of independent subchannels, according to the central limit theorem [54], the expression

$$\sum_{j=1}^{n_R}\sum_{i=1}^{n_T} \lambda_i |\beta_{j,i}|^2 \tag{2.53}$$

approaches a Gaussian random variable D with the mean

$$\mu_D = \sum_{j=1}^{n_R} \sum_{i=1}^{n_T} \lambda_i (1 + K^{j,i}) \tag{2.54}$$

and the variance

$$\sigma_D^2 = \sum_{j=1}^{n_R} \sum_{i=1}^{n_T} \lambda_i^2 (1 + 2K^{j,i}) \tag{2.55}$$

The pairwise error probability can then be upper-bounded by

$$P(\mathbf{X}, \hat{\mathbf{X}}) \le \int_{D=0}^{+\infty} \frac{1}{2} \exp\left(-\frac{E_s}{4N_0} D\right) p(D) dD \tag{2.56}$$

where $p(D)$ is the pdf of the Gaussian random variable D. Using the equation

$$\int_{D=0}^{+\infty} \exp(-\gamma D) p(D) dD = \exp\left(\frac{1}{2}\gamma^2 \sigma_D^2 - \gamma \mu_D\right) Q\left(\frac{\gamma \sigma_D^2 - \mu_D}{\sigma_D}\right), \quad \gamma > 0 \tag{2.57}$$

the upper bound in (2.56) can be further expressed as

$$P(\mathbf{X}, \hat{\mathbf{X}}) \le \frac{1}{2} \exp\left(\frac{1}{2}\left(\frac{E_s}{4N_0}\right)^2 \sigma_D^2 - \frac{E_s}{4N_0}\mu_D\right) Q\left(\frac{E_s}{4N_0}\sigma_D - \frac{\mu_D}{\sigma_D}\right) \tag{2.58}$$

Let us now consider the special case of Rayleigh fading. In this case, $\mu_a^{j,i} = 0$ and thus $K^{j,i} = 0$. The mean and the variance of the Gaussian random variable D become

$$\mu_D = n_R \sum_{i=1}^{r} \lambda_i \tag{2.59}$$

and

$$\sigma_D^2 = n_R \sum_{i=1}^{r} \lambda_i^2, \tag{2.60}$$

respectively. Substituting (2.59) and (2.60) into (2.58), we obtain the pairwise error probability upper bound on Rayleigh fading channels as

$$P(\mathbf{X}, \hat{\mathbf{X}}) \le \frac{1}{2} \exp\left(\frac{1}{2}\left(\frac{E_s}{4N_0}\right)^2 n_R \sum_{i=1}^{r} \lambda_i^2 - \frac{E_s}{4N_0} n_R \sum_{i=1}^{r} \lambda_i\right)$$

$$\cdot Q\left(\frac{E_s}{4N_0}\sqrt{n_R \sum_{i=1}^{r} \lambda_i^2} - \frac{\sqrt{n_R} \sum_{i=1}^{r} \lambda_i}{\sqrt{\sum_{i=1}^{r} \lambda_i^2}}\right) \tag{2.61}$$

The Pairwise Error Probability Upper Bound for Small Values of rn_R

When the number of independent subchannels rn_R is small, e.g. $rn_R < 4$, the Gaussian assumption is no longer valid and the pairwise error probability can be expressed as

$$P(\mathbf{X}, \hat{\mathbf{X}}) \leq \int \cdots \int_{|\beta_{j,i}|=0}^{\infty} P(\mathbf{X}, \hat{\mathbf{X}}|\mathbf{H}) p(|\beta_{1,1}|) p(|\beta_{1,2}|) \cdots p(|\beta_{n_R,n_T}|)$$

$$\cdot\, d|\beta_{1,1}| d|\beta_{1,2}| \cdots d|\beta_{n_R,n_T}| \tag{2.62}$$

where $|\beta_{j,i}|$ are independent Rician-distributed random variables with pdf in (2.50). This expression can be integrated analytically term by term. By substituting (2.50) into (2.62), the pairwise error probability is upper bounded by [6]

$$P(\mathbf{X}, \hat{\mathbf{X}}) \leq \prod_{j=1}^{n_R} \left(\prod_{i=1}^{n_T} \frac{1}{1 + \frac{E_s}{4N_0}\lambda_i} \exp\left(-\frac{K^{j,i} \frac{E_s}{4N_0}\lambda_i}{1 + \frac{E_s}{4N_0}\lambda_i} \right) \right) \tag{2.63}$$

In the case of Rayleigh fading, the upper bound of the pairwise error probability becomes [6]

$$P(\mathbf{X}, \hat{\mathbf{X}}) \leq \left(\prod_{i=1}^{n_T} \frac{1}{1 + \frac{E_s}{4N_0}\lambda_i} \right)^{n_R} \tag{2.64}$$

At high SNR's, the above upper bound can be simplified as [6]

$$P(\mathbf{X}, \hat{\mathbf{X}}) \leq \left(\prod_{i=1}^{r} \lambda_i \right)^{-n_R} \left(\frac{E_s}{4N_0} \right)^{-rn_R} \tag{2.65}$$

where r denotes the rank of matrix $\mathbf{A}(\mathbf{X}, \hat{\mathbf{X}})$, and $\lambda_1, \lambda_2, \ldots, \lambda_r$ are the nonzero eigenvalues of matrix $\mathbf{A}(\mathbf{X}, \hat{\mathbf{X}})$.

Using a union bound technique, we can compute an upper bound of the code frame error probability, which sums the contributions of the pairwise error probabilities over all error events. Note that the pairwise error probability in (2.65) decreases exponentially with the increasing SNR. The frame error probability at high SNR's is dominated by the pairwise error probability with the minimum product rn_R. The exponent of the SNR term, rn_R, is called the *diversity gain* and

$$G_c = \frac{(\lambda_1\lambda_2 \ldots \lambda_r)^{1/r}}{d_u^2} \tag{2.66}$$

is called the *coding gain*, where d_u^2 is the squared Euclidean distance of the reference uncoded system. Note that both diversity and coding gains are obtained as the minimum rn_R and $(\lambda_1\lambda_2 \ldots \lambda_r)^{1/r}$ over all pairs of distinct codewords. The diversity gain is an approximate measure of the power gain of the system with space diversity over the system without diversity at the same error probability. The coding gain measures the power gain of the coded system over an uncoded system with the same diversity gain, at the same error probability. The diversity gain determines the slope of an error rate curve plotted as a function of SNR, while the coding gain determines the horizontal shift of the uncoded system error rate curve to the space-time coded error rate curve obtained for the same diversity order.

In general, to minimize the error probability, it is desirable to make both diversity and coding gain as large as possible. Since the diversity gain is an exponent in the error probability upper bound (2.65), it is clear that achieving a large diversity gain is more important than achieving a high coding gain for systems with a small value of rn_R.

Example 2.2: A Time-Switched Space-Time Code

Let us consider a time-switched space-time code (TS STC)

$$\mathbf{X} = \begin{bmatrix} x_t & 0 \\ 0 & x_t \end{bmatrix} \tag{2.67}$$

where 0 means erasure. In this scheme, only one antenna is active in each time slot and the modulated symbol x_t is transmitted from antenna one and two at time $2t$ and $2t + 1$, respectively. Since each modulated symbol is transmitted in two time slots, the code rate is 1/2. Let

$$\hat{\mathbf{X}} = \begin{bmatrix} \hat{x}_t & 0 \\ 0 & \hat{x}_t \end{bmatrix} \tag{2.68}$$

be another codeword where $x_t \neq \hat{x}_t$. The codeword difference matrix between the codewords is

$$\mathbf{B}(\mathbf{X}, \hat{\mathbf{X}}) = \begin{bmatrix} x_t - \hat{x}_t & 0 \\ 0 & x_t - \hat{x}_t \end{bmatrix} \tag{2.69}$$

Since $x_t \neq \hat{x}_t$, the rank of matrix $\mathbf{B}(\mathbf{X}, \hat{\mathbf{X}})$ is $r = 2$. Note that matrices $\mathbf{A}(\mathbf{X}, \hat{\mathbf{X}})$ and $\mathbf{B}(\mathbf{X}, \hat{\mathbf{X}})$ have the same rank, as $\mathbf{A}(\mathbf{X}, \hat{\mathbf{X}}) = \mathbf{B}(\mathbf{X}, \hat{\mathbf{X}}) \cdot \mathbf{B}^H(\mathbf{X}, \hat{\mathbf{X}})$. This scheme achieves a diversity gain of two, if the number of receive antennas is one.

For a single receive antenna, the received signals at times $2t$ and $2t + 1$, denoted by r_t^1 and r_t^2, respectively, are given by

$$r_t^1 = h_t^1 x_t + n_t^1$$
$$r_t^2 = h_t^2 x_t + n_t^2 \tag{2.70}$$

where h_t^1 and h_t^2 are the channel fading coefficients between transmit antenna one and two and the receive antenna, respectively, and n_t^1 and n_t^2 are the noise samples at times $2t$ and $2t + 1$, respectively. Let us assume that the fading coefficients are perfectly known at the receiver. The maximum likelihood decoder chooses a signal x_t' from the modulation signal set to minimize the squared Euclidean distance metric

$$d^2(r_t^1, h_t^1 x_t') + d^2(r_t^2, h_t^2 x_t') \tag{2.71}$$

Based on the MRC method, the receiver constructs a decision statistics signal, denoted by $\tilde{x}_t$, as

$$\tilde{x}_t = (h_t^1)^* r_t^1 + (h_t^2)^* r_t^2$$
$$= \left(|h_t^1|^2 + |h_t^2|^2 \right) x_t + (h_t^1)^* n_t^1 + (h_t^2)^* n_t^2 \tag{2.72}$$

The maximum likelihood decoding rule can be rewritten as

$$x_t' = \arg\min \left(|h_t^1|^2 + |h_t^2|^2 - 1 \right) + d^2(\tilde{x}_t, x_t') \tag{2.73}$$

The time-switched space-time code with a single receive antenna can achieve the same diversity gain of two as a two-branch MRC receive diversity scheme. However, the time-switched space-time code is of half rate and there is a 3 dB performance penalty relative to the MRC receive diversity scheme.

Example 2.3: Repetition Code

We consider a space-time code by transmitting the same modulated symbols from two antennas. The space-time codeword matrix is given by

$$\mathbf{X} = \begin{bmatrix} x_t \\ x_t \end{bmatrix} \tag{2.74}$$

The code rate is one. For any two distinct codewords with $x_t \neq \hat{x}_t$, the codeword difference matrix is given by

$$\mathbf{B}(\mathbf{X}, \hat{\mathbf{X}}) = \begin{bmatrix} x_t - \hat{x}_t \\ x_t - \hat{x}_t \end{bmatrix} \tag{2.75}$$

Obviously, the rank of the matrix is only one. Therefore, the repetition code has the same performance as a no diversity scheme ($n_T = n_R = 1$).

2.5.2 Error Probability on Fast Fading Channels

The analysis for slow fading channels in the previous section can be directly applied to fast fading channels.

At each time t, we define a *space-time symbol difference vector* $\mathbf{F}(\mathbf{x}_t, \hat{\mathbf{x}}_t)$ as

$$\mathbf{F}(\mathbf{x}_t, \hat{\mathbf{x}}_t) = [x_t^1 - \hat{x}_t^1, x_t^2 - \hat{x}_t^2, \ldots, x_t^{n_T} - \hat{x}_t^{n_T}]^T \tag{2.76}$$

Let us consider an $n_T \times n_T$ matrix $\mathbf{C}(\mathbf{x}_t, \hat{\mathbf{x}}_t)$, defined as

$$\mathbf{C}(\mathbf{x}_t, \hat{\mathbf{x}}_t) = \mathbf{F}(\mathbf{x}_t, \hat{\mathbf{x}}_t) \cdot \mathbf{F}^H(\mathbf{x}_t, \hat{\mathbf{x}}_t) \tag{2.77}$$

It is clear that the matrix $\mathbf{C}(\mathbf{x}_t, \hat{\mathbf{x}}_t)$ is Hermitian. Therefore, there exist a unitary matrix $\mathbf{V}_t$ and a real diagonal matrix $\mathbf{D}_t$, such that

$$\mathbf{V}_t \cdot \mathbf{C}(\mathbf{x}_t, \hat{\mathbf{x}}_t) \cdot \mathbf{V}_t^H = \mathbf{D}_t \tag{2.78}$$

The diagonal elements of $\mathbf{D}_t$ are the eigenvalues, D_t^i, $i = 1, 2, \ldots, n_T$, and the rows of $\mathbf{V}_t$, $\{\mathbf{v}_t^1, \mathbf{v}_t^2, \ldots, \mathbf{v}_t^{n_T}\}$, are the eigenvectors of $\mathbf{C}(\mathbf{x}_t, \hat{\mathbf{x}}_t)$, forming a complete orthonormal basis of an N-dimensional vector space.

In the case that $\mathbf{x}_t = \hat{\mathbf{x}}_t$, $\mathbf{C}(\mathbf{x}_t, \hat{\mathbf{x}}_t)$ is an all-zero matrix and all the eigenvalues D_t^i, $i = 1, 2, \ldots, n_T$, are zero. On the other hand, if $\mathbf{x}_t \neq \hat{\mathbf{x}}_t$, the matrix $\mathbf{C}(\mathbf{x}_t, \hat{\mathbf{x}}_t)$ has only one nonzero eigenvalue and the other $n_T - 1$ eigenvalues are zero. Let D_t^1 be the nonzero eigenvalue element which is equal to the squared Euclidean distance between the two space-time symbols $\mathbf{x}_t$ and $\hat{\mathbf{x}}_t$.

$$D_t^1 = |\mathbf{x}_t - \hat{\mathbf{x}}_t|^2 = \sum_{i=1}^{n_T} |x_t^i - \hat{x}_t^i|^2 \tag{2.79}$$

The eigenvector of $\mathbf{C}(\mathbf{x}_t, \hat{\mathbf{x}}_t)$ corresponding to the nonzero eigenvalue D_t^1 is denoted by $\mathbf{v}_t^1$.
Let us define $\mathbf{h}_t^j$ as

$$\mathbf{h}_t^j = (h_{j,1}^t, h_{j,2}^t, \ldots, h_{j,n_T}^t) \tag{2.80}$$

Equation (2.35) can be rewritten as

$$d_h^2(\mathbf{X}, \hat{\mathbf{X}}) = \sum_{t=1}^{L} \sum_{j=1}^{n_R} \sum_{i=1}^{n_T} |\beta_{j,i}^t|^2 \cdot D_t^i \tag{2.81}$$

where

$$\beta_{j,i}^t = \mathbf{h}_t^j \cdot \mathbf{v}_t^i \tag{2.82}$$

Since at each time t there is at most only one nonzero eigenvalue, D_t^1, the expression (2.81) can be represented by

$$d_h^2(\mathbf{X}, \hat{\mathbf{X}}) = \sum_{t \in \rho(\mathbf{x}, \hat{\mathbf{x}})} \sum_{j=1}^{n_R} |\beta_{j,1}^t|^2 \cdot D_t^1$$

$$= \sum_{t \in \rho(\mathbf{x}, \hat{\mathbf{x}})} \sum_{j=1}^{n_R} |\beta_{j,1}^t|^2 \cdot |\mathbf{x}_t - \hat{\mathbf{X}}_t|^2 \tag{2.83}$$

where $\rho(\mathbf{x}, \hat{\mathbf{x}})$ denotes the set of time instances $t = 1, 2, \ldots, L$, such that $|\mathbf{x}_t - \hat{\mathbf{x}}_t| \neq 0$.
Substituting (2.83) into (2.39), we get

$$P(\mathbf{X}, \hat{\mathbf{X}}|\mathbf{H}) \leq \frac{1}{2} \exp\left(- \sum_{t \in \rho(\mathbf{x}, \hat{\mathbf{x}})} \sum_{j=1}^{n_R} |\beta_{j,1}^t|^2 |\mathbf{x}_t - \hat{\mathbf{x}}_t|^2 \frac{E_s}{4N_0} \right) \tag{2.84}$$

Comparing (2.82) with (2.47), it is obvious that $\beta_{j,1}^t$ are also independent complex Gaussian random variables with variance 1/2 per dimension and $|\beta_{j,1}^t|$ follows a Rician distribution with the pdf

$$p(|\beta_{j,1}^t|) = 2|\beta_{j,1}^t| \exp\left(-|\beta_{j,1}^t|^2 - K_t^{j,1}\right) I_0\left(2|\beta_{j,1}^t|\sqrt{K_t^{j,1}}\right) \tag{2.85}$$

where

$$K_t^{j,1} = \left| [\mu_{h_t}^{j,1}, \mu_{h_t}^{j,2}, \ldots, \mu_{h_t}^{j,n_T}] \cdot \mathbf{v}_t^1 \right|^2 \tag{2.86}$$

The conditional pairwise error probability upper bound (2.84) can be averaged over independent Rician-distributed variables $|\beta_{j,1}^t|$. If we define δ_H as the number of space-time symbols in which the two codewords $\mathbf{X}$ and $\hat{\mathbf{X}}$ differ, then at the right hand side of inequality (2.84), there are $\delta_H n_R$ independent random variables. As before, we will distinguish two cases in the analysis, depending on the value of $\delta_H n_R$. The term δ_H is also called the *space-time symbol-wise Hamming distance* between the two codewords.

The Pairwise Error Probability Upper Bound for Large $\delta_H n_R$

Provided that the value of $\delta_H n_R$ for a given code is large, e.g., $\delta_H n_R \geq 4$, according to the central limit theorem, the expression $d_h^2(\mathbf{X}, \hat{\mathbf{X}})$ in (2.83) can be approximated by a Gaussian random variable with the mean

$$\mu_d = \sum_{t \in \rho(\mathbf{x},\hat{\mathbf{x}})} \sum_{j=1}^{n_R} |\mathbf{x}_t - \hat{\mathbf{x}}_t|^2 (1 + K_t^{j,1}) \tag{2.87}$$

and the variance

$$\sigma_d^2 = \sum_{t \in \rho(\mathbf{x},\hat{\mathbf{x}})} \sum_{j=1}^{n_R} |\mathbf{x}_t - \hat{\mathbf{x}}_t|^4 (1 + 2K_t^{j,1}) \tag{2.88}$$

By averaging (2.84) over the Gaussian random variable and using Eq. (2.57), the pairwise error probability can be upper-bounded by

$$P(\mathbf{X}, \hat{\mathbf{X}}) \leq \frac{1}{2} \exp\left(\frac{1}{2}\left(\frac{E_s}{4N_0}\right)^2 \sigma_d^2 - \frac{E_s}{4N_0}\mu_d \right) Q\left(\frac{E_s}{4N_0}\sigma_d - \frac{\mu_d}{\sigma_d} \right) \tag{2.89}$$

For Rayleigh fading channels, the pairwise error probability upper bound can be approximated by

$$P(\mathbf{X}, \hat{\mathbf{X}}) \leq \frac{1}{2} \exp\left(\frac{1}{2}\left(\frac{E_s}{4N_0}\right)^2 n_R D^4 - \frac{E_s}{4N_0}n_R d_E^2 \right) Q\left(\frac{E_s}{4N_0}\sqrt{n_R D^4} - \frac{\sqrt{n_R}d_E^2}{\sqrt{D^4}} \right) \tag{2.90}$$

where d_E^2 is the accumulated squared Euclidean distance between the two space-time symbol sequences, given by

$$d_E^2 = \sum_{t \in \rho(\mathbf{x},\hat{\mathbf{x}})} |\mathbf{x}_t - \hat{\mathbf{x}}_t|^2 \tag{2.91}$$

and D^4 is defined as

$$D^4 = \sum_{t \in \rho(\mathbf{x},\hat{\mathbf{x}})} |\mathbf{x}_t - \hat{\mathbf{x}}_t|^4 \tag{2.92}$$

The Pairwise Error Probability Upper Bound for Small $\delta_H n_R$

When the value of $\delta_H n_R$ is small, e.g., $\delta_H n_R < 4$, the central limit theorem argument is not valid and the average pairwise error probability can be expressed as

$$P(\mathbf{X}, \hat{\mathbf{X}}) \leq \int \cdots \int_{|\beta_{j,1}^t|=0}^{\infty} P(\mathbf{X}, \hat{\mathbf{X}}|\mathbf{H}) p(|\beta_{1,1}^1|) p(|\beta_{2,1}^1|) \cdots p(|\beta_{n_R,1}^L|)$$

$$\cdot \, d|\beta_{1,1}^1| d|\beta_{2,1}^1| \cdots d|\beta_{n_R,1}^L| \tag{2.93}$$

where $|\beta_{j,1}^t|$, $t = 1, 2, \ldots, L$, and $j = 1, 2, \ldots, n_R$, are independent Rician-distributed random variables with the pdf given by (2.85). By integrating (2.93) term by term, the pairwise error probability becomes [6]

$$P(\mathbf{X}, \hat{\mathbf{X}}) \le \prod_{t \in \rho(\mathbf{x}, \hat{\mathbf{x}})} \prod_{j=1}^{n_R} \frac{1}{1 + \frac{E_s}{4N_0}|\mathbf{x}_t - \hat{\mathbf{x}}_t|^2} \exp\left(-\frac{K_t^{j,1} \frac{E_s}{4N_0}|\mathbf{x}_t - \hat{\mathbf{x}}_t|^2}{1 + \frac{E_s}{4N_0}|\mathbf{x}_t - \hat{\mathbf{x}}_t|^2} \right) \qquad (2.94)$$

For a special case where $|\beta_{j,1}^t|$ are Rayleigh distributed, the upper bound of the pairwise error probability at high SNR's becomes [6]

$$P(\mathbf{X}, \hat{\mathbf{X}}) \le \prod_{t \in \rho(\mathbf{x}, \hat{\mathbf{x}})} \left(\frac{1}{1 + |\mathbf{x}_t - \hat{\mathbf{x}}_t|^2 \frac{E_s}{4N_0}} \right)^{n_R}$$

$$\le \left(d_p^2 \right)^{-n_R} \left(\frac{E_s}{4N_0} \right)^{-\delta_H n_R} \qquad (2.95)$$

where d_p^2 is the product of the squared Euclidean distances between the two space-time symbol sequences, given by

$$d_p^2 = \prod_{t \in \rho(\mathbf{x}, \hat{\mathbf{x}})} |\mathbf{x}_t - \hat{\mathbf{x}}_t|^2 \qquad (2.96)$$

By using a union bound technique, we can compute an upper bound of the code frame error probability, which sums the contributions of the pairwise error probabilities over all error events. As the pairwise error probability in (2.95) decreases exponentially with the increasing SNR, the frame error probability at high SNR's is dominated by the pairwise error probability with the minimum product $\delta_H n_R$. The exponent of the SNR term, $\delta_H n_R$, is called the *diversity gain* for fast Rayleigh fading channels and

$$G_c = \frac{(d_p^2)^{1/\delta_H}}{d_u^2} \qquad (2.97)$$

is called the *coding gain* for fast Rayleigh fading channels, where d_u^2 is the squared Euclidean distance of the reference uncoded system. Note that both diversity and coding gains are obtained as the minimum $\delta_H n_R$ and $(d_p^2)^{1/\delta_H}$ over all pairs of distinct codewords.

2.6 Space-Time Code Design Criteria

2.6.1 Code Design Criteria for Slow Rayleigh Fading Channels

As the error performance upper bounds (2.61) and (2.65) indicate, the design criteria for slow Rayleigh fading channels depend on the value of $r n_R$. The maximum possible value of $r n_R$ is $n_T n_R$. For small values of $n_T n_R$, corresponding to a small number of independent subchannels, the error probability at high SNR's is dominated by the minimum rank r of matrix $\mathbf{A}(\mathbf{X}, \hat{\mathbf{X}})$ over all possible codeword pairs. The product of the minimum rank and the number of receive antennas, $r n_R$, is called the minimum diversity. In addition, in order to minimize the error probability, the minimum product of nonzero eigenvalues,

$\prod_{i=1}^{r} \lambda_i$, of matrix $\mathbf{A}(\mathbf{X}, \hat{\mathbf{X}})$ along the pairs of codewords with the minimum rank should be maximized. Therefore, if the value of $n_T n_R$ is small, the space-time code design criteria for slow Rayleigh fading channels can be summarized as [6]:

Design Criteria Set I

[I-a] Maximize the minimum rank r of matrix $\mathbf{A}(\mathbf{X}, \hat{\mathbf{X}})$ over all pairs of distinct codewords

[I-b] Maximize the minimum product, $\prod_{i=1}^{r} \lambda_i$, of matrix $\mathbf{A}(\mathbf{X}, \hat{\mathbf{X}})$ along the pairs of distinct codewords with the minimum rank

Note that $\prod_{i=1}^{r} \lambda_i$ is the absolute value of the sum of determinants of all the principal $r \times r$ cofactors of matrix $\mathbf{A}(\mathbf{X}, \hat{\mathbf{X}})$ [6]. This criteria set is referred to as *rank & determinant criteria*. It is also called Tarokh/Seshadri/Calderbank (TSC) criteria. The minimum rank of matrix $\mathbf{A}(\mathbf{X}, \hat{\mathbf{X}})$ over all pairs of distinct codewords is called the minimum rank of the space-time code.

To maximize the minimum rank r means to find a space-time code with the full rank of matrix $\mathbf{A}(\mathbf{X}, \hat{\mathbf{X}})$, e.g., $r = n_T$. However, the full rank is not always achievable due to the restriction of the code structure. We discuss in detail how to design optimum space-time codes in Chapters 3 and 4.

For large values of $n_T n_R$, corresponding to a large number of independent subchannels, the pairwise error probability is upper-bounded by (2.61). In order to get an insight into the code design for systems of practical interest, we assume that the space-time code operates at a reasonably high SNR, which can be represented as[1]

$$\frac{E_s}{4N_0} \geq \frac{\sum_{i=1}^{r} \lambda_i}{\sum_{i=1}^{r} \lambda_i^2} \tag{2.98}$$

By using the inequality

$$Q(x) \leq \frac{1}{2} e^{-x^2/2}, x \geq 0 \tag{2.99}$$

the bound in (2.61) can be further approximated as

$$P(\mathbf{X}, \hat{\mathbf{X}}) \leq \frac{1}{4} \exp\left(-n_R \frac{E_s}{4N_0} \sum_{i=1}^{r} \lambda_i\right) \tag{2.100}$$

The bound in (2.100) shows that the error probability is dominated by the codewords with the minimum sum of the eigenvalues of $\mathbf{A}(\mathbf{X}, \hat{\mathbf{X}})$. In order to minimize the error probability, the minimum sum of all eigenvalues of matrix $\mathbf{A}(\mathbf{X}, \hat{\mathbf{X}})$ among all the pairs of distinct codewords should be maximized. For a square matrix the sum of all the eigenvalues is equal to the sum of all the elements on the matrix main diagonal, which is called the *trace of the matrix* [53]. It can be expressed as

$$tr(\mathbf{A}(\mathbf{X}, \hat{\mathbf{X}})) = \sum_{i=1}^{r} \lambda_i = \sum_{i=1}^{n_T} A^{i,i} \tag{2.101}$$

---

[1] The value of $\frac{\sum_{i=1}^{r} \lambda_i}{\sum_{i=1}^{r} \lambda_i^2}$ is usually small. For example, its value for the 4-state QPSK space-time code in [6], [25] and [30] is 0.5, 0.19 and 0.11, respectively.

where $A^{i,i}$ are the elements on the main diagonal of matrix $\mathbf{A}(\mathbf{X}, \hat{\mathbf{X}})$. Since

$$A^{i,j} = \sum_{t=1}^{L} (x_t^i - \hat{x}_t^i)(x_t^j - \hat{x}_t^j)^* \tag{2.102}$$

substituting (2.102) into (2.101), we get

$$tr(\mathbf{A}(\mathbf{X}, \hat{\mathbf{X}})) = \sum_{i=1}^{n_T} \sum_{t=1}^{L} |x_t^i - \hat{x}_t^i|^2 \tag{2.103}$$

Equation (2.103) indicates that the trace of matrix $\mathbf{A}(\mathbf{X}, \hat{\mathbf{X}})$ is equivalent to the squared Euclidean distance between the codewords $\mathbf{X}$ and $\hat{\mathbf{X}}$. Therefore, maximizing the minimum sum of all eigenvalues of matrix $\mathbf{A}(\mathbf{X}, \hat{\mathbf{X}})$ among the pairs of distinct codewords, or the minimum trace of matrix $\mathbf{A}(\mathbf{X}, \hat{\mathbf{X}})$, is equivalent to maximizing the minimum Euclidean distance between all pairs of distinct codewords. This design criterion is called the *trace criterion*.

It should be pointed out that formula (2.100) is valid for a large number of independent subchannels under the condition that the minimum value of rn_R is high. In this case, the space-time code design criteria for slow fading channels can be summarized as

Design Criteria Set II

[II-a] Make sure that the minimum rank r of matrix $\mathbf{A}(\mathbf{X}, \hat{\mathbf{X}})$ over all pairs of distinct codewords is such that $rn_R \geq 4$

[II-b] Maximize the minimum trace $\sum_{i=1}^{r} \lambda_i$ of matrix $\mathbf{A}(\mathbf{X}, \hat{\mathbf{X}})$ among all pairs of distinct codewords

It is important to note that the proposed design criteria are consistent with those for trellis codes over fading channels with a large number of diversity branches [38] [37]. A large number of diversity branches reduces the effect of fading and consequently, the channel approaches an AWGN model. Therefore, the trellis code design criteria derived for AWGN channels [36], which is maximizing the minimum code Euclidean distance, apply to fading channels with a large number of diversity. In a similar way, in space-time code design, when the number of independent subchannels rn_R is large, the channel converges to an AWGN channel. Thus, the code design is the same as that for AWGN channels.

From the above discussion, we can conclude that either the rank & determinant criteria or the trace criterion should be applied for design of space-time codes, depending on the diversity order rn_R. When $rn_R < 4$, the rank & determinant criteria should be applied and when $rn_R \geq 4$, the trace criterion should be applied.

The boundary value of rn_R between the two design criteria sets was chosen to be 4. This boundary is determined by the required number of random variables rn_R in (2.53) to satisfy the central limit theorem. In general, for random variables with smooth pdf's, the central limit theorem can be applied if the number of random variables in the sum is larger than 4 [54]. In the application of the central limit theorem in (2.53), the choice of 4 as the boundary has been further justified by the code design and performance simulation, as it was found that as long as $rn_R \geq 4$, the best codes based on the trace criterion outperform the best codes based on the rank and determinant criteria [31] [34].

2.6.2 Code Design Criteria for Fast Rayleigh Fading Channels

As the error performance upper bounds (2.90) and (2.95) indicate, the code design criteria for fast Rayleigh fading channels depend on the value of $\delta_H n_R$. For small values of $\delta_H n_R$, the error probability at high SNR's is dominated by the minimum space-time symbol-wise Hamming distance δ_H over all distinct codeword pairs. In addition, in order to minimize the error probability, the minimum product distance, d_p^2, along the path of the pairs of codewords with the minimum symbol-wise Hamming distance δ_H, should be maximized. Therefore, if the value of $\delta_H n_R$ is small, the space-time code design criteria for fast fading channels can be summarized as [6]:

Design Criteria Set III

[III-a] Maximize the minimum space-time symbol-wise Hamming distance δ_H between all pairs of distinct codewords

[III-b] Maximize the minimum product distance, d_p^2, along the path with the minimum symbol-wise Hamming distance δ_H

For large values of $\delta_H n_R$ the pairwise error probability is upper-bounded by (2.90). As before, we assume the space-time code works at a reasonably high SNR, which corresponds to

$$\frac{E_s}{4N_0} \geq \frac{d_E^2}{D^4} \qquad (2.104)$$

where d_E^2 and D^4 are given by (2.91) and (2.92), respectively. By using the inequality (2.99), the bound (2.90) can be further approximated by

$$P(\mathbf{X}, \hat{\mathbf{X}}) \leq \exp\left(-n_R \frac{E_s}{4N_0} \sum_{t=1}^{L} \sum_{i=1}^{n_T} |x_t^i - \hat{x}_t^i|^2\right)$$

$$= \exp\left(-n_R \frac{E_s}{4N_0} d_E^2\right) \qquad (2.105)$$

From (2.105), it is clear that the frame error probability at high SNR's is dominated by the pairwise error probability with the minimum squared Euclidean distance d_E^2. To minimize the error probability on fading channels, the codes should satisfy

Design Criteria Set IV

[IV-a] Make sure that the product of the minimum space-time symbol-wise Hamming distance and the number of receive antennas, $\delta_H n_R$, is large enough (larger than or equal to 4)

[IV-b] Maximize the minimum Euclidean distance among all pairs of distinct codewords.

It is interesting to note that this design criterion is the same as the trace criterion for space-time code on slow fading channels if the value of $r n_R$ is large. It is also consistent with the design criterion for trellis coded modulation on fading channels if the symbol-wise Hamming distance is large [38].

Based on the previous discussion, we can conclude that code design on fading channels is very much dependent on the possible diversity order of the space-time coded system. For codes on slow fading channels, the total diversity is the product of the receive diversity, n_R, and the transmit diversity provided by the code scheme, r. On the other hand, for codes on fast fading channels, the total diversity is the product of the receive diversity, n_R, and the time diversity achieved by the code scheme, δ_H. If the total diversity is small, in the code design for slow fading channels one should attempt to maximize the diversity and the coding gain by choosing a code with the largest minimum rank and the determinant; while for fast fading channels one should attempt to choose a code with the largest minimum symbol-wise Hamming distance and the product distance. In this case, the diversity gain dominates the code performance and it has much more influence on error probability than the coding gain. However, when the total diversity is getting larger, increasing the diversity order cannot achieve a substantial performance improvement. In contrast, the coding gain becomes more important. Since a high order of diversity drives the fading channel towards an AWGN channel as shown in Fig. 2.9, the error probability is dominated by the minimum Euclidean distance. Thus, the code design criterion for AWGN channels, which is maximizing the minimum Euclidean distance, is valid for both slow and fast fading channels provided that the diversity is large.

Example 2.4

To illustrate the design criteria and evaluate the importance of the rank, determinant and trace in determining the code performance for systems with various numbers of the transmit and receive antennas on slow Rayleigh fading channels, we consider the following example.

Let us consider three QPSK space-time trellis codes with 4 states and 2 transmit antennas. The three codes are denoted by A, B and C, respectively. The code trellis structures are shown in Fig. 2.11. These codes have the same bandwidth efficiency of 2 bits/s/Hz. The minimum rank, determinant and trace of the codes are also listed in Fig. 2.11. It is shown that codes A and B have a full rank and the same determinant of 4, while code C is not of full rank and therefore, its determinant is 0. On the other hand, the minimum trace for codes B and C is 10 while code A has a smaller minimum trace of 4.

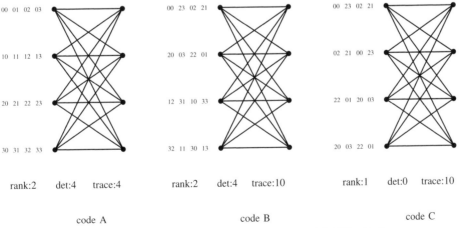

Figure 2.11 Trellis structures for 4-state space-time coded QPSK with 2 antennas

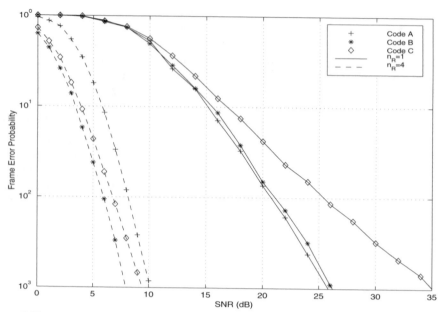

Figure 2.12   FER performance of the 4-state space-time trellis coded QPSK with 2 transmit antennas, Solid: 1 receive antenna, Dash: 4 receive antennas

The performance of the codes with various numbers of the receive antennas on slow Rayleigh fading channels is evaluated by simulation. The frame length was 130 symbols. The frame error rate (FER) performance versus the SNR per receive antenna, e.g., SNR = $n_T E_s/N_0$, is shown in Fig. 2.12.

From Fig. 2.12, it can be observed that codes A and B outperform code C if one receive antenna is employed. This is explained as follows. When the number of independent subchannels $n_T n_R$ is small, the minimum rank of the code dominates the code performance. Since both code A and code B are of full rank ($r = 2$) and code C is not ($r = 1$), codes A and B achieve a better performance relative to code C. It can also be seen that the performance curves for codes A and B have an asymptotic slope of -2 while the slope for code C is -1, consistent with the diversity order of 2 for codes A and B, and 1 for code C. At a FER of 10^{-2}, codes A and B outperform code C by about 5 dB due to a larger diversity order. It clearly indicates that the minimum rank is much more important in determining the code performance for systems with a small number of independent subchannels.

However, when the number of receive antennas is 4, code C performs better than code A as shown in Fig. 2.12, which means the code with a full rank is worse than the code with a smaller rank. This occurs as the diversity gain $r n_R$ in this case is 8 and 4 for codes A and C, respectively. According to Design Criteria II, code C is superior to code A due to a larger minimum trace value. At a FER of 10^{-2}, the advantage of the code C relative to code A is about 1.3 dB.

From Fig. 2.12 we can also see that code B is about 0.8 dB better than code C at the FER of 10^{-2}, although they have the same minimum trace. This is due to the fact that code B has the same minimum trace and a larger rank than code C. Therefore, code B can achieve a larger diversity, which is manifested by a steeper error rate slope for code B than for code C.

When the number of receive antennas increases further to 6, it is shown in [31] that the performance of code B on slow Rayleigh fading channels is very close to its performance on AWGN channels, which verifies the convergence of Rayleigh fading channels to AWGN channels, provided a large diversity is available.

This example clearly verified the code design criteria for slow fading channels.

2.6.3   Code Performance at Low to Medium SNR Ranges

The code design criteria are derived based on the asymptotic code performance at very high SNR's. However, in practical communication systems, given the number of transmit and receive antennas to be used and the FER performance requirement, the code may work at a low or a medium SNR range. Let us denote $\frac{E_s}{4N_0}$ by γ. We assume that $\gamma \gg 1$ refers to a high SNR range, $\gamma \ll 1$ as a low SNR range and $\gamma \approx 1$ as a medium SNR range. For example, for two transmit and two receive antennas, to achieve the FER of 10^{-2} and bandwidth efficiency of 2 bits/s/Hz, the space-time coded QPSK will work around the SNR of about 10 dB. This SNR corresponds to $\gamma \approx 1$, so the previous code design criteria derived for very high SNR's may not be very accurate.

In [29], modified design criteria for space-time codes under various SNR conditions are discussed. Recall the pairwise error probability upper bound in (2.64). The bound can be rewritten as

$$P(\mathbf{X}, \hat{\mathbf{X}}) \leq \left(\prod_{i=1}^{n_T} \frac{1}{1 + \gamma \lambda_i} \right)^{n_R} \tag{2.106}$$

We will consider three different cases.

CASE 1: $\gamma \ll 1$

This case refers to a low SNR range. Since $\gamma \ll 1$, when the denominator in the upper bound (2.106) is expanded, one can ignore the contribution of the high order terms of γ, so that the upper bound becomes

$$P(\mathbf{X}, \hat{\mathbf{X}}) \leq \left(1 + \gamma \sum_{i=1}^{r} \lambda_i \right)^{-n_R} \tag{2.107}$$

Recall that the sum of the eigenvalues is equal to the trace. From this bound, we can see that in the code design for a low SNR range, the minimum trace should be maximized. This means that in order to achieve a specified FER at a low SNR, n_T and/or n_R need to be large. In other words, this case falls into the situation of large values of $n_T n_R$ as we discussed previously.

CASE 2: $\gamma \gg 1$

This case refers to a very high SNR range. Under this condition, the space-time code performance is upper bounded by [6]

$$P(\mathbf{X}, \hat{\mathbf{X}}) \leq \left(\prod_{i=1}^{r} \lambda_i \right)^{-n_R} \gamma^{-r n_R} \tag{2.108}$$

It is obvious that the rank & determinant criteria should be used in code design. For a specified FER, if $n_T n_R$ is small, the SNR is typically large ($\gamma \gg 1$). Therefore, this case falls into the domain of small values of $n_T n_R$ as we discussed previously.

CASE 3: $\gamma \approx 1$

The case refers to a moderate SNR range. As we predicted by analysis and confirmed by simulation, the typical SNR for a space-time code with two transmit and two receive antennas to achieve the FER of 10^{-2} and bandwidth efficiency of 2 bits/s/Hz is around 10 dB. For this SNR value, $\gamma = \frac{E_s}{4N_0} = \frac{SNR}{4n_T} = \frac{10}{8} \approx 1$. In this range, for $\gamma \approx 1$, the pairwise upper bound becomes

$$P(\mathbf{X}, \hat{\mathbf{X}}) \le \left(\prod_{i=1}^{n_T} \frac{1}{1 + \lambda_i} \right)^{n_R} \tag{2.109}$$

We can formulate the code design criterion as maximizing the minimum determinant of the matrix $\mathbf{I} + \mathbf{A}(\mathbf{X}, \hat{\mathbf{X}})$, where $\mathbf{I}$ is an $n_T \times n_T$ identical matrix. This criterion is derived for a very specific SNR. In practical systems, space-time codes may work at a range of SNR's. Good codes designed by this criterion may not be optimal in the whole range of SNR's. Therefore, this criterion is less practical relative to other design criteria.

2.7   Exact Evaluation of Code Performance

In the previous code performance analysis and code design, we considered only the worst case pairwise error probability upper bound. In order to get the accurate performance evaluation, one possible method is to compute the code distance spectrum and apply the union bound technique to calculate the average pairwise error probability. The obtained upper bound is asymptotically tight at high SNR's for a small number of receive antennas but loose for other scenarios [41]. A more accurate performance evaluation can be obtained with exact evaluation of the pairwise error probability, rather than evaluating the bounds. This can be done by using residue methods based on the characteristic function technique [42] [43] or on the moment generating function method [44] [45].

Recall that the pairwise error probability conditioned on the MIMO fading coefficients is given by

$$P(\mathbf{X}, \hat{\mathbf{X}} | \mathbf{H}) = Q \left(\sqrt{\frac{E_s}{2N_0} \sum_{t=1}^{L} \| \mathbf{H}_t (\mathbf{x}_t - \hat{\mathbf{x}}_t) \|^2} \right) \tag{2.110}$$

Let

$$\Gamma = \frac{E_s}{2N_0} \sum_{t=1}^{L} \| \mathbf{H}_t (\mathbf{x}_t - \hat{\mathbf{x}}_t) \|^2 \tag{2.111}$$

By using Graig's formula for the Gaussian Q function [48]

$$Q(x) = \frac{1}{\pi} \int_0^{\pi/2} \exp \left(-\frac{x^2}{2 \sin^2 \theta} \right) d\theta \tag{2.112}$$

we can rewrite for the conditional pairwise error probability

$$P(\mathbf{X}, \hat{\mathbf{X}}|\mathbf{H}) = \frac{1}{\pi} \int_0^{\pi/2} \exp\left(-\frac{\Gamma^2}{2\sin^2\theta}\right) d\theta \tag{2.113}$$

In order to compute the average pairwise error probability, we average (2.113) with respect to the distribution of Γ. The average pairwise error probability can be expressed in terms of the moment generating function (MGF) of Γ, denoted by $M_\Gamma(s)$, which is given by

$$M_\Gamma(s) = \int_0^\infty e^{s\Gamma} P_\Gamma(\Gamma) \, d\Gamma \tag{2.114}$$

The average pairwise error probability can be represented as [44]

$$\begin{aligned}
P(\mathbf{X}, \hat{\mathbf{X}}) &= \frac{1}{\pi} \int_0^{\pi/2} E\left[\exp\left(-\frac{\Gamma^2}{2\sin^2\theta}\right)\right] d\theta \\
&= \frac{1}{\pi} \int_0^{\pi/2} \int_0^\infty \exp\left(-\frac{\Gamma^2}{2\sin^2\theta}\right) P_\Gamma(\Gamma) d\Gamma \, d\theta \\
&= \frac{1}{\pi} \int_0^{\pi/2} M_\Gamma\left(-\frac{1}{2\sin^2\theta}\right) d\theta
\end{aligned} \tag{2.115}$$

For fast Rayleigh fading MIMO channels, the MGF can be expressed in a closed form as [44]

$$M_\Gamma(s) = \prod_{t=1}^{L}\left[1 - \frac{sE_s}{2N_0}\sum_{i=1}^{n_T}|x_t^i - \hat{x}_t^i|^2\right]^{-n_R} \tag{2.116}$$

where

$$s = -\frac{1}{2\sin^2\theta}. \tag{2.117}$$

Substituting (2.116) into (2.115), we get for the average pairwise error probability

$$P(\mathbf{X}, \hat{\mathbf{X}}) = \frac{1}{\pi} \int_0^{\pi/2} \prod_{t=1}^{L}\left[1 + \frac{E_s}{4N_0\sin^2\theta}\sum_{i=1}^{n_T}|x_t^i - \hat{x}_t^i|^2\right]^{-n_R} d\theta \tag{2.118}$$

For slow Rayleigh fading channels, the MGF can be represented as [44]

$$M_\Gamma(s) = \left(det\left[\mathbf{I}_{n_T} - s(\mathbf{X} - \hat{\mathbf{X}})(\mathbf{X} - \hat{\mathbf{X}})^H \frac{E_s}{2N_0}\right]\right)^{-n_R}. \tag{2.119}$$

If the rows of matrix $(\mathbf{X} - \hat{\mathbf{X}})$ are orthogonal, this MGF can be simplified as

$$M_\Gamma(s) = \prod_{i=1}^{n_T}\left[1 - \frac{sE_s}{2N_0}\sum_{t=1}^{L}|x_t^i - \hat{x}_t^i|^2\right]^{-n_R}. \tag{2.120}$$

In this case, the average pairwise error probability is given by

$$P(\mathbf{X}, \hat{\mathbf{X}}) = \frac{1}{\pi} \int_0^{\pi/2} \prod_{i=1}^{n_T}\left[1 + \frac{E_s}{4N_0\sin^2\theta}\sum_{t=1}^{L}|x_t^i - \hat{x}_t^i|^2\right]^{-n_R} d\theta. \tag{2.121}$$

Example 2.5

Let us consider a 4-state QPSK space-time trellis code with two transmit antennas. The code trellis structure is shown in Fig. 2.13. For the pairwise error event of length 2, illustrated by the thick lines in Fig. 2.13, matrices $\mathbf{X}$ and $\hat{\mathbf{X}}$ are

$$\mathbf{X} = \begin{bmatrix} 1 & 1 \\ 1 & 1 \end{bmatrix}, \qquad \hat{\mathbf{X}} = \begin{bmatrix} 1 & -1 \\ -1 & 1 \end{bmatrix} \tag{2.122}$$

The codeword distance matrix for the error event is given by

$$\begin{aligned} \mathbf{A}(\mathbf{X}, \hat{\mathbf{X}}) &= (\mathbf{X} - \hat{\mathbf{X}})(\mathbf{X} - \hat{\mathbf{X}})^H \\ &= \begin{bmatrix} 4 & 0 \\ 0 & 4 \end{bmatrix} \end{aligned} \tag{2.123}$$

Substituting (2.123) into (2.119) gives

$$M_{\Gamma}(s) = \left(1 - 2s \frac{E_s}{N_0} \right)^{-2n_R} \tag{2.124}$$

The pairwise error probability for slow Rayleigh fading channels can be expressed as

$$P(\mathbf{X}, \hat{\mathbf{X}}) = \frac{1}{\pi} \int_0^{\pi/2} \left[1 + \frac{E_s}{N_0 \sin^2 \theta} \right]^{-2n_R} d\theta \tag{2.125}$$

$$= \frac{1}{2} \left[1 - \sqrt{\frac{E_s/N_0}{1 + E_s/N_0}} \sum_{k=0}^{2n_R-1} \binom{2k}{k} \left(\frac{1}{4(1 + E_s/N_0)} \right)^k \right] \tag{2.126}$$

The pairwise error probability (2.126) is plotted in Figs. 2.14 and 2.15 for $n_R = 1$ and $n_R = 2$, respectively. The pairwise error probability upper bounds in (2.61), (2.64), and (2.65) are also shown in these figures for comparison.

The exact evaluation of the pairwise error probability based on the transfer function technique gives a transfer function upper bound on the average frame error probability or

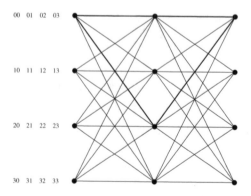

Figure 2.13   Trellis structure for a 4-state QPSK space-time code with two antennas

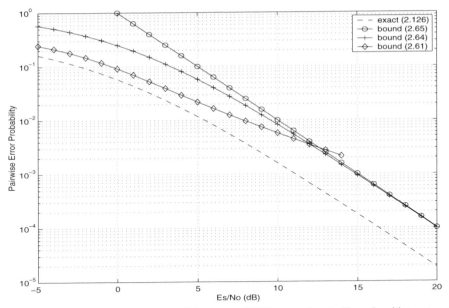

Figure 2.14   Pairwise error probability of the 4-state QPSK space-time trellis code with two transmit and one receive antenna

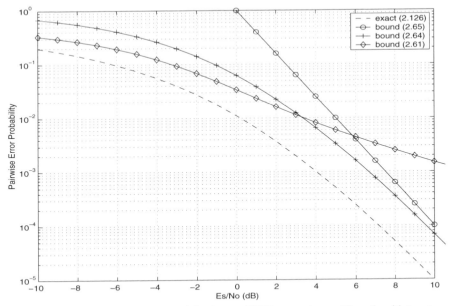

Figure 2.15   Pairwise error probability of the 4-state QPSK space-time trellis code with two transmit and two receive antennas

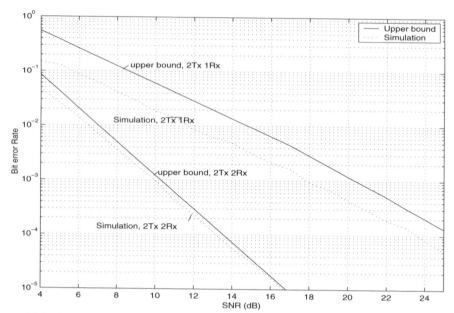

Figure 2.16   Average bit error rate of the 4-state QPSK space-time trellis code with two transmit antennas and one and two receive antennas

bit error probability for space-time codes, which is asymptotically tight at high SNR's and tighter than a Chernoff bound [44].

Figure 2.16 illustrates the performance comparison between the analytical and simulated average bit error rate of the 4-state QPSK space-time trellis code with two transmit antennas and one and two receive antennas on slow Rayleigh fading channels. In the analytical evaluating the average bit error rate, only the error events of lengths two and three are considered for simplicity.

Bibliography

[1] G. J. Foschini and M. Gans, "On the limits of wireless communication in a fading environment when using multiple antennas", *Wireless Personal Communication,* vol. 6, pp. 311–335, Mar. 1998.

[2] G. J. Foschini, "Layered space-time architecture for wireless communication in fading environments when using multiple antennas", *Bell Labs Tech. J.,* 2, Autumn 1996.

[3] E. Telatar, "Capacity of multi-antenna Gaussian channels", *Technical Report, AT&T-Bell Labs,* June 1995, and *ETT,* vol. 10, no. 6, pp. 585–595, Nov. 1999.

[4] B. Hochwald and T. L. Marzetta, "Capacity of a mobile multiple-antenna communication link in a Rayleigh flat fading environment", *IEEE Trans. Inform. Theory,* vol. 45, no. 1, pp. 139–157, Jan. 1999.

[5] G. G. Raleigh and J.M. Cioffi, "Spatio-temporal coding for wireless communication", *IEEE Trans. Communi.,* vol. 46, no. 3, pp. 357–366, Mar. 1998.

[6] V. Tarokh, N. Seshadri and A. R. Calderbank, "Space-time codes for high data rate wireless communication: performance criterion and code construction", *IEEE Trans. Inform. Theory,* vol. 44, no. 2, pp. 744–765, Mar. 1998.

[7] J.-C. Guey, M. R. Bell, M. P. Fitz and W. Y. Kuo, "Signal design for transmitter diversity wireless communication systems over Rayleigh fading channels", in *Proc. IEEE Vehicular Technology Conference,* pp. 136–140, Atlanta, US, 1996, and *IEEE Trans. Commun.,* vol. 47, no. 4, pp. 527–537, Apr. 1998.

[8] A. Naguib, V. Tarokh, N. Seshadri and A. Calderbank, "A space-time coding modem for high-data-rate wireless communications", *IEEE J. Select. Areas Commun.,* vol. 16, no. 10, pp. 1459–1478, Oct. 1998.

[9] S. M. Alamouti, "A simple transmit diversity technique for wireless communications", *IEEE J. Select. Areas Commun.,* vol. 16, no. 10, pp. 1451–1458, Oct. 1998.

[10] V. Tarokh, H. Jafarkhani and A. R. Calderbank, "Space-time block codes from orthogonal designs", *IEEE Trans. Inform. Theory,* vol. 45, no. 5, pp. 1456–1467, July 1999.

[11] V. Tarokh, H. Jafarkhani and A. R. Calderbank, "Space-time block coding for wireless communications: performance results", *IEEE J. Select. Areas Commun.,* vol. 17, no. 3, pp. 451–460, Mar. 1999.

[12] J. Grimm, M. P. Fitz and J. V. Krogmeier, "Further results in space-time coding for Rayleigh fading", in *Proc. 36th Allerton Conference on Communications, Control and Computing,* Sep. 1998.

[13] V. Tarokh, A. Naguib, N. Seshadri and A. R. Calderbank, "Combined array processing and space-time coding", *IEEE Trans. Inform. Theory,* vol. 45, no. 4, pp. 1121–1128, May 1999.

[14] V. Tarokh, A. Naguib, N. Seshadri and A. R. Calderbank, "Space-time codes for high data rate wireless communication: performance criteria in the presence of channel estimation errors, mobility, and multiple paths", *IEEE Trans. Commun.,* vol. 47, no. 2, pp. 199–207, Feb. 1999.

[15] A. F. Naguib and R. Calderbank, "Space-time coding and signal processing for high data rate wireless communications", *IEEE Signal Processing Magazine,* vol. 17, no. 3, pp. 76–92, Mar. 2000.

[16] J. H. Winters, "Switched diversity with feedback for DPSK mobile radio systems", *IEEE Trans. Vehicular Technology,* vol. 32, pp. 134–150, 1983.

[17] A. Wittneben, "Base station modulation diversity for digital SIMULCAST", in *Proc. IEEE Vehicular Technology Conf.,* vol. 1, pp. 848–853, May 1991.

[18] A. Winttneben, "A new bandwidth efficient transmit antenna modulation diversity scheme for linear digital modulation", in *Proc. 1993 IEEE International Conf. Communications (ICC93),* pp. 1630–1634, May 1993.

[19] N. Seshadri and J. H. Winters, "Two signaling schemes for improving the error performance of FDD transmission systems using transmitter antenna diversity", in *Proc. 1993 IEEE Vehicular Technology Conf.,* pp. 508–511, May 1993.

[20] J. H. Winters, "The diversity gain of transmit diversity in wireless systems with Rayleigh fading", in *Proc. 1994 IEEE International Conf. Communications (ICC94),* vol. 2, pp. 1121–1125, New Orleans, LA, May 1994.

[21] J. H. Winters, "The diversity gain of transmit diversity in wireless systems with Rayleigh fading", *IEEE Trans. Vehicular Technology,* vol. 47, pp. 119–123, 1998.

[22] A. Hiroike, F. Adachi and N. Nakajima, "Combined effects of phase sweeping transmitter diversity and channel coding", *IEEE Trans. Vehicular Technology,* vol. 42, pp. 170–176, 1992.

[23] A. Hiroike and K. Hirade, "Multitransmitter simulcast digital signal transmission by using frequency offset strategy in land mobile radio telephone", *IEEE Trans. Vehicular Technology,* vol. 27, pp. 231–238, 1978.

[24] A. R. Hammons and H. E. Gammal, "On the theory of space-time codes for PSK modulation", *IEEE Trans. on. Inform. Theory,* vol. 46, no. 2, pp. 524–542, Mar. 2000.

[25] S. Baro, G. Bauch and A. Hansmann, "Improved codes for space-time trellis coded modulation", *IEEE Commun. Lett.,* vol. 4, no. 1, pp. 20–22, Jan. 2000.

[26] R. S. Blum, "New analytical tools for designing space-time convolutional codes", Invited paper at CISS, Princeton University, Princeton, NJ, Mar. 2000, also submitted to *IEEE Trans. on Inform. Theory.*

[27] Q. Yan and R. S. Blum, "Optimum space-time convolutional codes", in *Proc. IEEE WCNC'00,* Chicago, IL., pp. 1351–1355, Sep. 2000.

[28] D. M. Ionescu, K. K. Mukkavilli, Z. Yan and J. Lilleberg, "Improved 8- and 16-state space-time codes for 4PSK with two transmit antennas", *IEEE Commun. Lett.,* vol. 5, pp. 301–303, Jul. 2001.

[29] M. Tao and R. S. Cheng, "Improved design criteria and new trellis codes for space-time coded modulation in slow flat fading channels", *IEEE Comm. Lett.,* vol. 5, pp. 313–315, Jul. 2001.

[30] Z. Chen, J. Yuan and B. Vucetic, "Improved space-time trellis coded modulation scheme on slow Rayleigh fading channels", *Electron. Lett.,* vol. 37, no. 7, pp. 440–441, Mar. 2001.

[31] Z. Chen, J. Yuan and B. Vucetic, "An improved space-time trellis coded modulation scheme on slow Rayleigh fading channels", in *Proc. IEEE ICC'01,* Helsinki, Finland, pp. 1110–1116, Jun. 2001.

[32] J. Yuan, B. Vucetic, B. Xu and Z. Chen, "Design of space-time codes and its performance in CDMA systems", in *Proc. IEEE Vehicular Technology Conf.,* pp. 1292–1296, May 2001.

[33] J. Yuan, Z. Chen, B. Vucetic and W. Firmanto, "Performance analysis of space-time coding on fading channels", in *Proc. IEEE ISIT'01,* Washington D.C., pp. 153, Jun. 2001.

[34] J. Yuan, Z. Chen, B. Vucetic and W. Firmanto, "Performance analysis and design of space-time coding on fading channels", submitted to *IEEE Trans. Commun..*

[35] W. Firmanto, B. Vucetic and J. Yuan, "Space-time TCM with improved performance on fast fading channels", *IEEE Commun. Lett.,* vol. 5, no. 4, pp. 154–156, Apr. 2001.

[36] G. Ungerboeck, "Channel coding for multilevel/phase signals", *IEEE Trans. Inform. Theory,* vol. 28, no. 1, pp. 55–67, Jan. 1982.

[37] J. Ventura-Traveset, G. Caire, E. Biglieri and G. Taricco, "Impact of diversity reception on fading channels with coded modulation—Part I: coherent detection", *IEEE Trans. Commun.,* vol. 45, no. 5, pp. 563–572, May 1997.

[38] B. Vucetic and J. Nicolas, "Performance of M-PSK trellis codes over nonlinear fading mobile satellite channels", *IEE Proceedings I,* vol. 139, pp. 462–471, Aug. 1992.

[39] B. Vucetic, "Bandwidth efficient concatenated coding schemes for fading channels", *IEEE Trans. Commun.,* vol. 41, no. 1, pp. 50–61, Jan. 1993.

[40] Z. Liu, G. B. Giannakis, S. Zhuo and B. Muquet, "Space-time coding for broadband wireless communications", *Wireless Communications and Mobile Computing*, vol. 1, no. 1, pp. 35–53, Jan. 2001.

[41] D. K. Aktas and M. P. Fitz, "Computing the distance spectrum of space-time trellis codes", in *Proc. Wireless Commun. and Networking Conf.*, Chicago, IL, pp. 51–55, Sept. 2000.

[42] M. Uysal and C. N. Georghiades, "Error performance analysis of space-time codes over Rayleigh fading channels", *Journal of Commun. and Networks*, vol. 2, no. 4, pp. 351–355, Dec. 2000.

[43] J. K. Cavers and P. Ho, "Analysis of error performance of trellis coded modulation in Rayleigh fading channels", *IEEE Trans. Commun.*, vol. 40, no. 1, pp. 74–83, Jan. 1992.

[44] M. K. Simon, "Evaluation of average bit error probability for space-time coding based on a simpler exact evaluation of pairwise error probability", *Journal of Commun. and Networks*, vol. 3, no. 3, pp. 257–264, Sep. 2001.

[45] G. Taricco and E. Biglieri, "Exact pairwise error probability of space-time codes", *IEEE Trans. Inform. Theory*, vol. 48, no. 2, pp. 510–513, Feb. 2002.

[46] S. Chennakeshu and J. B. Anderson, "Error rates for Rayleigh fading multichannel reception of MPSK signals", *IEEE Trans. Communications*, vol. 43, no. 2/3/4, pp. 338–346, Feb/Mar/Apr. 1995.

[47] J. G. Proakis, *Digital Communications,* 4th Ed., McGraw-Hill, New York, 2001.

[48] M. K. Simon and M.-S. Alouini, *Digital Communication over Fading Channels: A Unified Approach to Performance Analysis*, John Wiley & Sons, 2000.

[49] T. S. Rappaport, *Wireless Communications: Principles and Practice*, Prentice Hall, 1996.

[50] G. L. Stuber, *Principles of Mobile Communications*, Norwell, Kluwer Academic Publishers, 1996.

[51] H. V. Poor and G. W. Wornell, *Wireless Communications: Signal Processing Perspectives*, Prentice Hall, 1998.

[52] V. K. Garg and J. E. Wilkes, *Wireless and Personal Communications Systems*, Prentice Hall, 1996.

[53] R. A. Horn and C. R. Johnson, *Matrix Analysis*, New York, Cambridge Univ. Press, 1988.

[54] A. Papoulis, *Probability, Random Variables, and Stochastic Processes*, Third Edition, McGraw Hill.

3

Space-Time Block Codes

3.1 Introduction

In Chapter 2 we discussed the performance analysis and design criteria for space-time codes in general. In this chapter, we present space-time block codes and evaluate their performance on MIMO fading channels. We first introduce the Alamouti code [1], which is a simple two-branch transmit diversity scheme. The key feature of the scheme is that it achieves a full diversity gain with a simple maximum-likelihood decoding algorithm. In this chapter, we also present space-time block codes with a large number of transmit antennas based on orthogonal designs [3]. The decoding algorithms for space-time block codes with both real and complex signal constellations are discussed. The performance of the schemes on MIMO fading channels under various channel conditions is evaluated by simulations.

3.2 Alamouti Space-Time Code

The Alamouti scheme is historically the first space-time block code to provide full transmit diversity for systems with two transmit antennas [1]. It is worthwhile to mention that delay diversity schemes [2] can also achieve a full diversity, but they introduce interference between symbols and complex detectors are required at the receiver. In this section, we present Alamouti's transmit diversity technique, including encoding and decoding algorithms and its performance.

3.2.1 Alamouti Space-Time Encoding

Figure 3.1 shows the block diagram of the Alamouti space-time encoder.

Let us assume that an M-ary modulation scheme is used. In the Alamouti space-time encoder, each group of m information bits is first modulated, where $m = \log_2 M$. Then, the encoder takes a block of two modulated symbols x_1 and x_2 in each encoding operation and maps them to the transmit antennas according to a code matrix given by

$$\mathbf{X} = \begin{bmatrix} x_1 & -x_2^* \\ x_2 & x_1^* \end{bmatrix} \tag{3.1}$$

Space-Time Coding   Branka Vucetic and Jinhong Yuan
© 2003 John Wiley & Sons, Ltd   ISBN: 0-470-84757-3

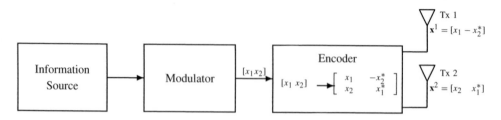

Figure 3.1 A block diagram of the Alamouti space-time encoder

The encoder outputs are transmitted in two consecutive transmission periods from two transmit antennas. During the first transmission period, two signals x_1 and x_2 are transmitted simultaneously from antenna one and antenna two, respectively. In the second transmission period, signal $-x_2^*$ is transmitted from transmit antenna one and signal x_1^* from transmit antenna two, where x_1^* is the complex conjugate of x_1.

It is clear that the encoding is done in both the space and time domains. Let us denote the transmit sequence from antennas one and two by $\mathbf{x}^1$ and $\mathbf{x}^2$, respectively.

$$\mathbf{x}^1 = [x_1, -x_2^*]$$

$$\mathbf{x}^2 = [x_2, \quad x_1^*] \tag{3.2}$$

The key feature of the Alamouti scheme is that the transmit sequences from the two transmit antennas are orthogonal, since the inner product of the sequences $\mathbf{x}^1$ and $\mathbf{x}^2$ is zero, i.e.

$$\mathbf{x}^1 \cdot \mathbf{x}^2 = x_1 x_2^* - x_2^* x_1 = 0 \tag{3.3}$$

The code matrix has the following property

$$\mathbf{X} \cdot \mathbf{X}^H = \begin{bmatrix} |x_1|^2 + |x_2|^2 & 0 \\ 0 & |x_1|^2 + |x_2|^2 \end{bmatrix}$$

$$= (|x_1|^2 + |x_2|^2)\mathbf{I}_2 \tag{3.4}$$

where $\mathbf{I}_2$ is a 2×2 identity matrix.

Let us assume that one receive antenna is used at the receiver. The block diagram of the receiver for the Alamouti scheme is shown in Fig. 3.2. The fading channel coefficients from the first and second transmit antennas to the receive antenna at time t are denoted by $h_1(t)$ and $h_2(t)$, respectively. Assuming that the fading coefficients are constant across two consecutive symbol transmission periods, they can be expressed as follows

$$h_1(t) = h_1(t + T) = h_1 = |h_1|e^{j\theta_1} \tag{3.5}$$

and

$$h_2(t) = h_2(t + T) = h_2 = |h_2|e^{j\theta_2} \tag{3.6}$$

where $|h_i|$ and θ_i, $i = 0, 1$, are the amplitude gain and phase shift for the path from transmit antenna i to the receive antenna, and T is the symbol duration.

At the receive antenna, the received signals over two consecutive symbol periods, denoted by r_1 and r_2 for time t and $t + T$, respectively, can be expressed as

$$r_1 = h_1 x_1 + h_2 x_2 + n_1 \tag{3.7}$$

$$r_2 = -h_1 x_2^* + h_2 x_1^* + n_2 \tag{3.8}$$

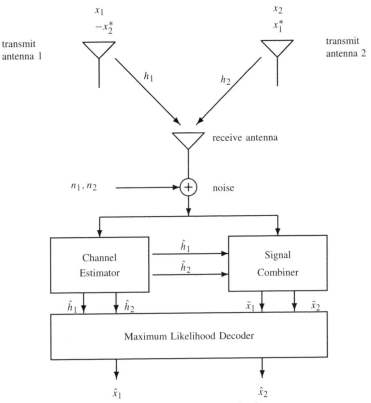

Figure 3.2   Receiver for the Alamouti scheme

where n_1 and n_2 are independent complex variables with zero mean and power spectral density $N_0/2$ per dimension, representing additive white Gaussian noise samples at time t and $t + T$, respectively.

3.2.2   Combining and Maximum Likelihood Decoding

If the channel fading coefficients, h_1 and h_2, can be perfectly recovered at the receiver, the decoder will use them as the channel state information (CSI). Assuming that all the signals in the modulation constellation are equiprobable, a maximum likelihood decoder chooses a pair of signals $(\hat{x}_1, \hat{x}_2)$ from the signal modulation constellation to minimize the distance metric

$$d^2(r_1, h_1\hat{x}_1 + h_2\hat{x}_2) + d^2(r_2, -h_1\hat{x}_2^* + h_2\hat{x}_1^*)$$
$$= |r_1 - h_1\hat{x}_1 - h_2\hat{x}_2|^2 + |r_2 + h_1\hat{x}_2^* - h_2\hat{x}_1^*|^2 \qquad (3.9)$$

over all possible values of $\hat{x}_1$ and $\hat{x}_2$. Substituting (3.7) and (3.8) into (3.9), the maximum likelihood decoding can be represented as

$$(\hat{x}_1, \hat{x}_2) = \arg \min_{(\hat{x}_1, \hat{x}_2) \in C} (|h_1|^2 + |h_2|^2 - 1)(|\hat{x}_1|^2 + |\hat{x}_2|^2) + d^2(\tilde{x}_1, \hat{x}_1) + d^2(\tilde{x}_2, \hat{x}_2) \quad (3.10)$$

where C is the set of all possible modulated symbol pairs $(\hat{x}_1, \hat{x}_2)$, $\tilde{x}_1$ and $\tilde{x}_2$ are two decision statistics constructed by combining the received signals with channel state information. The decision statistics are given by

$$\tilde{x}_1 = h_1^* r_1 + h_2 r_2^*$$
$$\tilde{x}_2 = h_2^* r_1 - h_1 r_2^* \tag{3.11}$$

Substituting r_1 and r_2 from (3.7) and (3.8), respectively, into (3.11), the decision statistics can be written as,

$$\tilde{x}_1 = (|h_1|^2 + |h_2|^2) x_1 + h_1^* n_1 + h_2 n_2^*$$
$$\tilde{x}_2 = (|h_1|^2 + |h_2|^2) x_2 - h_1 n_2^* + h_2^* n_1 \tag{3.12}$$

For a given channel realization h_1 and h_2, the decision statistics $\tilde{x}_i$, $i = 1, 2$, is only a function of x_i, $i = 1, 2$. Thus, the maximum likelihood decoding rule (3.10) can be separated into two independent decoding rules for x_1 and x_2, given by

$$\hat{x}_1 = \arg \min_{\hat{x}_1 \in S} (|h_1|^2 + |h_2|^2 - 1)|\hat{x}_1|^2 + d^2(\tilde{x}_1, \hat{x}_1) \tag{3.13}$$

and

$$\hat{x}_2 = \arg \min_{\hat{x}_2 \in S} (|h_1|^2 + |h_2|^2 - 1)|\hat{x}_2|^2 + d^2(\tilde{x}_2, \hat{x}_2) \tag{3.14}$$

respectively. For M-PSK signal constellations, $(|h_1|^2 + |h_2|^2 - 1)|\hat{x}_i|^2$, $i = 1, 2$, are constant for all signal points, given the channel fading coefficients. Therefore, the decision rules in (3.13) and (3.14) can be further simplified to

$$\hat{x}_1 = \arg \min_{\hat{x}_1 \in S} d^2(\tilde{x}_1, \hat{x}_1)$$
$$\hat{x}_2 = \arg \min_{\hat{x}_2 \in S} d^2(\tilde{x}_2, \hat{x}_2) \tag{3.15}$$

3.2.3 The Alamouti Scheme with Multiple Receive Antennas

The Alamouti scheme can be applied for a system with two transmit and n_R receive antennas. The encoding and transmission for this configuration is identical to the case of a single receive antenna. Let us denote by r_1^j and r_2^j the received signals at the jth receive antenna at time t and $t + T$, respectively.

$$r_1^j = h_{j,1} x_1 + h_{j,2} x_2 + n_1^j$$
$$r_2^j = -h_{j,1} x_2^* + h_{j,2} x_1^* + n_2^j \tag{3.16}$$

where $h_{j,i}$, $i = 1, 2$, $j = 1, 2, \ldots, n_R$, is the fading coefficient for the path from transmit antenna i to receive antenna j, and n_1^j and n_2^j are the noise signals for receive antenna j at time t and $t + T$, respectively.

The receiver constructs two decision statistics based on the linear combination of the received signals. The decision statistics, denoted by $\tilde{x}_1$ and $\tilde{x}_2$, are given by

$$\tilde{x}_1 = \sum_{j=1}^{n_R} h_{j,1}^* r_1^j + h_{j,2}(r_2^j)^*$$

$$= \sum_{i=1}^{2} \sum_{j=1}^{n_R} |h_{j,i}|^2 x_1 + \sum_{j=1}^{n_R} h_{j,1}^* n_1^j + h_{j,2}(n_2^j)^*$$

$$\tilde{x}_2 = \sum_{j=1}^{n_R} h_{j,2}^* r_1^j - h_{j,1}(r_2^j)^*$$

$$= \sum_{i=1}^{2} \sum_{j=1}^{n_R} |h_{j,i}|^2 x_2 + \sum_{j=1}^{n_R} h_{j,2}^* n_1^j - h_{j,1}(n_2^j)^* \qquad (3.17)$$

The maximum likelihood decoding rules for the two independent signals x_1 and x_2 are given by

$$\hat{x}_1 = \arg\min_{\hat{x}_1 \in S} \left[\left(\sum_{j=1}^{n_R} (|h_{j,1}|^2 + |h_{j,2}|^2) - 1 \right) |\hat{x}_1|^2 + d^2(\tilde{x}_1, \hat{x}_1) \right] \qquad (3.18)$$

$$\hat{x}_2 = \arg\min_{\hat{x}_2 \in S} \left[\left(\sum_{j=1}^{n_R} (|h_{j,1}|^2 + |h_{j,2}|^2) - 1 \right) |\hat{x}_2|^2 + d^2(\tilde{x}_2, \hat{x}_2) \right] \qquad (3.19)$$

For M-PSK modulation, all the signals in the constellation have equal energy. The maximum likelihood decoding rules are equivalent to the case of a single receive antennas, shown in (3.15).

3.2.4 Performance of the Alamouti Scheme

Now we show that due to the orthogonality between the sequences coming from the two transmit antennas, the Alamouti scheme can achieve the full transmit diversity of $n_T = 2$. Let us consider any two distinct code sequences $\mathbf{X}$ and $\hat{\mathbf{X}}$ generated by the inputs (x_1, x_2) and $(\hat{x}_1, \hat{x}_2)$, respectively, where $(x_1, x_2) \neq (\hat{x}_1, \hat{x}_2)$. The codeword difference matrix is given by

$$\mathbf{B}(\mathbf{X}, \hat{\mathbf{X}}) = \begin{bmatrix} x_1 - \hat{x}_1 & -x_2^* + \hat{x}_2^* \\ x_2 - \hat{x}_2 & x_1^* - \hat{x}_1^* \end{bmatrix} \qquad (3.20)$$

Since the rows of the code matrix are orthogonal, the rows of the codeword difference matrix are orthogonal as well. The codeword distance matrix is given by

$$\mathbf{A}(\mathbf{X}, \hat{\mathbf{X}}) = \mathbf{B}(\mathbf{X}, \hat{\mathbf{X}})\mathbf{B}^H(\mathbf{X}, \hat{\mathbf{X}})$$

$$= \begin{bmatrix} |x_1 - \hat{x}_1|^2 + |x_2 - \hat{x}_2|^2 & 0 \\ 0 & |x_1 - \hat{x}_1|^2 + |x_2 - \hat{x}_2|^2 \end{bmatrix} \qquad (3.21)$$

Since $(x_1, x_2) \neq (\hat{x}_1, \hat{x}_2)$, it is clear that the distance matrices of any two distinct codewords have a full rank of two. In other words, the Alamouti scheme can achieve a full transmit diversity of $n_T = 2$. The determinant of matrix $\mathbf{A}(\mathbf{X}, \hat{\mathbf{X}})$ is given by

$$\det(\mathbf{A}(\mathbf{X}, \hat{\mathbf{X}})) = (|x_1 - \hat{x}_1|^2 + |x_2 - \hat{x}_2|^2)^2 \qquad (3.22)$$

It is obvious from (3.21) that for the Alamouti scheme, the codeword distance matrix has two identical eigenvalues. The minimum eigenvalue is equal to the minimum squared Euclidean distance in the signal constellation. This means for the Alamouti scheme, the minimum distance between any two transmitted code sequences remains the same as in the uncoded system. Therefore, the Alamouti scheme does not provide any coding gain relative to the uncoded modulation scheme, i.e.

$$G_c = \frac{(\lambda_1 \lambda_2)^{1/2}}{d_u^2} = 1 \qquad (3.23)$$

For a pair of codewords $\mathbf{X}$ and $\hat{\mathbf{X}}$, let us define the squared Euclidean distance between the codewords, denoted by $d_E^2(\mathbf{X}, \hat{\mathbf{X}})$, as

$$d_E^2(\mathbf{X}, \hat{\mathbf{X}}) = |x_1 - \hat{x}_1|^2 + |x_2 - \hat{x}_2|^2 \qquad (3.24)$$

To obtain the average pairwise error probability, we compute the moment generating function (MGF) for the Alamouti scheme on slow Rayleigh fading channels based on (2.120) as

$$M_\Gamma(s) = \left(1 - sd_E^2(\mathbf{X}, \hat{\mathbf{X}}) \frac{E_s}{2N_0}\right)^{-2n_R} \qquad (3.25)$$

where

$$s = -\frac{1}{2\sin^2\theta}$$

The pairwise error probability in this case can be given by

$$P(\mathbf{X}, \hat{\mathbf{X}}) = \frac{1}{\pi} \int_0^{\pi/2} \left[1 + d_E^2(\mathbf{X}, \hat{\mathbf{X}}) \frac{E_s}{4N_0 \sin^2\theta}\right]^{-2n_R} d\theta \qquad (3.26)$$

$$= \frac{1}{2}\left[1 - \sqrt{\frac{d_E^2(\mathbf{X}, \hat{\mathbf{X}})E_s/4N_0}{1 + d_E^2(\mathbf{X}, \hat{\mathbf{X}})E_s/4N_0}}\right]$$

$$\cdot \sum_{k=0}^{2n_R-1} \binom{2k}{k} \left(\frac{1}{4(1 + d_E^2(\mathbf{X}, \hat{\mathbf{X}})E_s/4N_0)}\right)^k\right] \qquad (3.27)$$

The performance of the Alamouti transmit diversity scheme on slow Rayleigh fading channels is evaluated by simulation. In the simulations, it is assumed that fading from each transmit antenna to each receive antenna is mutually independent and that the receiver has the perfect knowledge of the channel coefficients. Figure 3.3 shows the bit error rate (BER)

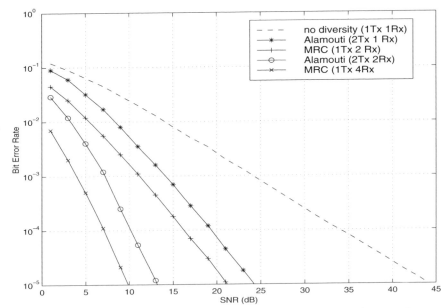

Figure 3.3 The BER performance of the BPSK Alamouti scheme with one and two receive antennas on slow Rayleigh fading channels

performance of the Alamouti scheme with coherent BPSK modulation against the signal-to-noise ratio (SNR) per receive antenna. The BER performance of two and four-branch receive diversity schemes with single transmit antenna and maximal ratio combining (MRC) is also shown in the figure for comparison. Furthermore, we assume that the total transmit power from two antennas for the Alamouti scheme is the same as the transmit power from the single transmit antenna for the MRC receiver diversity scheme and that it is normalized to one.

The simulation results show that the Alamouti scheme with two transmit antennas and a single receive antenna achieves the same diversity order as a two-branch MRC receive diversity scheme as the slopes of the two curves are the same. However, the performance of the Alamouti scheme is 3 dB worse. The 3-dB performance penalty is due to the fact that the energy radiated from each transmit antenna in the Alamouti scheme is half of that radiated from the single antenna in the MRC receive diversity scheme in order that the two schemes have the same total transmitted power. If each transmit antenna in the Alamouti scheme was to radiate the same energy as the single transmit antenna in the MRC receive diversity scheme, the Alamouti scheme would be equivalent to the MRC receive diversity scheme. Similarly, from the figure, we can see that the Alamouti scheme with two receive antennas achieves the same diversity as a four-branch MRC receive diversity scheme but its performance is 3 dB worse. In general, the Alamouti scheme with two transmit and n_R receive antennas has the same diversity gain as an MRC receive diversity scheme with one transmit and $2n_R$ receive antennas.

Figures 3.4 and 3.5 show the frame error rate (FER) performance of the Alamouti scheme with coherent BPSK and QPSK, respectively, on slow Rayleigh fading channels. The frame size is 130 symbols. One and two receive antennas are employed in the simulations.

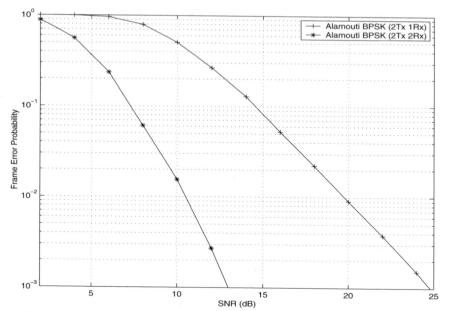

Figure 3.4   The FER performance of the BPSK Alamouti scheme with one and two receive antennas on slow Rayleigh fading channels

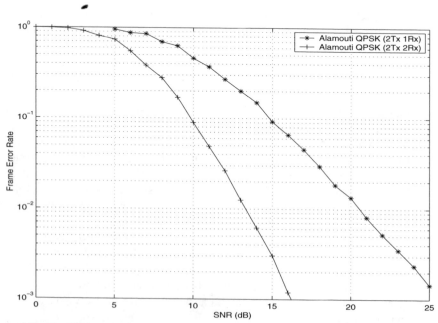

Figure 3.5   The FER performance of the QPSK Alamouti scheme with one and two receive antennas on slow Rayleigh fading channels

3.3   Space-Time Block Codes (STBC)

The Alamouti scheme achieves the full diversity with a very simple maximum-likelihood decoding algorithm. The key feature of the scheme is orthogonality between the sequences generated by the two transmit antennas. This scheme was generalized to an arbitrary number of transmit antennas by applying the theory of *orthogonal designs*. The generalized schemes are referred to as *space-time block codes* (STBCs) [3]. The space-time block codes can achieve the full transmit diversity specified by the number of the transmit antennas n_T, while allowing a very simple maximum-likelihood decoding algorithm, based only on linear processing of the received signals [3].

3.3.1   Space-Time Block Encoder

Figure 3.6 shows an encoder structure for space-time block codes. In general, a space-time block code is defined by an $n_T \times p$ transmission matrix $\mathbf{X}$. Here n_T represents the number of transmit antennas and p represents the number of time periods for transmission of one block of coded symbols.

Let us assume that the signal constellation consists of 2^m points. At each encoding operation, a block of km information bits are mapped into the signal constellation to select k modulated signals $x_1, x_2, \ldots, x_k$, where each group of m bits selects a constellation signal. The k modulated signals are encoded by a space-time block encoder to generate n_T parallel signal sequences of length p according to the transmission matrix $\mathbf{X}$. These sequences are transmitted through n_T transmit antennas simultaneously in p time periods.

In the space-time block code, the number of symbols the encoder takes as its input in each encoding operation is k. The number of transmission periods required to transmit the space-time coded symbols through the multiple transmit antennas is p. In other words, there are p space-time symbols transmitted from each antennas for each block of k input symbols. The *rate* of a space-time block code is defined as the ratio between the number of symbols the encoder takes as its input and the number of space-time coded symbols transmitted from each antenna. It is given by

$$R = k/p \tag{3.28}$$

The spectral efficiency of the space-time block code is given by

$$\eta = \frac{r_b}{B} = \frac{r_s m R}{r_s} = \frac{km}{p} \text{ bits/s/Hz} \tag{3.29}$$

where r_b and r_s are the bit and symbol rate, respectively, and B is the bandwidth.

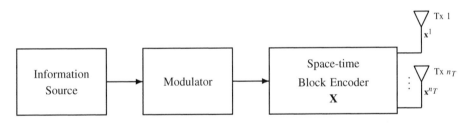

Figure 3.6   Encoder for STBC

The entries of the transmission matrix $\mathbf{X}$ are linear combinations of the k modulated symbols $x_1, x_2, \ldots, x_k$ and their conjugates $x_1^*, x_2^*, \ldots, x_k^*$. In order to achieve the full transmit diversity of n_T, the transmission matrix $\mathbf{X}$ is constructed based on orthogonal designs such that [3]

$$\mathbf{X} \cdot \mathbf{X}^H = c(|x_1|^2 + |x_2|^2 + \cdots + |x_k|^2)\mathbf{I}_{n_T} \tag{3.30}$$

where c is a constant, $\mathbf{X}^H$ is the Hermitian of $\mathbf{X}$ and $\mathbf{I}_{n_T}$ is an $n_T \times n_T$ identity matrix. The ith row of $\mathbf{X}$ represents the symbols transmitted from the ith transmit antenna consecutively in p transmission periods, while the jth column of $\mathbf{X}$ represents the symbols transmitted simultaneously through n_T transmit antennas at time j. The jth column of $\mathbf{X}$ is regarded as a space-time symbol transmitted at time j. The element of $\mathbf{X}$ in the ith row and jth column, $x_{i,j}$, $i = 1, 2, \ldots, n_T$, $j = 1, 2, \ldots, p$, represents the signal transmitted from the antenna i at time j.

It has been shown that the rate of a space-time block code with full transmit diversity is less than or equal to one, $R \leq 1$ [3]. The code with a full rate $R = 1$ requires no bandwidth expansion, while the code with rate $R < 1$ requires a bandwidth expansion of $1/R$. For space-time block codes with n_T transmit antennas, the transmission matrix is denoted by $\mathbf{X}_{n_T}$. The code is called the space-time block code with size n_T.

Note that orthogonal designs are applied to construct space-time block codes. The rows of the transmission matrix $\mathbf{X}_{n_T}$ are orthogonal to each other. This means that in each block, the signal sequences from any two transmit antennas are orthogonal. For example, if we assume that $\mathbf{x}_i = (x_{i,1}, x_{i,2}, \ldots, x_{i,p})$ is the transmitted sequence from the ith antenna, $i = 1, 2, \ldots, n_T$, we have

$$\mathbf{x}_i \cdot \mathbf{x}_j = \sum_{t=1}^{p} x_{i,t} \cdot x_{j,t}^* = 0, \quad i \neq j, \ i, j \in \{1, 2, \ldots, n_T\} \tag{3.31}$$

where $\mathbf{x}_i \cdot \mathbf{x}_j$ denotes the inner product of the sequences $\mathbf{x}_i$ and $\mathbf{x}_j$. The orthogonality enables to achieve the full transmit diversity for a given number of transmit antennas. In addition, it allows the receiver to decouple the signals transmitted from different antennas and consequently, a simple maximum likelihood decoding, based only on linear processing of the received signals.

3.4 STBC for Real Signal Constellations

Based on the type of the signal constellations, space-time block codes can be classified into space-time block codes with real signals and space-time block codes with complex signals.

In general, if an $n_T \times p$ real transmission matrix $\mathbf{X}_{n_T}$ with variables $x_1, x_2, \ldots, x_k$ satisfies

$$\mathbf{X}_{n_T} \cdot \mathbf{X}_{n_T}^T = c(|x_1|^2 + |x_2|^2 + \cdots + |x_k|^2)\mathbf{I}_{n_T} \tag{3.32}$$

the space-time block code can provide the full transmit diversity of n_T with a code rate of k/p.

For simplicity, we begin by considering the space-time block codes with a square transmission matrix $\mathbf{X}_{n_T}$. For any arbitrary real signal constellation, such as M-ASK, space-time block codes with $n_T \times n_T$ square transmission matrix $\mathbf{X}_{n_T}$ exist if and only if the number of transmit antennas $n_T = 2$, 4, or 8 [3]. These codes are of full rate $R = 1$ and offer the full transmit diversity of n_T. The transmission matrices are given by

$$\mathbf{X}_2 = \begin{bmatrix} x_1 & -x_2 \\ x_2 & x_1 \end{bmatrix} \tag{3.33}$$

for $n_T = 2$ transmit antennas,

$$\mathbf{X}_4 = \begin{bmatrix} x_1 & -x_2 & -x_3 & -x_4 \\ x_2 & x_1 & x_4 & -x_3 \\ x_3 & -x_4 & x_1 & x_2 \\ x_4 & x_3 & -x_2 & x_1 \end{bmatrix} \tag{3.34}$$

for $n_T = 4$ transmit antennas, and

$$\mathbf{X}_8 = \begin{bmatrix} x_1 & -x_2 & -x_3 & -x_4 & -x_5 & -x_6 & -x_7 & -x_8 \\ x_2 & x_1 & -x_4 & x_3 & -x_6 & x_5 & x_8 & -x_7 \\ x_3 & x_4 & x_1 & -x_2 & -x_7 & -x_8 & x_5 & x_6 \\ x_4 & -x_3 & x_2 & x_1 & -x_8 & x_7 & -x_6 & x_5 \\ x_5 & x_6 & x_7 & x_8 & x_1 & -x_2 & -x_3 & -x_4 \\ x_6 & -x_5 & x_8 & -x_7 & x_2 & x_1 & x_4 & -x_3 \\ x_7 & -x_8 & -x_5 & x_6 & x_3 & -x_4 & x_1 & x_2 \\ x_8 & x_7 & -x_6 & -x_5 & x_4 & x_3 & -x_2 & x_1 \end{bmatrix} \tag{3.35}$$

for $n_T = 8$ transmit antennas.

The square transmission matrices have orthogonal rows with entries $\pm x_1, \pm x_2, \ldots, \pm x_k$. From the matrices, it can be observed that for a block of k modulated message symbols, both the number of transmit antennas n_T and the number of time periods p required to transmit the block code are equal to the message block length k. For example, consider a space-time block code, specified by $\mathbf{X}_4$, with four transmit antennas. The encoder takes $k = 4$ real modulated symbols x_1, x_2, x_3, and x_4 as its input and generates the code sequences. At time $t = 1$, signals x_1, x_2, x_3, and x_4 are transmitted from antenna 1 through 4, respectively. At time $t = 2$, signals $-x_2$, x_1, $-x_4$, and x_3 are transmitted from antenna 1 through 4, respectively, and so on. For this example, four transmit antennas and four time periods are needed to transmit four message symbols. Therefore, no bandwidth expansion is required for this code, or in other words the code can achieve the full code rate of 1.

It is desirable to construct the full code rate $R = 1$ transmission schemes for any number of transmit antennas, since full rate codes are bandwidth efficient. In general, for n_T transmit antennas, the minimum value of transmission periods p to achieve the full rate is given by [3]

$$\min(2^{4c+d}) \tag{3.36}$$

where the minimization is taken over the set

$$c, d \mid 0 \leq c, \; 0 \leq d \leq 4, \text{ and } 8c + 2^d \geq n_T \tag{3.37}$$

For $n_T \le 8$, the minimum value of p is given by

$$n_T = 2, \quad p = 2 \tag{3.38}$$

$$n_T = 3, \quad p = 4$$

$$n_T = 4, \quad p = 4$$

$$n_T = 5, \quad p = 8$$

$$n_T = 6, \quad p = 8$$

$$n_T = 7, \quad p = 8$$

$$n_T = 8, \quad p = 8$$

These values provide guidelines to construct full rate space-time block codes. According to these values, non-square transmission matrices $\mathbf{X}_3$, $\mathbf{X}_5$, $\mathbf{X}_6$, and $\mathbf{X}_7$ were constructed based on real orthogonal designs for full rate and full diversity space-time block codes with sizes of 3, 5, 6 and 7, respectively. These matrices are given as follows [3]

$$\mathbf{X}_3 = \begin{bmatrix} x_1 & -x_2 & -x_3 & -x_4 \\ x_2 & x_1 & x_4 & -x_3 \\ x_3 & -x_4 & x_1 & x_2 \end{bmatrix} \tag{3.39}$$

$$\mathbf{X}_5 = \begin{bmatrix} x_1 & -x_2 & -x_3 & -x_4 & -x_5 & -x_6 & -x_7 & -x_8 \\ x_2 & x_1 & -x_4 & x_3 & -x_6 & x_5 & x_8 & -x_7 \\ x_3 & x_4 & x_1 & -x_2 & -x_7 & -x_8 & x_5 & x_6 \\ x_4 & -x_3 & x_2 & x_1 & -x_8 & x_7 & -x_6 & x_5 \\ x_5 & x_6 & x_7 & x_8 & x_1 & -x_2 & -x_3 & -x_4 \end{bmatrix} \tag{3.40}$$

$$\mathbf{X}_6 = \begin{bmatrix} x_1 & -x_2 & -x_3 & -x_4 & -x_5 & -x_6 & -x_7 & -x_8 \\ x_2 & x_1 & -x_4 & x_3 & -x_6 & x_5 & x_8 & -x_7 \\ x_3 & x_4 & x_1 & -x_2 & -x_7 & -x_8 & x_5 & x_6 \\ x_4 & -x_3 & x_2 & x_1 & -x_8 & x_7 & -x_6 & x_5 \\ x_5 & x_6 & x_7 & x_8 & x_1 & -x_2 & -x_3 & -x_4 \\ x_6 & -x_5 & x_8 & -x_7 & x_2 & x_1 & x_4 & -x_3 \end{bmatrix} \tag{3.41}$$

$$\mathbf{X}_7 = \begin{bmatrix} x_1 & -x_2 & -x_3 & -x_4 & -x_5 & -x_6 & -x_7 & -x_8 \\ x_2 & x_1 & -x_4 & x_3 & -x_6 & x_5 & x_8 & -x_7 \\ x_3 & x_4 & x_1 & -x_2 & -x_7 & -x_8 & x_5 & x_6 \\ x_4 & -x_3 & x_2 & x_1 & -x_8 & x_7 & -x_6 & x_5 \\ x_5 & x_6 & x_7 & x_8 & x_1 & -x_2 & -x_3 & -x_4 \\ x_6 & -x_5 & x_8 & -x_7 & x_2 & x_1 & x_4 & -x_3 \\ x_7 & -x_8 & -x_5 & x_6 & x_3 & -x_4 & x_1 & x_2 \end{bmatrix} \tag{3.42}$$

To explain the concepts involved, let us consider X_6, the matrix for space-time block codes with six transmit antennas. The input to the space-time block encoder is a block of eight symbols, $x_1, x_2, \ldots, x_8$, from a real signal constellation. After the encoding, the coded symbols are transmitted through six transmit antennas in eight transmission periods, e.g., from the third antenna, signals x_3, x_4, x_1, $-x_2$, $-x_7$, $-x_8$, x_5 and x_6 are transmitted in the first, second, third, etc., up to the eight transmission period, successively. It is obvious that the number of symbols that the encoder takes as its input is equal to the number of time periods required to transmit these symbols. Thus, this scheme requires no bandwidth expansion. This property holds true for all above examples.

3.5   STBC for Complex Signal Constellations

In general, if an $n_T \times p$ complex transmission matrix $\mathbf{X}_{n_T}$ with complex entries x_1, $x_2, \ldots, x_k$ satisfies

$$\mathbf{X}_{n_T} \cdot \mathbf{X}_{n_T}^H = c(|x_1|^2 + |x_2|^2 + \cdots + |x_k|^2)\mathbf{I}_{n_T} \tag{3.43}$$

the space-time block code can provide the full transmit diversity of n_T with a code rate of k/p.

The Alamouti scheme can be regarded as a space-time block code with complex signals for two transmit antennas. The transmission matrix is represented by

$$\mathbf{X}_2^c = \begin{bmatrix} x_1 & -x_2^* \\ x_2 & x_1^* \end{bmatrix} \tag{3.44}$$

This scheme provides the full diversity of 2 and the full rate of 1.

The Alamouti scheme is unique in that it is the only space-time block code with an $n_T \times n_T$ complex transmission matrix to achieve the full rate [3]. If the number of the transmit antennas is larger than two, the code design goal is to construct high-rate complex transmission matrices $\mathbf{G}_{n_T}^c$ with low decoding complexity that achieve the full diversity. In addition, similar to real orthogonal designs, the value of p must be minimized in order to minimize the decoding delay.

For an arbitrary complex signal constellation, there are space-time block codes that can achieve a rate of 1/2 for any given number of transmit antennas. For example, complex transmission matrices $\mathbf{X}_3^c$ and $\mathbf{X}_4^c$ are orthogonal designs for space-time block codes with three and four transmit antennas, respectively. These codes have the rate 1/2. Matrices $\mathbf{X}_3^c$ and $\mathbf{X}_4^c$ are given below [3]

$$\mathbf{X}_3^c = \begin{bmatrix} x_1 & -x_2 & -x_3 & -x_4 & x_1^* & -x_2^* & -x_3^* & -x_4^* \\ x_2 & x_1 & x_4 & -x_3 & x_2^* & x_1^* & x_4^* & -x_3^* \\ x_3 & -x_4 & x_1 & x_2 & x_3^* & -x_4^* & x_1^* & x_2^* \end{bmatrix} \tag{3.45}$$

$$\mathbf{X}_4^c = \begin{bmatrix} x_1 & -x_2 & -x_3 & -x_4 & x_1^* & -x_2^* & -x_3^* & -x_4^* \\ x_2 & x_1 & x_4 & -x_3 & x_2^* & x_1^* & x_4^* & -x_3^* \\ x_3 & -x_4 & x_1 & x_2 & x_3^* & -x_4^* & x_1^* & x_2^* \\ x_4 & x_3 & -x_2 & x_1 & x_4^* & x_3^* & -x_2^* & x_1^* \end{bmatrix} \tag{3.46}$$

It can be shown that the inner product of any two rows of these matrices is zero, which proves the orthogonality of these structures. With matrix $\mathbf{X}_3^c$, four complex symbols are taken at a time, and transmitted via three transmit antennas in eight symbol periods; hence the transmission rate is 1/2. With regard to matrix $\mathbf{X}_4^c$, four symbols from a complex constellation are taken at a time and transmitted via four transmit antennas in eight symbol periods, resulting in a transmission rate of 1/2 as well.

A more involved linear processing results in a higher rate for space-time block codes with a complex constellation and more than two antennas. The following two matrices $\mathbf{X}_3^h$

and $\mathbf{X}_4^h$ are complex generalized orthogonal designs for space-time block codes with rate 3/4 [3]

$$
\mathbf{X}_3^h = \begin{bmatrix} x_1 & -x_2^* & \dfrac{x_3^*}{\sqrt{2}} & \dfrac{x_3^*}{\sqrt{2}} \\[2mm] x_2 & x_1^* & \dfrac{x_3^*}{\sqrt{2}} & \dfrac{-x_3^*}{\sqrt{2}} \\[2mm] \dfrac{x_3}{\sqrt{2}} & \dfrac{x_3}{\sqrt{2}} & \dfrac{(-x_1 - x_1^* + x_2 - x_2^*)}{2} & \dfrac{(x_2 + x_2^* + x_1 - x_1^*)}{2} \end{bmatrix}
\tag{3.47}
$$

$$
\mathbf{X}_4^h = \begin{bmatrix} x_1 & -x_2 & \dfrac{x_3^*}{\sqrt{2}} & \dfrac{x_3^*}{\sqrt{2}} \\[2mm] x_2 & x_1 & \dfrac{x_3^*}{\sqrt{2}} & \dfrac{-x_3^*}{\sqrt{2}} \\[2mm] \dfrac{x_3}{\sqrt{2}} & \dfrac{x_3}{\sqrt{2}} & \dfrac{(-x_1 - x_1^* + x_2 - x_2^*)}{2} & \dfrac{(x_2 + x_2^* + x_1 - x_1^*)}{2} \\[2mm] \dfrac{x_3}{\sqrt{2}} & \dfrac{-x_3}{\sqrt{2}} & \dfrac{(-x_2 - x_2^* + x_1 - x_1^*)}{2} & \dfrac{-(x_1 + x_1^* + x_2 - x_2^*)}{2} \end{bmatrix}
\tag{3.48}
$$

Another rate 3/4 space-time block code with three antennas over complex signal constellations shown in [12] is given by

$$
\mathbf{X}_3^{h'} = \begin{bmatrix} x_1 & x_2^* & x_3^* & 0 \\ -x_2 & x_1^* & 0 & -x_3^* \\ -x_3 & 0 & x_1^* & x_2^* \end{bmatrix}
\tag{3.49}
$$

3.6 Decoding of STBC

Now let us consider the decoding algorithm. For simplicity, we start with a STBC described by a square transmission matrix over a real signal constellation, such as $\mathbf{X}_2$, $\mathbf{X}_4$ and $\mathbf{X}_8$. In this case, the first column of the transmission matrix is a vector $[x_1, x_2, \ldots, x_{n_T}]^T$. The other columns of $\mathbf{X}_{n_T}$ are all permutations of the first column with possible different signs. Let ϵ_t denote the permutations of the symbols from the first column to the t-th column. The row position of x_i in the t-th column is represented by $\epsilon_t(i)$ and the sign of x_i in the t-th column is denoted by $\mathrm{sgn}_t(i)$.

We assume that the channel coefficients $h_{j,i}(t)$ are constant over p symbol periods.

$$
h_{j,i}(t) = h_{j,i}, \quad t = 1, 2, \ldots, p
\tag{3.50}
$$

In deriving the maximum likelihood decoding, similar to the one for the Alamouti scheme, we can construct the decision statistics for the transmitted signal x_i as

$$
\tilde{x}_i = \sum_{t=1}^{n_T} \sum_{j=1}^{n_R} \mathrm{sgn}_t(i) \cdot r_t^j \cdot h_{j,\epsilon_t(i)}^*
\tag{3.51}
$$

where $i = 1, 2, \ldots, n_T$. Because of the orthogonality of pairwise rows of the transmission matrix, minimizing the maximum likelihood metric

$$\sum_{t=1}^{n_T} \sum_{j=1}^{n_R} \left| r_t^j - \sum_{i=1}^{n_T} h_{j,i} x_t^i \right|^2 \tag{3.52}$$

is equivalent to minimizing the joint decision metric

$$\sum_{i=1}^{n_T} \left[|\tilde{x}_i - x_i|^2 + \left(\sum_{t=1}^{n_T} \sum_{j=1}^{n_R} |h_{j,t}|^2 - 1 \right) |x_i|^2 \right] \tag{3.53}$$

Since the value of $\tilde{x}_i$ only depends on the code symbol x_i, given the received signals, the path coefficients and the structure of the orthogonal transmission matrix, minimizing the joint decision metric is further equivalent to minimizing each individual decision metric

$$|\tilde{x}_i - x_i|^2 + \left(\sum_{t=1}^{n_T} \sum_{j=1}^{n_R} |h_{j,t}|^2 - 1 \right) |x_i|^2 \tag{3.54}$$

This algorithm simplifies the joint decoding significantly by performing separate decoding for each transmitted signal. Due to the orthogonality, the decision statistics for the desired transmitted signal x_i is independent of the other transmitted signals x_j, $j = 1, 2, \ldots, n_T$, $j \neq i$. The decoding metric for each signal x_i is based on linear processing of its decision statistics $\tilde{x}_i$.

For the STBC with a non-square transmission matrix over real signal constellations, such as $\mathbf{X}_3$, $\mathbf{X}_5$, $\mathbf{X}_6$ and $\mathbf{X}_7$, the decision statistics at the receiver can be constructed as

$$\tilde{x}_i = \sum_{t \in \eta(i)} \sum_{j=1}^{n_R} \mathrm{sgn}_t(i) \cdot r_t^j \cdot h_{j,\epsilon_t(i)}^* \tag{3.55}$$

where $i = 1, 2, \ldots, p$, and $\eta(i)$ is the set of columns of the transmission matrix, in which x_i appears. For example, consider the transmission matrix $\mathbf{X}_3$ with three transmit antennas

$$\begin{aligned} \eta(1) &= \{1, 2, 3\}; & \eta(2) &= \{1, 2, 4\}; \\ \eta(3) &= \{1, 3, 4\}; & \eta(4) &= \{2, 3, 4\} \end{aligned} \tag{3.56}$$

The decision metric for each individual signal x_i is given by

$$|\tilde{x}_i - x_i|^2 + \left(\sum_{t=1}^{n_T} \sum_{j=1}^{n_R} |h_{j,t}|^2 - 1 \right) |x_i|^2 \tag{3.57}$$

Similar decoding algorithms can be derived for STBC with complex signal constellations. For the rate 1/2 STBC $\mathbf{X}_3^c$ and $\mathbf{X}_4^c$, the decision statistics $\tilde{x}_i$ can be represented by

$$\tilde{x}_i = \sum_{t \in \eta(i)} \sum_{j=1}^{n_R} \mathrm{sgn}_t(i) \cdot \tilde{r}_t^j(i) \cdot \tilde{h}_{j,\epsilon_t(i)} \tag{3.58}$$

where

$$\tilde{r}_t^j(i) = \begin{cases} r_t^j & \text{if } x_i \text{ belongs to the } t\text{-th column of } \mathbf{X}_{n_T}^c \\ (r_t^j)^* & \text{if } x_i^* \text{ belongs to the } t\text{-th column of } \mathbf{X}_{n_T}^c \end{cases} \tag{3.59}$$

and

$$\tilde{h}_{j,\epsilon_t(i)} = \begin{cases} h_{j,\epsilon_t(i)}^* & \text{if } x_i \text{ belongs to the } t\text{-th column of } \mathbf{X}_{n_T}^c \\ h_{j,\epsilon_t(i)} & \text{if } x_i^* \text{ belongs to the } t\text{-th column of } \mathbf{X}_{n_T}^c \end{cases} \tag{3.60}$$

The decision metric is given by

$$|\tilde{x}_i - x_i|^2 + \left(2 \sum_{t=1}^{n_T} \sum_{j=1}^{n_R} |h_{j,t}|^2 - 1 \right) |x_i|^2 \tag{3.61}$$

As an example, let us calculate the decision statistics for the STBC $\mathbf{X}_3^c$ and $\mathbf{X}_4^c$.
According to (3.58), the decision statistics for $\mathbf{X}_3^c$ can be expressed as

$$\tilde{x}_1 = \sum_{j=1}^{n_R} (r_1^j h_{j,1}^* + r_2^j h_{j,2}^* + r_3^j h_{j,3}^* + (r_5^j)^* h_{j,1} + (r_6^j)^* h_{j,2} + (r_7^j)^* h_{j,3})$$

$$= \rho_3 x_1 + \sum_{j=1}^{n_R} (n_1^j h_{j,1}^* + n_2^j h_{j,2}^* + n_3^j h_{j,3}^* + (n_5^j)^* h_{j,1} + (n_6^j)^* h_{j,2} + (n_7^j)^* h_{j,3})$$

$$\tilde{x}_2 = \sum_{j=1}^{n_R} (r_1^j h_{j,2}^* - r_2^j h_{j,1}^* + r_4^j h_{j,3}^* + (r_5^j)^* h_{j,2} - (r_6^j)^* h_{j,1} + (r_8^j)^* h_{j,3})$$

$$= \rho_3 x_2 + \sum_{j=1}^{n_R} (n_1^j h_{j,2}^* - n_2^j h_{j,1}^* + n_4^j h_{j,3}^* + (n_5^j)^* h_{j,2} - (n_6^j)^* h_{j,1} + (n_8^j)^* h_{j,3})$$

$$\tilde{x}_3 = \sum_{j=1}^{n_R} (r_1^j h_{j,3}^* - r_3^j h_{j,1}^* - r_4^j h_{j,2}^* + (r_5^j)^* h_{j,3} - (r_7^j)^* h_{j,1} - (r_8^j)^* h_{j,2})$$

$$= \rho_3 x_3 + \sum_{j=1}^{n_R} (n_1^j h_{j,3}^* - n_3^j h_{j,1}^* - n_4^j h_{j,2}^* + (n_5^j)^* h_{j,3} - (n_7^j)^* h_{j,1} - (n_8^j)^* h_{j,2})$$

$$\tilde{x}_4 = \sum_{j=1}^{n_R} (-r_2^j h_{j,3}^* + r_3^j h_{j,2}^* - r_4^j h_{j,1}^* - (r_6^j)^* h_{j,3} + (r_7^j)^* h_{j,2} - (r_8^j)^* h_{j,1})$$

$$= \rho_3 x_4 + \sum_{j=1}^{n_R} (-n_2^j h_{j,3}^* + n_3^j h_{j,2}^* - n_4^j h_{j,1}^* - (n_6^j)^* h_{j,3} + (n_7^j)^* h_{j,2} - (n_8^j)^* h_{j,1})$$

where

$$\rho_3 = 2 \sum_{i=1}^{3} \sum_{j=1}^{n_R} |h_{j,i}|^2. \tag{3.62}$$

The decision statistics for $\mathbf{X}_4^c$ can be expressed as

$$\tilde{x}_1 = \sum_{j=1}^{n_R}(r_1^j h_{j,1}^* + r_2^j h_{j,2}^* + r_3^j h_{j,3}^* + r_4^j h_{j,4}^* + (r_5^j)^* h_{j,1} + (r_6^j)^* h_{j,2} + (r_7^j)^* h_{j,3} + (r_8^j)^* h_{j,4})$$

$$= \rho_4 x_1 + \sum_{j=1}^{n_R}(n_1^j h_{j,1}^* + n_2^j h_{j,2}^* + n_3^j h_{j,3}^* + n_4^j h_{j,4}^*$$

$$+ (n_5^j)^* h_{j,1} + (n_6^j)^* h_{j,2} + (n_7^j)^* h_{j,3} + (n_8^j)^* h_{j,4})$$

$$\tilde{x}_2 = \sum_{j=1}^{n_R}(r_1^j h_{j,2}^* - r_2^j h_{j,1}^* - r_3^j h_{j,4}^* + r_4^j h_{j,3}^* + (r_5^j)^* h_{j,2} - (r_6^j)^* h_{j,1} - (r_7^j)^* h_{j,4} + (r_8^j)^* h_{j,3})$$

$$= \rho_4 x_2 + \sum_{j=1}^{n_R}(n_1^j h_{j,2}^* - n_2^j h_{j,1}^* - n_3^j h_{j,4}^* + n_4^j h_{j,3}^*$$

$$+ (n_5^j)^* h_{j,2} - (n_6^j)^* h_{j,1} - (n_7^j)^* h_{j,4} + (n_8^j)^* h_{j,3})$$

$$\tilde{x}_3 = \sum_{j=1}^{n_R}(r_1^j h_{j,3}^* + r_2^j h_{j,4}^* - r_3^j h_{j,1}^* - r_4^j h_{j,2}^* + (r_5^j)^* h_{j,3} + (r_6^j)^* h_{j,4} - (r_7^j)^* h_{j,1} - (r_8^j)^* h_{j,2})$$

$$= \rho_4 x_3 + \sum_{j=1}^{n_R}(n_1^j h_{j,3}^* + n_2^j h_{j,4}^* - n_3^j h_{j,1}^* - n_4^j h_{j,2}^* + (n_5^j)^* h_{j,3}$$

$$+ (n_6^j)^* h_{j,4} - (n_7^j)^* h_{j,1} - (n_8^j)^* h_{j,2})$$

$$\tilde{x}_4 = \sum_{j=1}^{n_R}(-r_1^j h_{j,4}^* - r_2^j h_{j,3}^* + r_3^j h_{j,2}^* - r_4^j h_{j,1}^* - (r_5^j)^* h_{j,4} - (r_6^j)^* h_{j,3} + (r_7^j)^* h_{j,2} - (r_8^j)^* h_{j,1})$$

$$= \rho_4 x_4 + \sum_{j=1}^{n_R}(-n_1^j h_{j,4}^* - n_2^j h_{j,3}^* + n_3^j h_{j,2}^* - n_4^j h_{j,1}^*$$

$$- (n_5^j)^* h_{j,4} - (n_6^j)^* h_{j,3} + (n_7^j)^* h_{j,2} - (n_8^j)^* h_{j,1})$$

where

$$\rho_4 = 2\sum_{i=1}^{4}\sum_{j=1}^{n_R}|h_{j,i}|^2. \tag{3.63}$$

To decode the rate 3/4 code $\mathbf{X}_3^h$, the receiver constructs the decision statistics as follows

$$\tilde{x}_1 = \sum_{j=1}^{n_R}\left(r_1^j h_{j,1}^* + (r_2^j)^* h_{j,2} + \frac{(r_4^j - r_3^j)h_{j,3}^*}{2} - \frac{(r_4^j - r_3^j)^* h_{j,3}}{2}\right)$$

$$\tilde{x}_2 = \sum_{j=1}^{n_R}\left(r_1^j h_{j,2}^* - (r_2^j)^* h_{j,1} + \frac{(r_4^j + r_3^j)h_{j,3}^*}{2} + \frac{(-r_3^j + r_4^j)^* h_{j,3}}{2}\right)$$

$$\tilde{x}_3 = \sum_{j=1}^{n_R} \left(\frac{(r_1^j + r_2^j)h_{j,3}^*}{\sqrt{2}} + \frac{(r_3^j)^*(h_{j,1} + h_{j,2})}{\sqrt{2}} + \frac{(r_4^j)^*(h_{j,1} - h_{j,2})}{\sqrt{2}} \right)$$

Similarly, to decode the rate 3/4 code $\mathbf{X}_4^h$, the receiver constructs the decision statistics as follows

$$\tilde{x}_1 = \sum_{j=1}^{n_R}(r_1^j h_{j,1}^* + (r_2^j)^* h_{j,2} + \frac{(r_4^j - r_3^j)(h_{j,3}^* - h_{j,4}^*)}{2} - \frac{(r_3^j + r_4^j)^*(h_{j,3} + h_{j,4})}{2})$$

$$\tilde{x}_2 = \sum_{j=1}^{n_R}(r_1^j h_{j,2}^* - (r_2^j)^* h_{j,1} + \frac{(r_4^j + r_3^j)(h_{j,3}^* - h_{j,4}^*)}{2} + \frac{(-r_3^j + r_4^j)^*(h_{j,3} + h_{j,4})}{2})$$

$$\tilde{x}_3 = \sum_{j=1}^{n_R} \left(\frac{(r_1^j + r_2^j)h_{j,3}^*}{\sqrt{2}} \frac{(r_1^j - r_2^j)h_{j,4}^*}{\sqrt{2}} + \frac{(r_3^j)^*(h_{j,1} + h_{j,2})}{\sqrt{2}} + \frac{(r_4^j)^*(h_{j,1} - h_{j,2})}{\sqrt{2}} \right)$$

3.7 Performance of STBC

In this section, we show simulation results for the performance of STBC on Rayleigh fading channels. In the simulations, it is assumed that the receiver knows the perfect channel state information. The bit error rate (BER) and the symbol error rate (SER) for STBC with 3 bits/s/Hz and a variable number of transmit antennas are shown in Figs. 3.7 and 3.8, respectively. The performance of an uncoded 8-PSK is plotted in the figures for comparison.

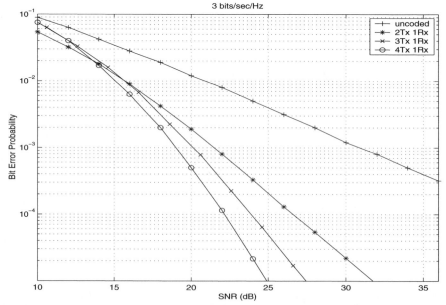

Figure 3.7 Bit error rate performance for STBC of 3 bits/s/Hz on Rayleigh fading channels with one receive antenna

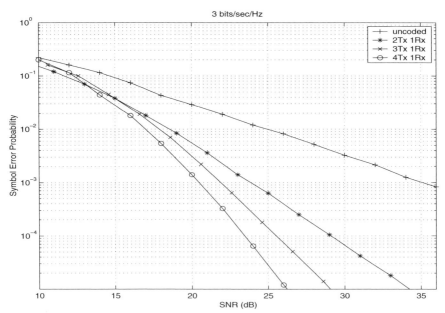

Figure 3.8 Symbol error rate performance for STBC of 3 bits/s/Hz on Rayleigh fading channels with one receive antenna

For transmission with two transmit antennas, the rate one code $\mathbf{X}_2^c$ and 8-PSK modulation are employed. For three and four transmit antennas, 16-QAM and the rate 3/4 codes $\mathbf{X}_3^h$ and $\mathbf{X}_4^h$ are used, respectively. Therefore, the transmission rate is 3 bits/s/Hz in all cases. For Fig. 3.7, we can see that at the BER of 10^{-5}, the code $\mathbf{X}_4^h$ is better by about 7dB and 2.5 dB than the code $\mathbf{X}_2^c$ and the code $\mathbf{X}_3^h$, respectively.

Figures 3.9 and 3.10 show BER and SER performance, respectively, for STBC of 2 bits/s/Hz with two, three, and four transmit antennas and one receive antenna on Rayleigh fading channels. The STBC with two transmit antennas is the rate one code $\mathbf{X}_2^c$ with QPSK modulation. The STBC with three and four transmit antennas are the rate 1/2 codes $\mathbf{X}_3^c$ and $\mathbf{X}_4^c$, respectively, with 16-QAM modulation. It can be observed that at the BER of 10^{-5}, the code with four transmit antennas gains about 5 dB and 3 dB relative to the codes with two and three transmit antennas, respectively.

The BER and SER performance for the codes with 1 bit/s/Hz, a variable number of the transmit antennas and a single receive antenna are illustrated in Figs. 3.11 and 3.12, respectively. The STBC with two transmit antennas is the rate one code $\mathbf{X}_2^c$ with BPSK modulation. The STBC with three and four transmit antennas are the rate 1/2 codes $\mathbf{X}_3^c$ and $\mathbf{X}_4^c$, respectively, with QPSK modulation. It can be observed that at the BER of 10^{-5}, the code with four transmit antennas is superior by about 8 dB and 2.5 dB to the codes with two and three transmit antennas, respectively.

The simulation results show that increasing the number of transmit antennas can provide a significant performance gain. The increase in decoding complexity for STBC with a large number of transmit antennas is very little due to the fact that only linear processing is required for decoding. In order to further improve the code performance, it is possible to concatenate an outer code, such as trellis or turbo code, with an STBC as an inner code.

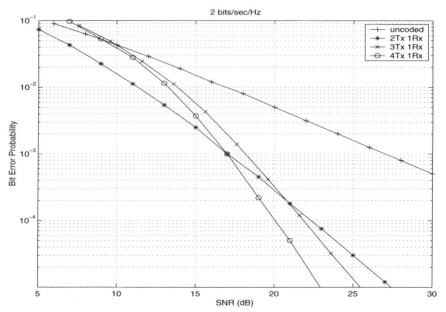

Figure 3.9 Bit error rate performance for STBC of 2 bits/s/Hz on Rayleigh fading channels with one receive antenna

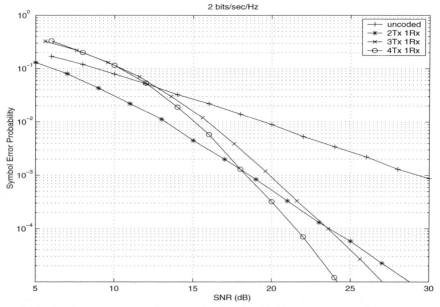

Figure 3.10 Symbol error rate performance for STBC of 2 bits/s/Hz on Rayleigh fading channels with one receive antenna

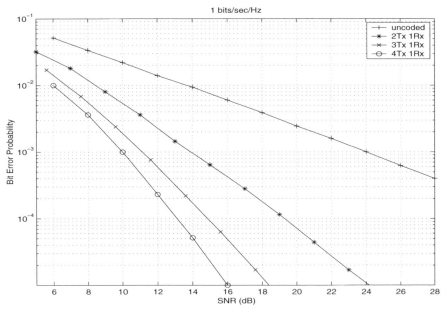

Figure 3.11 Bit error rate performance for STBC of 1 bits/s/Hz on Rayleigh fading channels with one receive antenna

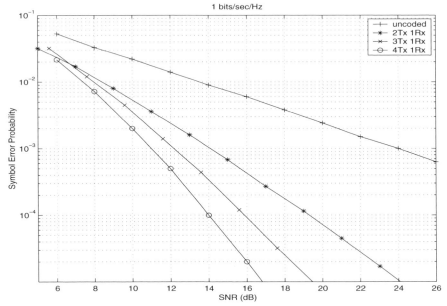

Figure 3.12 Symbol error rate performance for STBC of 1 bits/s/Hz on Rayleigh fading channels with one receive antenna

3.8 Effect of Imperfect Channel Estimation on Performance

In this section, the effect of imperfect channel state information on the code performance is discussed. We start with the description of the channel estimation method used in the simulations [6]. The channel fading coefficients are estimated by inserting pilot sequences in the transmitted signals. It is assumed that the channel is constant over the duration of a frame and independent between the frames. In general, with n_T transmit antennas we need to have n_T different pilot sequences $P_1, P_2, \ldots, P_{n_T}$. At the beginning of each frame transmitted from antennas i, a pilot sequence P_i consisting of k symbols

$$P_i = (P_{i,1}, P_{i,2}, \ldots, P_{i,k}) \tag{3.64}$$

is appended. Since the signals at the receive antennas are linear superpositions of all transmitted signals, the pilot sequences $P_1, P_2, \ldots, P_{n_T}$ are designed to be orthogonal to each other.

During the channel estimation, the received signal at antenna j and time t can be represented by

$$r_t^j = \sum_{i=1}^{n_T} h_{j,i} P_{i,t} + n_t^j \tag{3.65}$$

where $h_{j,i}$ is the fading coefficient for the path from transmit antenna i to the receive antenna j and n_t^j is the noise sample at receive antenna j and time t. The received signal and noise sequence at antenna j can be represented as

$$\mathbf{r}^j = (r_1^j, r_2^j, \ldots, r_k^j)$$

$$\mathbf{n}^j = (n_1^j, n_2^j, \ldots, n_k^j) \tag{3.66}$$

The receiver estimates the channel fading coefficients $h_{j,i}$ by using the observed sequences $\mathbf{r}^j$. Since the pilot sequences $P_1, P_2, \ldots, P_{n_T}$ are orthogonal, the minimum mean square error (MMSE) estimate of $h_{j,i}$ is given by [6]

$$\tilde{h}_{j,i} = \frac{\mathbf{r}^j \cdot P_i}{||P_i||^2}$$

$$= h_{j,i} + \frac{\mathbf{n}^j \cdot P_i}{||P_i||^2}$$

$$= h_{j,i} + e_{j,i} \tag{3.67}$$

where $e_{j,i}$ is the estimation error due to the noise, given by

$$e_{j,i} = \frac{\mathbf{n}^j \cdot P_i}{P_i \cdot P_i} \tag{3.68}$$

Since n_t^j is a zero-mean complex Gaussian random variable with single-sided power spectral density N_0, the estimation error $e_{j,i}$ has a zero mean and single-sided power spectral density N_0/k [6].

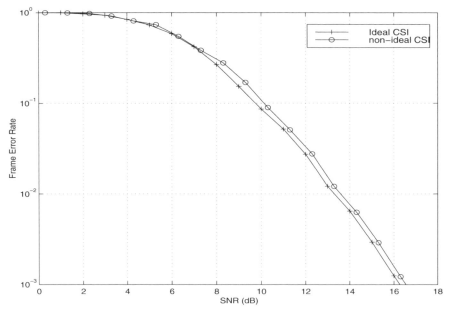

Figure 3.13   Performance of the STBC with 2 bits/s/Hz on correlated slow Rayleigh fading channels with two transmit and two receive antennas

The performance of the STBC with imperfect channel state information at the receiver is shown in Fig. 3.13. In the simulation, QPSK modulation and the rate one code $\mathbf{X}_2^c$ with two transmit and two receive antennas are employed. It is assumed that the channel is described as a slow Rayleigh fading model with constant coefficients over a frame of 130 symbols. The pilot sequence inserted in each frame has a length of 10 symbols. The simulation results show that due to imperfect channel estimation, the code performance is degraded by about 0.3 dB compared to the case of ideal channel state information. Note that the degradation in code performance also accounts for the loss of the signal energy by appending the pilot sequences.

If the number of transmit antennas is small, the performance degradation due to the channel estimation error is small. However, as the number of transmit antennas increases, the sensitivity of the system to channel estimation error increases [6].

3.9   Effect of Antenna Correlation on Performance

Figure 3.14 shows the performance of the STBC with 2 bits/s/Hz on correlated slow Rayleigh fading channels with two transmit and two receive antennas. We assume that the transmit antennas are not correlated but the receive antennas are correlated. The receive antenna correlation matrix is given by

$$\Theta_R = \begin{bmatrix} 1 & \theta \\ \theta & 1 \end{bmatrix} \tag{3.69}$$

where θ is the correlation factor between the receive antennas. In the simulation, the correlation factor is chosen to be 0.25, 0.5, 0.75 and 1. It can be observed that the code performance

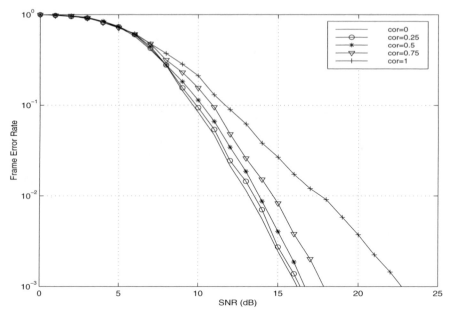

Figure 3.14 Performance of the STBC with 2 bits/s/Hz on correlated slow Rayleigh fading channels with two transmit and two receive antennas

is slightly degraded when the correlation factor is 0.25. However, relative to the case with uncorrelated antennas, the code is getting worse by 0.7 dB and 1.6 dB at a FER of 10^{-2} for the correlation factors of 0.5 and 0.75, respectively. When the channels are fully correlated, the penalty on the code performance is about 4.2 dB at the same FER.

Bibliography

[1] S. M. Alamouti, "A simple transmit diversity technique for wireless communications", *IEEE Journal Select. Areas Commun.*, vol. 16, no. 8, pp. 1451–1458, Oct. 1998.

[2] A. Wittneben, "A new bandwidth efficient transmit antenna modulation diversity scheme for linear digital modulation", in *Proc. IEEE ICC93*, pp. 1630–1634, 1993.

[3] V. Tarokh, H. Jafarkhani and A. R. Calderbank, "Space-time block codes from orthogonal designs", *IEEE Trans. Inform. Theory*, vol. 45, no. 5, pp. 1456–1467, July 1999.

[4] V. Tarokh, H. Jafarkhani and A. R. Calderbank, "Space-time block coding for wireless communications: performance results", *IEEE J. Select. Areas Commun.*, vol. 17, no. 3, pp. 451–460, Mar. 1999.

[5] V. Tarokh, A. Naguib, N. Seshadri and A. R. Calderbank, "Combined array processing and space-time coding", *IEEE Trans. Inform. Theory*, vol. 45, no. 4, pp. 1121–1128, May 1999.

[6] V. Tarokh, A. Naguib, N. Seshadri and A. R. Calderbank, "Space-time codes for high data rate wireless communication: Performance criteria in the presence of channel estimation errors, mobility, and multiple paths", *IEEE Trans. Commun.*, vol. 47, no. 2, pp. 199–207, Feb. 1999.

[7] V. Tarokh and H. Jafarkhani, "A differential detection scheme for transmit diversity", *IEEE J. Select. Areas Commun.,* vol. 18, pp. 1169–1174, July 2000.

[8] H. Jafarkhani and V. Tarokh, "Multiple transmit antenna differential detection from generalized orthogonal designs", *IEEE Trans. Inform. Theory,* vol. 47, no. 6, pp. 2626–2631, Sep. 2001.

[9] B. L. Hughes, "Differential space-time modulation", *IEEE Trans. Inform. Theory,* vol. 46, no. 7, pp. 2567–2578, Nov. 2000.

[10] B. M. Hochwald and T. L. Marzetta, "Unitary space-time modulation for multiple-antenna communications in Rayleigh flat fading", *IEEE Trans. Inform. Theory,* vol. 46, no. 2, pp. 543–564, Mar. 2000.

[11] B. M. Hochwald and W. Sweldens, "Differential unitary space-time modulation", *IEEE Trans. Communi.,* vol. 48, no. 12, Dec. 2000.

[12] B. Hochwald, T. L. Marzetta and C. B. Papadias, "A transmitter diversity scheme for wideband CDMA systems based on space-time spreading", *IEEE Journal on Selected Areas in Commun.,* vol. 19, no. 1, Jan. 2001, pp. 48–60.

[13] T. S. Rappaport, *Wireless Communications: Principles and Practice,* Prentice Hall, 1996.

4

Space-Time Trellis Codes

4.1 Introduction

Space-time block codes can achieve a maximum possible diversity advantage with a simple decoding algorithm. It is very attractive because of its simplicity. However, no coding gain can be provided by space-time block codes, while non-full rate space-time block codes can introduce bandwidth expansion. In this chapter, we consider a joint design of error control coding, modulation, transmit and receive diversity to develop an effective signalling scheme, space-time trellis codes (STTC), which is able to combat the effects of fading. STTC was first introduced by Tarokh, Seshadri and Calderbank [4]. It was widely discussed and explored in the literature as STTC can simultaneously offer a substantial coding gain, spectral efficiency, and diversity improvement on flat fading channels.

In this chapter, we introduce an encoder structure of space-time trellis codes. By applying the space-time code design criteria, optimum space-time trellis coded M-PSK schemes for various numbers of transmit antennas and spectral efficiencies are constructed for slow and fast fading channels. The code performance is evaluated by simulations and compared against the capacity limit. The effects of imperfect channel estimation and correlated antenna elements on the code performance are also presented.

4.2 Encoder Structure for STTC

For space-time trellis codes, the encoder maps binary data to modulation symbols, where the mapping function is described by a trellis diagram.

Let us consider an encoder of space-time trellis coded M-PSK modulation with n_T transmit antennas as shown in Fig. 4.1. The input message stream, denoted by $\mathbf{c}$, is given by

$$\mathbf{c} = (\mathbf{c}_0, \mathbf{c}_1, \mathbf{c}_2, \ldots, \mathbf{c}_t, \ldots) \tag{4.1}$$

where $\mathbf{c}_t$ is a group of $m = \log_2 M$ information bits at time t and given by

$$\mathbf{c}_t = (c_t^1, c_t^2, \ldots, c_t^m) \tag{4.2}$$

Space-Time Coding   Branka Vucetic and Jinhong Yuan
© 2003 John Wiley & Sons, Ltd   ISBN: 0-470-84757-3

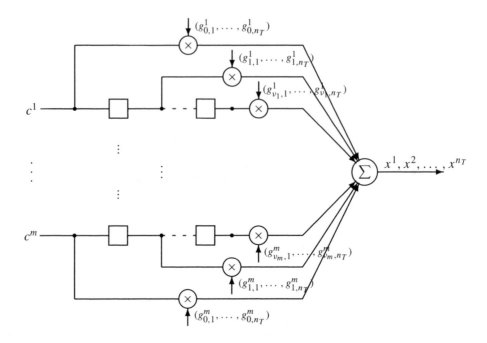

Figure 4.1   Encoder for STTC

The encoder maps the input sequence into an M-PSK modulated signal sequence, which is given by

$$\mathbf{x} = (\mathbf{x}_0, \mathbf{x}_1, \mathbf{x}_2, \ldots, \mathbf{x}_t, \ldots) \tag{4.3}$$

where $\mathbf{x}_t$ is a *space-time symbol* at time t and given by

$$\mathbf{x}_t = (x_t^1, x_t^2, \ldots, x_t^{n_T})^T \tag{4.4}$$

The modulated signals, $x_t^1, x_t^2, \ldots, x_t^{n_T}$, are transmitted simultaneously through n_T transmit antennas.

4.2.1   Generator Description

In the STTC encoder as shown in Fig. 4.1, m binary input sequences $\mathbf{c}^1, \mathbf{c}^2, \ldots, \mathbf{c}^m$ are fed into the encoder, which consists of m feedforward shift registers. The k-th input sequence $\mathbf{c}^k = (c_0^k, c_1^k, c_2^k, \ldots, c_t^k, \ldots)$, $k = 1, 2, \ldots m$, is passed to the k-th shift register and multiplied by an encoder coefficient set. The multiplier outputs from all shift registers are added modulo M, giving the encoder output $\mathbf{x} = (\mathbf{x}^1, \mathbf{x}^2, \ldots, \mathbf{x}^{n_T})$. The connections between the shift register elements and the modulo M adder can be described by the following m multiplication coefficient set sequences

$$\mathbf{g}^1 = [(g_{0,1}^1, g_{0,2}^1, \ldots, g_{0,n_T}^1), (g_{1,1}^1, g_{1,2}^1, \ldots, g_{1,n_T}^1), \ldots, (g_{v_1,1}^1, g_{v_1,2}^1, \ldots, g_{v_1,n_T}^1)]$$

$$\mathbf{g}^2 = [(g_{0,1}^2, g_{0,2}^2, \ldots, g_{0,n_T}^2), (g_{1,1}^2, g_{1,2}^2, \ldots, g_{1,n_T}^2), \ldots, (g_{v_2,1}^2, g_{v_2,2}^2, \ldots, g_{v_2,n_T}^2)]$$

$$\vdots$$

$$\mathbf{g}^m = [(g_{0,1}^m, g_{0,2}^m, \ldots, g_{0,n_T}^m), (g_{1,1}^m, g_{1,2}^m, \ldots, g_{1,n_T}^m), \ldots, (g_{v_m,1}^m, g_{v_m,2}^m, \ldots, g_{v_m,n_T}^m)]$$

where $g_{j,i}^k$, $k = 1, 2, \ldots, m$, $j = 1, 2, \ldots, v_k$, $i = 1, 2, \ldots, n_T$, is an element of the M-PSK constellation set, and v_k is the memory order of the k-th shift register.

The encoder output at time t for transmit antenna i, denoted by x_t^i, can be computed as

$$x_t^i = \sum_{k=1}^{m} \sum_{j=0}^{v_k} g_{j,i}^k c_{t-j}^k \quad \text{mod } M, \quad i = 1, 2, \ldots, n_T \tag{4.5}$$

These outputs are elements of an M-PSK signal set. Modulated signals form the space-time symbol transmitted at time t

$$\mathbf{x}_t = (x_t^1, x_t^2, \ldots, x_t^{n_T})^T. \tag{4.6}$$

The space-time trellis coded M-PSK can achieve a bandwidth efficiency of m bits/s/Hz. The total memory order of the encoder, denoted by v, is given by

$$v = \sum_{k=1}^{m} v_k \tag{4.7}$$

where v_k, $k = 1, 2, \ldots, m$, is the memory order for the k-th encoder branch. The value of v_k for M-PSK constellations is determined by

$$v_k = \left\lfloor \frac{v + k - 1}{\log_2 M} \right\rfloor \tag{4.8}$$

The total number of states for the trellis encoder is 2^v. The m multiplication coefficient set sequences are also called the *generator sequences*, since they can fully describe the encoder structure.

For example, let us consider a simple space-time trellis coded QPSK with two transmit antennas. The encoder consists of two feedforward shift registers. The encoder structure for the scheme with memory order of v is shown in Fig. 4.2.

Two binary input streams $\mathbf{c}^1 = (c_0^1, c_1^1, \ldots, c_t^1, \ldots)$ and $\mathbf{c}^2 = (c_0^2, c_1^2, \ldots, c_t^2, \ldots)$ are fed into the upper and lower encoder registers. The memory orders of the upper and lower encoder registers are v_1 and v_2, respectively, where $v = v_1 + v_2$. The two input streams are delayed and multiplied by the coefficient pairs

$$\mathbf{g}^1 = [(g_{0,1}^1, g_{0,2}^1), (g_{1,1}^1, g_{1,2}^1), \ldots, (g_{v_1,1}^1, g_{v_1,2}^1)]$$
$$\mathbf{g}^2 = [(g_{0,1}^2, g_{0,2}^2), (g_{1,1}^2, g_{1,2}^2), \ldots, (g_{v_2,1}^2, g_{v_2,2}^2)] \tag{4.9}$$

respectively, where $g_{j,i}^k \in \{0, 1, 2, 3\}$, $k = 1, 2$; $i = 1, 2$; $j = 0, 1, \ldots, v_k$. The multiplier outputs are added modulo 4, giving the output

$$x_t^i = \sum_{k=1}^{2} \sum_{j=0}^{v_k} g_{j,i}^k c_{t-j}^k \quad \text{mod } 4, \quad i = 1, 2 \tag{4.10}$$

The adder outputs x_t^1 and x_t^2 are points from a QPSK constellation. They are transmitted simultaneously through the first and second antenna, respectively.

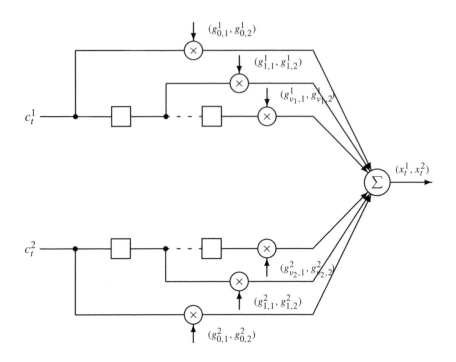

Figure 4.2   STTC encoder for two transmit antennas

4.2.2   Generator Polynomial Description

The STTC encoder can also be described in generator polynomial format. Let us consider a space-time encoder with two transmit antennas as shown in Fig. 4.2. The input binary sequence to the upper shift register can be represented as

$$\mathbf{c}^1(D) = c_0^1 + c_1^1 D + c_2^1 D^2 + c_3^1 D^3 + \cdots \tag{4.11}$$

Similarly, the binary input sequence to the lower shift register can be written as

$$\mathbf{c}^2(D) = c_0^2 + c_1^2 D + c_2^2 D^2 + c_3^2 D^3 + \cdots \tag{4.12}$$

where c_j^k, $j = 0, 1, 2, 3, \ldots$, $k = 1, 2$, are binary symbols 0, 1. The feedforward generator polynomial for the upper encoder and transmit antenna i, where $i = 1, 2$, can be written as

$$\mathbf{G}_i^1(D) = g_{0,i}^1 + g_{1,i}^1 D + \cdots + g_{\nu_1,i}^1 D^{\nu_1} \tag{4.13}$$

where $g_{j,i}^1$, $j = 0, 1, \ldots, \nu_1$ are non-binary coefficients that can take values 0, 1, 2, 3 for QPSK modulation and ν_1 is the memory order of the upper encoder. Similarly, the feedforward generator polynomial for the lower encoder and transmit antenna i, where $i = 1, 2$, can be written as

$$\mathbf{G}_i^2(D) = g_{0,i}^2 + g_{1,i}^2 D + \cdots + g_{\nu_2,i}^2 D^{\nu_1} \tag{4.14}$$

where $g_{j,i}^2$, $j = 1, 2, \ldots, \nu_2$, are non-binary coefficients that can take values 0, 1, 2, 3 for QPSK modulation and ν_2 is the memory order of the lower encoder. The encoded symbol sequence transmitted from antenna i is given by

$$\mathbf{x}^i(D) = \mathbf{c}^1(D)\mathbf{G}_i^1(D) + \mathbf{c}^2(D)\mathbf{G}_i^2(D) \quad \text{mod } 4 \tag{4.15}$$

The relationship in (4.15) can be written in the following form

$$\mathbf{x}^i(D) = \begin{bmatrix} \mathbf{c}^1(D) & \mathbf{c}^2(D) \end{bmatrix} \begin{bmatrix} \mathbf{G}_i^1(D) \\ \mathbf{G}_i^2(D) \end{bmatrix} \quad \text{mod } 4 \tag{4.16}$$

A systematic recursive STTC can be obtained by setting

$$\mathbf{G}_1(D) = \begin{bmatrix} 2 \\ 1 \end{bmatrix}$$

which means that the output of the first antenna is obtained by directly mapping the input sequences $\mathbf{c}^1$ and $\mathbf{c}^2$ into a QPSK sequence.

4.2.3   Example

Let us assume that the generator sequences of a 4-state space-time trellis coded QPSK scheme with 2 transmit antennas are

$$\mathbf{g}^1 = [(02), (20)]$$
$$\mathbf{g}^2 = [(01), (10)]$$

The trellis structure for the code is shown in Fig. 4.3. The trellis consists of $2^\nu = 4$ states, represented by state nodes. The encoder takes $m = 2$ bits as its input at each time. There are $2^m = 4$ branches leaving from each state corresponding to four different input patterns. Each branch is labelled by $c_t^1 \, c_t^2/x_t^1 \, x_t^2$, where c_t^1 and c_t^2 are a pair of encoder input bits, and x_t^1 and x_t^2 represent two coded QPSK symbols transmitted through antennas 1 and 2, respectively. The row listed next to a state node in Fig. 4.3 indicates the branch labels for transitions from that state corresponding to the encoder inputs 00, 01, 10, and 11, respectively.

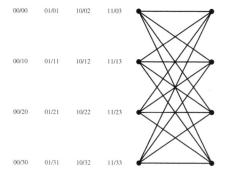

00/00	01/01	10/02	11/03
00/10	01/11	10/12	11/13
00/20	01/21	10/22	11/23
00/30	01/31	10/32	11/33

Figure 4.3   Trellis structure for a 4-state space-time coded QPSK with 2 antennas

Assume that the input sequence is

$$\mathbf{c} = (10, 01, 11, 00, 01, \ldots)$$

The output sequence generated by the space-time trellis encoder is given by

$$\mathbf{x} = (02, 21, 13, 30, 01, \ldots)$$

The transmitted signal sequences from the two transmit antennas are

$$\mathbf{x}^1 = (0, 2, 1, 3, 0, \ldots)$$
$$\mathbf{x}^2 = (2, 1, 3, 0, 1, \ldots)$$

Note that this example is actually a delay diversity scheme since the signal sequence transmitted from the first antenna is a delayed version of the signal sequence from the second antenna.

For STTC, the decoder employs the Viterbi algorithm to perform maximum likelihood decoding. Assuming that perfect CSI is available at the receiver, for a branch labelled by $(x_t^1, x_t^2, \ldots, x_t^{n_T})$, the branch metric is computed as the squared Euclidean distance between the hypothesised received symbols and the actual received signals as

$$\sum_{j=1}^{n_R} \left| r_t^j - \sum_{i=1}^{n_T} h_{j,i}^t x_t^i \right|^2 \tag{4.17}$$

The Viterbi algorithm selects the path with the minimum path metric as the decoded sequence.

4.3   Design of Space-Time Trellis Codes on Slow Fading Channels

Optimum space-time trellis coded M-PSK schemes for a given number of transmit antennas and memory order are designed by applying the design criteria introduced in Chapter 2.

For a given encoder structure, a set of encoder coefficients is determined by minimizing the error probability. It is important to note that the STTC encoder structure cannot guarantee geometrical uniformity of the code [19]. Therefore, the search was conducted over all possible pairs of paths in the code trellis.

As discussed in Chapter 2, the code design depends on the code parameter r and the number of receive antennas n_R in the system. If $r n_R < 4$, the rank & determinant criteria are applicable, while the trace criterion is used if $r n_R \geq 4$.

To maximize the minimum rank r for matrix $\mathbf{A}(\mathbf{X}, \hat{\mathbf{X}})$ means to make the matrix full rank such as $r = n_T$. However, the full rank is not always achievable due to the restriction of the trellis structure. For a space-time trellis code with the memory order of v, the length of an error event, denoted by l, can be lower-bounded as [13]

$$l \geq \lfloor v/2 \rfloor + 1 \tag{4.18}$$

As we know the rank of $\mathbf{A}(\mathbf{X}, \hat{\mathbf{X}})$ is the same as the rank of $\mathbf{B}(\mathbf{X}, \hat{\mathbf{X}})$. For an error event path of length l in the trellis, $\mathbf{B}(\mathbf{X}, \hat{\mathbf{X}})$ is a matrix of size $n_T \times l$, which results in the maximum

Table 4.1 Upper bound of the rank values for STTC

	$n_T = 2$	$n_T = 3$	$n_T = 4$	$n_T = 5$	$n_T \geq 6$
$\nu = 2$	2	2	2	2	2
$\nu = 3$	2	2	2	2	2
$\nu = 4$	2	3	3	3	3
$\nu = 5$	2	3	3	3	3
$\nu = 6$	2	3	4	4	4

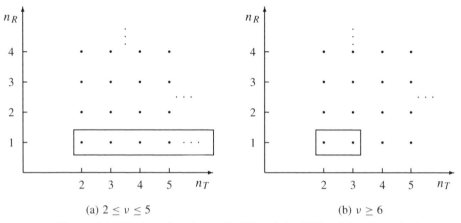

(a) $2 \leq \nu \leq 5$ (b) $\nu \geq 6$

Figure 4.4 The boundary for applicability of the TSC and the trace criteria

achievable rank is $\min(n_T, l)$. Consider the constraint of the error event length in (4.18). The maximum achievable rank for the code is determined by the value of $\min(n_T, \lfloor \nu/2 \rfloor + 1)$. The upper bound of the rank values for STTC with various numbers of transmit antennas and memory orders is listed in Table 4.1. It is clear from the table that the full rank is achievable only for STTC with two transmit antennas. For STTC with three and four transmit antennas, in order to achieve the full rank, the memory order of the encoder is at least four and six, respectively.

In the code design the number of receive antennas is normally not considered as a design parameter. Considering the relationship between the maximum achievable rank, the number of the transmit antennas and the memory order of an STTC shown in Table 4.1, we can visualize the cases in which each criteria set is applicable in the code design. The boundary between the rank & determinant criteria and the trace criterion is illustrated in Fig. 4.4. The points in the rectangular blocks are the cases where the rank & determinant criteria are to be employed. The trace criterion can be used for all other cases. The figure suggests that the rank & determinant criteria only apply to the systems with one receive antenna.

4.3.1 Optimal STTC Based on the Rank & Determinant Criteria

QPSK Codes with Two, Three and Four Transmit Antennas

For space-time codes with two or three transmit antennas and one receive antenna, the maximum possible diversity order rn_R is always less than 4. Following the rank & determinant

Table 4.2 Optimal QPSK STTC with two transmit antennas for slow fading channels based on rank & determinant criteria

code	ν	generator sequences	rank	det
TSC	2	$\mathbf{g}^1 = [(0, 2), (2, 0)]$ $\mathbf{g}^2 = [(0, 1), (1, 0)]$	2	4.0
BBH	2	$\mathbf{g}^1 = [(2, 2), (1, 0)]$ $\mathbf{g}^2 = [(0, 2), (3, 1)]$	2	8.0
optimum	2	$\mathbf{g}^1 = [(0, 2), (1, 0)]$ $\mathbf{g}^2 = [(2, 2), (0, 1)]$	2	8.0
TSC	3	$\mathbf{g}^1 = [(0, 2), (2, 0)]$ $\mathbf{g}^2 = [(0, 1), (1, 0), (2, 2)]$	2	12.0
BBH	3	$\mathbf{g}^1 = [(2, 2), (2, 0)]$ $\mathbf{g}^2 = [(0, 1), (1, 0), (2, 2)]$	2	12.0
optimum	3	$\mathbf{g}^1 = [(0, 2), (2, 0)]$ $\mathbf{g}^2 = [(2, 1), (1, 2), (0, 2)]$	2	16.0
TSC	4	$\mathbf{g}^1 = [(0, 2), (2, 0), (0, 2)]$ $\mathbf{g}^2 = [(0, 1), (1, 2), (2, 0)]$	2	12.0
BBH	4	$\mathbf{g}^1 = [(0, 2), (2, 0), (0, 2)]$ $\mathbf{g}^2 = [(2, 1), (1, 2), (2, 0)]$	2	20.0
optimum	4	$\mathbf{g}^1 = [(0, 2), (1, 2), (2, 2)]$ $\mathbf{g}^2 = [(2, 0), (1, 1), (0, 2)]$	2	32.0
TSC	5	$\mathbf{g}^1 = [(0, 2), (2, 2), (3, 3)]$ $\mathbf{g}^2 = [(0, 1), (1, 1), (2, 0), (2, 2)]$	2	12.0
optimum	5	$\mathbf{g}^1 = [(2, 0), (2, 3), (0, 2)]$ $\mathbf{g}^2 = [(2, 2), (1, 0), (1, 2), (2, 2)]$	2	36.0
optimum	6	$\mathbf{g}^1 = [(1, 2), (2, 2), (0, 3), (2, 0)]$ $\mathbf{g}^2 = [(2, 0), (2, 0), (1, 3), (0, 2)]$	2	48.0

criteria, optimum QPSK STTC were generated through computer search. The encoder coefficients and code parameters are listed in Table 4.2 for QPSK codes with memory orders 2 to 6.

The Tarokh/Seshadri/Calderbank (TSC) codes [4] and Baro/Bauch/Hansmann (BBH) codes [11] are considered as references. Their parameters are also shown in Table 4.2. It is clear from the table that for a given memory order, the proposed optimum STTC have the same minimum rank as the TSC and BBH codes, but a larger minimum determinant, which can result in a larger coding gain.

Optimal QPSK STTC with three and four transmit antennas based on the rank & determinant criteria are generated through systematic search. In order to achieve a full rank the memory order of an STTC is at least 4 for three transmit antennas and 6 for four transmit antennas, as shown in Table 4.1. The optimum codes with the full rank and the largest minimum determinant over all codeword distance matrices are presented in Table 4.3.

8-PSK Codes with Two Transmit Antennas

Optimal 8-PSK STTC with two transmit antennas and memory orders 3 to 5 based on the rank & determinant criteria are shown in Table 4.4. The TSC codes [4] are also considered

Table 4.3 Optimal QPSK STTC with three and four transmit antennas for slow fading channels based on rank & determinant criteria

n_T	ν	generator sequences	rank	det	tr
3	4	$\mathbf{g}^1 = [(0,0,2),(0,1,2),(2,3,1)]$ $\mathbf{g}^2 = [(2,0,0),(1,2,0),(2,3,3)]$	3	32	16
3	5	$\mathbf{g}^1 = [(0,2,1),(2,0,0),(0,0,2)]$ $\mathbf{g}^2 = [(3,1,0),(3,2,1),(3,2,2),(2,0,0)]$	3	64	14
3	6	$\mathbf{g}^1 = [(1,1,2),(2,1,2),(1,2,0),(2,0,0)]$ $\mathbf{g}^2 = [(0,3,0),(0,3,2),(2,2,1),(0,2,2)]$	3	96	18
4	6	$\mathbf{g}^1 = [(0,3,0,2),(2,3,0,2),(2,1,1,1),(2,2,2,0)]$ $\mathbf{g}^2 = [(3,0,2,0),(2,2,0,2),(0,0,3,2),(0,2,2,0)]$	4	64	26

Table 4.4 Optimal 8-PSK STTC with two transmit antennas for slow fading channels based on rank & determinant criteria

code	ν	generator sequences	r	det	tr
TSC	3	$\mathbf{g}^1 = [(0,4),(4,0)]$ $\mathbf{g}^2 = [(0,2),(2,0)]$ $\mathbf{g}^3 = [(0,1),(5,0)]$	2	2.0	4.0
optimum	3	$\mathbf{g}^1 = [(0,2),(2,0)]$ $\mathbf{g}^2 = [(0,4),(4,0)]$ $\mathbf{g}^3 = [(4,5),(1,4)]$	2	4.0	4.0
TSC	4	$\mathbf{g}^1 = [(0,4),(4,4)]$ $\mathbf{g}^2 = [(0,2),(2,2)]$ $\mathbf{g}^3 = [(0,1),(5,1),(1,5)]$	2	3.515	6.0
optimum	4	$\mathbf{g}^1 = [(0,4),(4,0)]$ $\mathbf{g}^2 = [(0,2),(2,0)]$ $\mathbf{g}^3 = [(2,0),(6,5),(1,4)]$	2	4.0	4.0
TSC	5	$\mathbf{g}^1 = [(0,4),(4,4)]$ $\mathbf{g}^2 = [(0,2),(2,2),(2,2)]$ $\mathbf{g}^3 = [(0,1),(5,1),(3,7)]$	2	3.515	8.0
optimum	5	$\mathbf{g}^1 = [(0,4),(4,4)]$ $\mathbf{g}^2 = [(0,2),(2,2),(2,0)]$ $\mathbf{g}^3 = [(3,5),(0,0),(4,0)]$	2	7.029	7.172

as references. It is clear from the table that for a given memory order, the proposed optimum codes have the same minimum rank as the TSC codes, but a larger minimum determinant which can result in a larger coding gain.

4.3.2 Optimal STTC Based on the Trace Criterion

QPSK Codes with Two, Three and Four Transmit Antennas

For space-time codes with $n_T n_R$ greater than or equal to 4, it is possible to find a code with a minimum diversity order $r n_R$, greater than or equal to 4. In this case, the code design

Table 4.5 Optimal QPSK STTC with two transmit antennas for slow fading channels based on trace criterion

n_T	ν	generator sequences	r	det	tr
2	2	$\mathbf{g}^1 = [(0, 2), (1, 2)]$ $\mathbf{g}^2 = [(2, 3), (2, 0)]$	2	4.0	10.0
2	3	$\mathbf{g}^1 = [(2, 2), (2, 1)]$ $\mathbf{g}^2 = [(2, 0), (1, 2), (0, 2)]$	2	8.0	12.0
2	4	$\mathbf{g}^1 = [(1, 2), (1, 3), (3, 2)]$ $\mathbf{g}^2 = [(2, 0), (2, 2), (2, 0)]$	2	8.0	16.0
2	5	$\mathbf{g}^1 = [(0, 2), (2, 3), (1, 2)]$ $\mathbf{g}^2 = [(2, 2), (1, 2), (2, 3), (2, 0)]$	2	20.0	16.0
2	6	$\mathbf{g}^1 = [(0, 2), (3, 1), (3, 3), (3, 2)]$ $\mathbf{g}^2 = [(2, 2), (2, 2), (0, 0), (2, 0)]$	2	16.0	18.0

Table 4.6 Optimal QPSK STTC with three transmit antennas for slow fading channels based on trace criterion

n_T	ν	generator sequences	r	det	tr
3	2	$\mathbf{g}^1 = [(0, 2, 2), (1, 2, 3)]$ $\mathbf{g}^2 = [(2, 3, 3), (2, 0, 2)]$	2	–	16.0
3	3	$\mathbf{g}^1 = [(2, 2, 2), (2, 1, 1)]$ $\mathbf{g}^2 = [(2, 0, 3), (1, 2, 0), (0, 2, 2)]$	2	–	20.0
3	4	$\mathbf{g}^1 = [(1, 2, 1), (1, 3, 2), (3, 2, 1)]$ $\mathbf{g}^2 = [(2, 0, 2), (2, 2, 0), (2, 0, 2)]$	2	–	24.0
3	5	$\mathbf{g}^1 = [(0, 2, 2), (2, 3, 3), (1, 2, 2)]$ $\mathbf{g}^2 = [(2, 2, 0), (1, 2, 2), (2, 3, 1), (2, 0, 0)]$	2	–	24.0
3	6	$\mathbf{g}^1 = [(0, 2, 2), (3, 1, 0), (3, 3, 2), (3, 2, 1)]$ $\mathbf{g}^2 = [(2, 2, 0), (2, 2, 2), (0, 0, 3), (2, 0, 1)]$	2	–	28.0

should be based on the criteria set II (the trace criterion), which means the minimum trace of the codeword distance matrix should be maximized. Table 4.5 shows the parameters of the optimum QPSK STTC with two transmit antennas found by code search based on this set of design criteria. These codes can achieve the full diversity and a larger coding gain relative to the TSC and the BBH codes with the same memory order, due to a larger trace value.

Optimal QPSK STTC with three and four transmit antennas based on the trace criterion are shown in Tables 4.6 and 4.7, respectively. These codes have a smaller minimum rank, but a larger minimum trace, relative to the codes presented in Table 4.3.

8-PSK Codes with Two, Three and Four Transmit Antennas

Optimal 8-PSK STTC with two, three and four transmit antennas, designed by the trace criterion, are shown in Tables 4.8, 4.9 and 4.10. respectively.

Table 4.7 Optimal QPSK STTC with four transmit antennas for slow fading channels based on trace criterion

n_T	ν	generator sequences	r	det	tr
4	2	$\mathbf{g}^1 = [(0, 2, 2, 0), (1, 2, 3, 2)]$ $\mathbf{g}^2 = [(2, 3, 3, 2), (2, 0, 2, 1)]$	2	–	20.0
4	3	$\mathbf{g}^1 = [(2, 2, 2, 2), (2, 1, 1, 2)]$ $\mathbf{g}^2 = [(2, 0, 3, 1), (1, 2, 0, 3), (0, 2, 2, 1)]$	2	–	26.0
4	4	$\mathbf{g}^1 = [(1, 2, 1, 1), (1, 3, 2, 2), (3, 2, 1, 3)]$ $\mathbf{g}^2 = [(2, 0, 2, 2), (2, 2, 0, 0), (2, 0, 2, 2)]$	≤ 3	–	32.0
4	5	$\mathbf{g}^1 = [(0, 2, 2, 2), (2, 3, 3, 2), (1, 2, 2, 1)]$ $\mathbf{g}^2 = [(2, 2, 0, 1), (1, 2, 2, 0), (2, 3, 1, 0), (2, 0, 0, 2)]$	≤ 3	–	36.0
4	6	$\mathbf{g}^1 = [(0, 2, 2, 1), (3, 1, 0, 2), (3, 3, 2, 2), (3, 2, 1, 3)]$ $\mathbf{g}^2 = [(2, 2, 0, 2), (2, 2, 2, 0), (0, 0, 3, 1), (2, 0, 1, 2)]$	3	–	38.0

Table 4.8 Optimal 8-PSK STTC with two transmit antennas for slow fading channels based on trace criterion

n_T	ν	generator sequences	r	det	tr
2	3	$\mathbf{g}^1 = [(2, 1), (3, 4)]$ $\mathbf{g}^2 = [(4, 6), (2, 0)]$ $\mathbf{g}^3 = [(0, 4), (4, 0)]$	2	2.0	7.172
2	4	$\mathbf{g}^1 = [(2, 4), (3, 7)]$ $\mathbf{g}^2 = [(4, 0), (6, 6)]$ $\mathbf{g}^3 = [(7, 2), (0, 7), (4, 4)]$	2	0.686	8.0
2	5	$\mathbf{g}^1 = [(0, 4), (4, 4)]$ $\mathbf{g}^2 = [(0, 2), (2, 3), (2, 2)]$ $\mathbf{g}^3 = [(4, 2), (4, 2), (3, 7)]$	2	2.343	8.586

Table 4.9 Optimal 8-PSK STTC codes with three transmit antennas for slow fading channels based on trace criterion

n_T	ν	generator sequences	r	det	tr
3	3	$\mathbf{g}^1 = [(2, 1, 3), (3, 4, 0)]$ $\mathbf{g}^2 = [(4, 6, 2), (2, 0, 4)]$ $\mathbf{g}^3 = [(0, 4, 4), (4, 0, 2)]$	2	–	12.0
3	4	$\mathbf{g}^1 = [(2, 4, 2), (3, 7, 2,)]$ $\mathbf{g}^2 = [(4, 0, 4), (6, 6, 4)]$ $\mathbf{g}^3 = [(7, 2, 2), (0, 7, 6), (4, 4, 0)]$	2	–	14.0
3	5	$\mathbf{g}^1 = [(0, 4, 0), (4, 4, 4)]$ $\mathbf{g}^2 = [(0, 2, 4), (2, 3, 7), (2, 2, 7)]$ $\mathbf{g}^3 = [(4, 2, 6), (4, 2, 0), (3, 7, 2)]$	2	–	16.0

Table 4.10 Optimal 8-PSK STTC codes with four transmit antennas for slow fading channels based on trace criterion

n_T	v	generator sequences	r	det	tr
4	3	$\mathbf{g}^1 = [(2, 1, 3, 7), (3, 4, 0, 5)]$ $\mathbf{g}^2 = [(4, 6, 2, 2), (2, 0, 4, 4)]$ $\mathbf{g}^3 = [(0, 4, 4, 4), (4, 0, 2, 0)]$	2	–	16.586
4	4	$\mathbf{g}^1 = [(2, 4, 2, 2), (3, 7, 2, 4)]$ $\mathbf{g}^2 = [(4, 0, 4, 4), (6, 6, 4, 0)]$ $\mathbf{g}^3 = [(7, 2, 2, 0), (0, 7, 6, 3), (4, 4, 0, 2)]$	2	–	20.0
4	5	$\mathbf{g}^1 = [(0, 4, 0, 3), (4, 4, 4, 3)]$ $\mathbf{g}^2 = [(0, 2, 4, 2), (2, 3, 7, 1), (2, 2, 7, 5)]$ $\mathbf{g}^3 = [(4, 2, 6, 5), (4, 2, 0, 7), (3, 7, 2, 6)]$	2	–	22.1

4.4 Performance Evaluation on Slow Fading Channels

The code FER performance is evaluated by simulations. In the simulations, each frame consisted of 130 symbols transmitted from each antenna. A maximum likelihood Viterbi decoder with perfect CSI is employed at the receiver. The performance curves are plotted against the signal-to-noise ratio (SNR) per receive antenna.

4.4.1 Performance of the Codes Based on the Rank & Determinant Criteria

The performance of the optimum QPSK codes with two transmit antennas and various numbers of states from Table 4.2 on slow Rayleigh fading channels is shown in Fig. 4.5. The number of receive antennas was one in the simulations. The figure shows that all the QPSK codes achieve the same diversity order, demonstrated by the same slope of the FER performance curves. The code performance is improved with increasing the number of states. However, the performance improvement is almost saturated when the number of states is above 16. The outage capacity limit for MIMO channels with two transmit and one receive antennas and spectral efficiency of 2 bits/s/Hz is also shown in the figure. We can observe that the 64-state QPSK code is within 2 dB away from the outage capacity.

The performance comparison between the optimum and reference QPSK codes with two transmit and one receive antennas on slow fading channels is shown in Fig. 4.6. From the figure, we can observe that the optimum 8-state code is slightly better than the corresponding TSC and BBH code. For 32 states, the optimum code is superior to the TSC code by 0.6 dB at a FER of 10^{-2}.

The performance of the optimum QPSK codes based on the rank & determinant criteria with three and four transmit antennas and various numbers of states on slow Rayleigh fading channels are shown in Figs. 4.7 and 4.8, respectively. The number of receive antennas was one in the simulations. The outage capacity limits for the corresponding MIMO channels and spectral efficiency of 2 bits/s/Hz are also plotted for comparison. We can see from the figures that the 64-state QPSK codes are about 2.8 dB and 3.3 dB away from the outage capacity for systems with three and four transmit antennas, respectively.

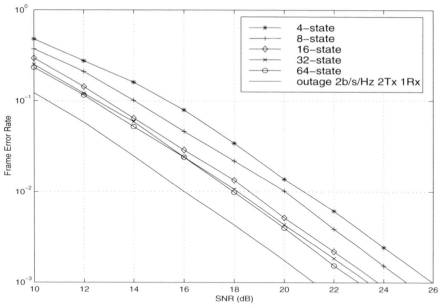

Figure 4.5 Performance comparison of the QPSK codes based on the rank & determinant criteria on slow fading channels with two transmit and one receive antennas

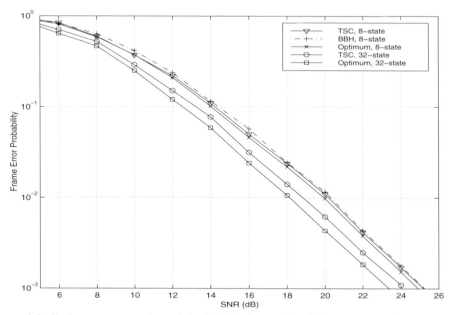

Figure 4.6 Performance comparison of the QPSK codes on slow fading channels with two transmit and one receive antennas

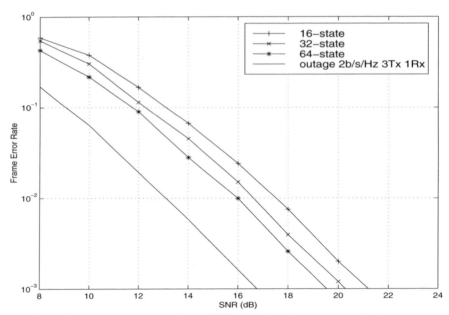

Figure 4.7 Performance comparison of the QPSK codes based on the rank & determinant criteria on slow fading channels with three transmit and one receive antennas

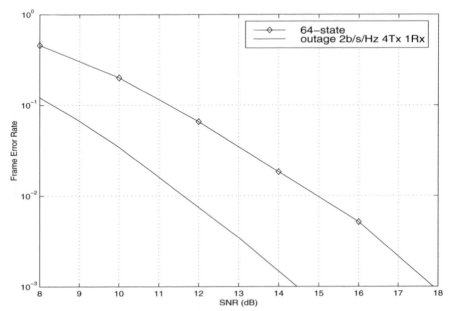

Figure 4.8 Performance comparison of the QPSK codes based on the rank & determinant criteria on slow fading channels with four transmit and one receive antennas

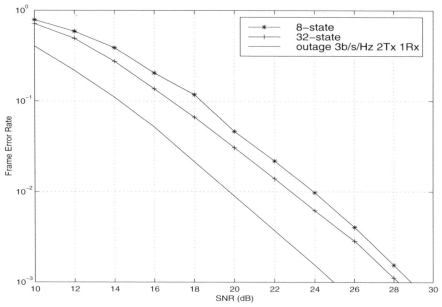

Figure 4.9 Performance comparison of the 8-PSK codes based on the rank & determinant criteria on slow fading channels with two transmit and one receive antennas

Figure 4.9 illustrates the performance of the optimum 8-PSK codes based on the rank & determinant criteria with two transmit and one receive antennas on slow Rayleigh fading channels along the corresponding outage capacity of 3 bits/s/Hz. The 32-state 8-PSK code is about 3.1 dB away from the capacity.

4.4.2 Performance of the Codes Based on the Trace Criterion

The performance of the optimum QPSK codes designed by the trace criterion with various numbers of states on slow Rayleigh fading channels is depicted in Figs. 4.10, 4.11 and 4.12 for two, three, and four transmit antennas, respectively, and two receive antennas.

The performance of the optimum 8-PSK codes with various numbers of states on slow Rayleigh fading channels is shown in Figs. 4.13, 4.14 and 4.15 for two, three, and four transmit antennas, respectively, and two receive antennas.

4.4.3 Performance Comparison for Codes Based on Different Design Criteria

Figure 4.16 illustrates the performance of the optimum 32-state QPSK codes with three transmit antennas based on the different design criteria. The code designed by using the trace criterion is superior by 0.3 dB and 0.6 dB at the FER of 10^{-3} for two and four receive antennas, respectively, and inferior by 0.6 dB for one receive antenna, compared with the code designed by using the rank & determinant criteria. Note that the 32-state code based on the trace criterion does not have a full rank. This is consistent with the previous discussion that, in general, the rank & determinant criteria are applicable for a single receive antenna and the trace criterion is applicable for two or more receive antennas.

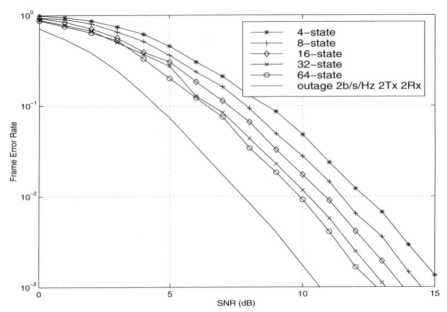

Figure 4.10   Performance comparison of the QPSK codes based on the trace criterion on slow fading channels with two transmit and two receive antennas

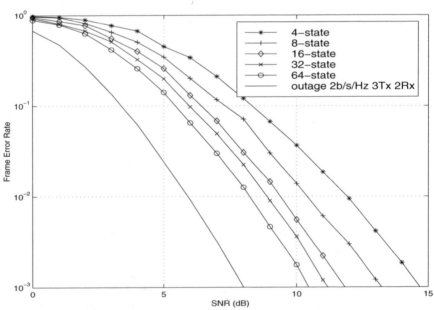

Figure 4.11   Performance comparison of the QPSK codes based on the trace criterion on slow fading channels with three transmit and two receive antennas

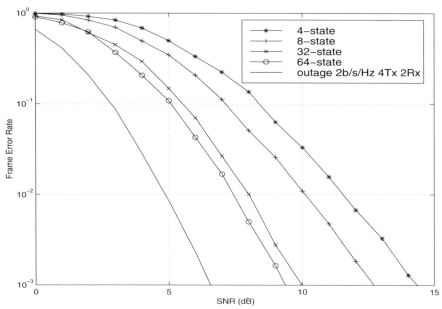

Figure 4.12   Performance comparison of the QPSK codes based on the trace criterion on slow fading channels with four transmit and two receive antennas

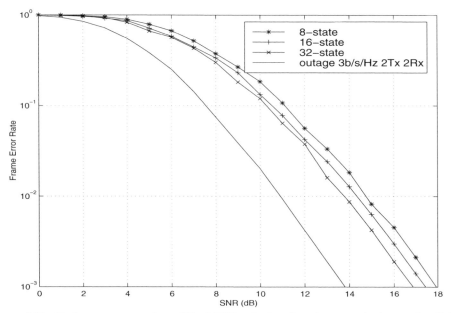

Figure 4.13   Performance comparison of the 8-PSK codes based on the trace criterion on slow fading channels with two transmit and two receive antennas

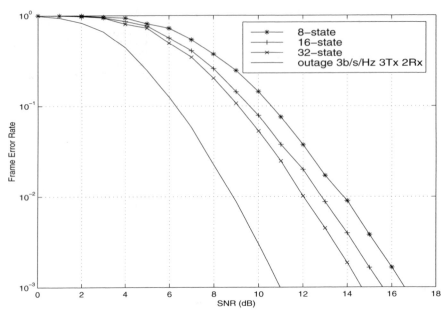

Figure 4.14   Performance comparison of the 8-PSK codes based on the trace criterion on slow fading channels with three transmit and two receive antennas

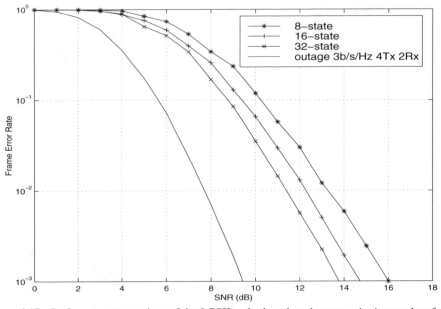

Figure 4.15   Performance comparison of the 8-PSK codes based on the trace criterion on slow fading channels with four transmit and two receive antennas

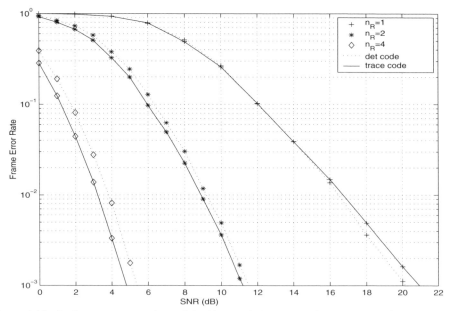

Figure 4.16   Performance comparison of the 32-state QPSK codes with three transmit antennas based on different criteria on slow fading channels

4.4.4   The Effect of the Number of Transmit Antennas on Code Performance

Figures 4.17 and 4.18 depict the performance comparison of the 32 and 64-state QPSK codes with various numbers of transmit antennas, respectively. We can see that a large performance improvement has been achieved by increasing the number of transmit antennas n_T. For the 32-state code with one receive antenna, a 2.9 dB and 5.0 dB improvement is observed at a FER of 10^{-3} when we increase n_T from two to three and four, respectively. In a system with two receive antennas, this code yields improvements of 2.0 dB and 3.1 dB, respectively. As the number of receive antennas gets larger, the performance gain from increasing the number of transmit antennas becomes smaller.

As Fig. 4.18 indicates, the 64-state code with one receive antenna brings a 3.9 dB and 5.6 dB improvement at a FER of 10^{-3} when we increase n_T from two to three and four, respectively. However, the same code gains only 2.0 dB and 3.4 dB, respectively, in a system with two receive antennas.

Figure 4.19 compares 8-state 8-PSK codes with two, three and four transmit antennas designed by the trace criterion. In systems with two receive antennas, the 8-PSK code with three transmit antennas is 1.5 dB better than the code with two transmit antennas at a FER of 10^{-3}. When the number of the receive antennas increases to four, the 8-PSK code with three transmit antennas is better than the code with two transmit antennas by 1.3 dB. However, in a system with a single receive antenna, the three codes have similar performance. This is because the code performance is dominated by the minimum rank if only one receive antenna is available and these codes have the same minimum rank of two.

The same comparison is also done for the 16-state 8-PSK codes with two, three and four transmit antennas. The results are shown in Fig. 4.20.

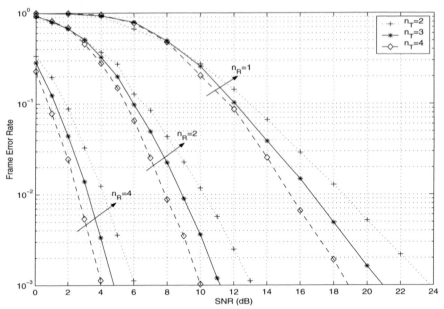

Figure 4.17 Performance comparison of the 32-state QPSK codes based on the trace criterion with two, three and four transmit antennas on slow fading channels

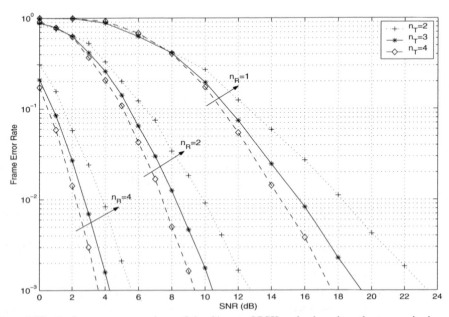

Figure 4.18 Performance comparison of the 64-state QPSK codes based on the trace criterion with two, three and four transmit antennas on slow fading channels

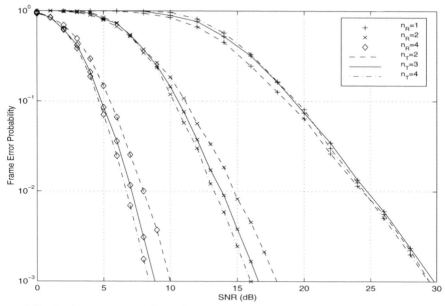

Figure 4.19 Performance comparison of the 8-state 8-PSK codes based on the trace criterion with two, three and four transmit antennas on slow fading channels

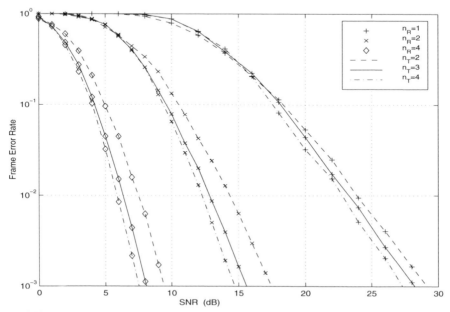

Figure 4.20 Performance comparison of the 16-state 8-PSK codes based on the trace criterion with two, three and four transmit antennas on slow fading channels

4.4.5 The Effect of the Number of Receive Antennas on Code Performance

Figure 4.21 illustrates the performance of the 4-state QPSK codes based on the trace criterion from Table 4.5 with various numbers of receive antennas on slow fading channels. It shows that for two receive antennas the optimum code outperforms the TSC and the BBH code by 0.6 and 0.2 dB, respectively, at a FER of 10^{-2}. For four receive antennas, the performance improvement gained by the optimum code relative to the TSC and the BBH code at a FER of 10^{-2} is 2.0 and 0.5 dB, respectively. Note that for the same number of receive antennas, the optimum code has the best performance due to the largest trace value and the TSC code has the worst performance due to the smallest trace value, although the optimum and the TSC code have the same minimum determinant. The trace criterion also explains why the optimum 4-state code, with a smaller minimum determinant, outperforms the 4-state BBH code. Note that for $n_R = 1$, the optimum code has approximately the same performance as the TSC and the BBH code.

Figure 4.22 shows the performance of the 8-state 8-PSK codes based on trace criterion from Table 4.8 with one, two, three and four receive antennas on slow fading channels. The dotted lines denote the TSC code and the solid lines represent the optimum code. The optimum code is superior to the TSC code by 0.6, 1.2 and 1.7 dB for two, three and four receive antennas, respectively. When only one receive antenna is available, the optimum code and the TSC code have approximately the same performance. As these results show, the advantage of the optimum codes relative to those designed by the rank & determinant criteria [4] increases with the increasing number of receive antennas. The presented simulation results agree with the performance analysis in Chapter 2.

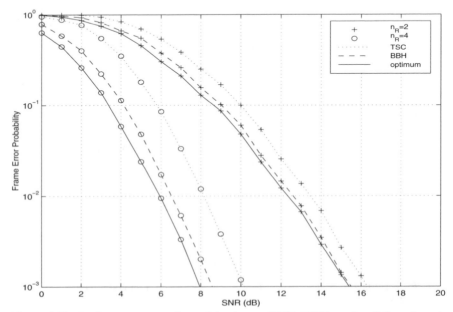

Figure 4.21 Performance comparison of the 4-state QPSK STTC on slow fading channels

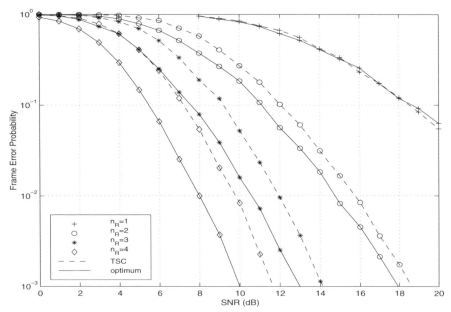

Figure 4.22   Performance comparison of the 8-state 8-PSK STTC on slow fading channels

4.4.6   The Effect of Channel Correlation on Code Performance

Figure 4.23 shows the 16-state QPSK code performance on correlated slow Rayleigh fading channels with two transmit and two receive antennas. The correlation factors between
receive antennas are chosen to be 0.25, 0.5, 0.75 and 1. It can be observed that the code
performance is slightly degraded, relative to the system with independent sub-channels,
when the correlation factor is 0.25. However, the code performance is getting worse by
0.5 dB and 1.3 dB compared to the case with uncorrelated antennas at a FER of 10^{-2} for
the correlation factors of 0.5 and 0.75, respectively. When the channels are fully correlated,
the penalty on the code performance is about 5.8 dB at the same FER.

4.4.7   The Effect of Imperfect Channel Estimation
on Code Performance

Figure 4.24 shows the 16-state QPSK code performance on slow Rayleigh fading channels
with two transmit and two receive antennas and imperfect channel estimation. In the simulation, ten orthogonal signals in each data frame are used as pilot sequence in order to
estimate the channel state information at the receiver. From the figure, we can see that the
deterioration due to the imperfect channel estimation is about 0.2 dB at a FER of 10^{-3}.

4.5   Design of Space-Time Trellis Codes on Fast
Fading Channels

Code Design Criteria Sets III and IV are applied to construct good STTC for fast Rayleigh
fading channels. For an STTC, the symbol-wise Hamming distance δ_H between two paths

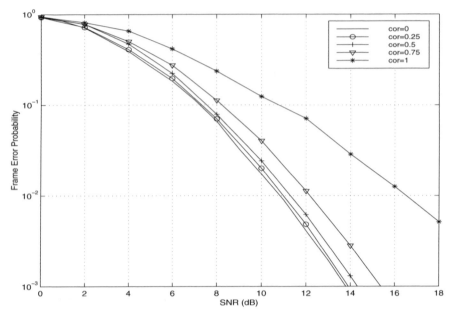

Figure 4.23 Performance of the 16-state QPSK code on correlated slow Rayleigh fading channels with two transmit and two receive antennas

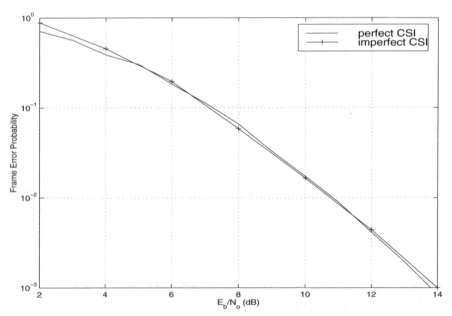

Figure 4.24 Performance of the 16-state QPSK code on slow Rayleigh fading channels with two transmit and two receive antennas and imperfect channel estimation

Table 4.11 Optimal QPSK STTC with two transmit antennas for fast fading channels

code	ν	generator sequences	δ_H	d_p^2
TSC	2	$\mathbf{g}^1 = [(0, 2), (2, 0)]$ $\mathbf{g}^2 = [(0, 1), (1, 0)]$	2	4.0
BBH	2	$\mathbf{g}^1 = [(2, 2), (1, 0)]$ $\mathbf{g}^2 = [(0, 2), (3, 1)]$	2	8.0
optimum	2	$\mathbf{g}^1 = [(3, 1), (2, 1)]$ $\mathbf{g}^2 = [(2, 2), (0, 2)]$	2	24.0
TSC	3	$\mathbf{g}^1 = [(0, 2), (2, 0)]$ $\mathbf{g}^2 = [(0, 1), (1, 0), (2, 2)]$	2	16.0
BBH	3	$\mathbf{g}^1 = [(2, 2), (2, 0)]$ $\mathbf{g}^2 = [(0, 1), (1, 0), (2, 2)]$	2	32.0
optimum	3	$\mathbf{g}^1 = [(0, 2), (1, 3), (2, 0)]$ $\mathbf{g}^2 = [(2, 2), (1, 2)]$	2	48.0
TSC	4	$\mathbf{g}^1 = [(0, 2), (2, 0), (0, 2)]$ $\mathbf{g}^2 = [(0, 1), (1, 2), (2, 0)]$	3	16.0
BBH	4	$\mathbf{g}^1 = [(0, 2), (2, 0), (0, 2)]$ $\mathbf{g}^2 = [(2, 1), (1, 2), (2, 0)]$	3	24.0
optimum	4	$\mathbf{g}^1 = [(0, 2), (0, 1), (2, 2)]$ $\mathbf{g}^2 = [(2, 0), (1, 2), (0, 2)]$	3	64.0
TSC	5	$\mathbf{g}^1 = [(0, 2), (2, 2), (3, 3)]$ $\mathbf{g}^2 = [(0, 1), (1, 1), (2, 0), (2, 2)]$	3	128.0
optimum	5	$\mathbf{g}^1 = [(1, 0), (0, 1), (2, 0), (0, 1)]$ $\mathbf{g}^2 = [(2, 2), (3, 1), (1, 2)]$	3	192.0

in an error event is upper-bounded by the error event length. Consider the constraint of the error event length in (4.18). The upper bound of the minimum symbol-wise Hamming distance δ_H for an STTC with the memory order of ν can be given by

$$\delta_H \leq \lfloor \nu/2 \rfloor + 1 \qquad (4.19)$$

For codes with memory order less than 6 and one receive antenna, the maximum possible diversity order $\delta_H n_R$ is less than 4. In this case, Set III should be used for code search. The good QPSK codes based on this criteria set are listed in Tables 4.11 and 4.12 for two and three transmit antennas, respectively. The good 8-PSK codes are shown in Tables 4.13, 4.14 and 4.15 for two, three and four transmit antennas, respectively. The minimum symbol-wise Hamming distance δ_H and the minimum product distance d_p^2 along the paths with minimum δ_H are shown in the tables. For two transmit antennas, the minimum δ_H and d_p^2 of the TSC and the BBH codes are listed in Table 4.11 for comparison. The data in the table indicate that, for a given memory order, the proposed optimum STTCs achieve the same minimum symbol-wise Hamming distance as the TSC and BBH codes, but a much larger minimum product distance. As a result, the proposed optimum codes achieve a larger coding gain compared to the TSC and the BBH codes.

Table 4.12 Optimal QPSK STTC with three transmit antennas for fast fading channels

n_T	ν	generator sequences	δ_H	d_p^2
3	2	$\mathbf{g}^1 = [(0, 2, 2), (1, 1, 2)]$ $\mathbf{g}^2 = [(2, 0, 2), (2, 2, 0)]$	2	64
3	3	$\mathbf{g}^1 = [(0, 2, 0), (1, 3, 0), (2, 0, 1)]$ $\mathbf{g}^2 = [(2, 2, 2), (1, 2, 2)]$	2	120
3	4	$\mathbf{g}^1 = [(0, 2, 2), (0, 1, 2), (2, 2, 0)]$ $\mathbf{g}^2 = [(2, 0, 2), (1, 2, 0), (0, 2, 2)]$	3	384

Table 4.13 Optimal 8-PSK STTC with two transmit antennas for fast fading channels

n_T	ν	generator sequences	δ_H	d_p^2
TSC	3	$\mathbf{g}^1 = [(0, 4), (4, 0)]$ $\mathbf{g}^2 = [(0, 2), (2, 0)]$ $\mathbf{g}^3 = [(0, 1), (5, 0)]$	2	2.0
optimum	3	$\mathbf{g}^1 = [(2, 1), (2, 4)]$ $\mathbf{g}^2 = [(0, 4), (4, 0)]$ $\mathbf{g}^3 = [(4, 6), (2, 1)]$	2	15.51
TSC	4	$\mathbf{g}^1 = [(0, 4), (4, 4)]$ $\mathbf{g}^2 = [(0, 2), (2, 2)]$ $\mathbf{g}^3 = [(0, 1), (5, 1), (1, 5)]$	2	8.0
optimum	4	$\mathbf{g}^1 = [(0, 4), (4, 2)]$ $\mathbf{g}^2 = [(1, 0), (2, 1), (0, 1)]$ $\mathbf{g}^3 = [(1, 1), (6, 4)]$	2	24.0
TSC	5	$\mathbf{g}^1 = [(0, 4), (4, 4)]$ $\mathbf{g}^2 = [(0, 2), (2, 2), (2, 2)]$ $\mathbf{g}^3 = [(0, 1), (5, 1), (3, 7)]$	2	13.66
optimum	5	$\mathbf{g}^1 = [(3, 4), (0, 4)]$ $\mathbf{g}^2 = [(1, 0), (0, 1), (6, 0)]$ $\mathbf{g}^3 = [(1, 1), (3, 1), (1, 1)]$	2	29.66

When the memory order of STTC is larger than 6, or more than 1 receive antenna is employed, it is always possible to achieve a minimum diversity order $\delta_H n_R$ greater than or equal to 4. In this case, the code design should be based on Criteria Set IV, which requires that the minimum squared Euclidean distance d_E^2 of the STTC should be maximized. This criteria set is equivalent to Criteria Set II. Thus, the codes in Tables 4.5–4.10, which have the largest minimum Euclidean distance, can also achieve an optimum performance on fast fading channels, when the number of the receive antennas is larger than one. In this sense, the codes in Tables 4.5–4.10 are robust, as they are optimum for both slow and fast fading channels.

Table 4.14 Optimal 8-PSK STTC codes with three transmit antennas for fast fading channels

n_T	ν	generator sequences	δ_H	d_p^2
4	3	$\mathbf{g}^1 = [(2, 1, 2), (2, 4, 2)]$ $\mathbf{g}^2 = [(0, 4, 4), (4, 0, 4)]$ $\mathbf{g}^3 = [(4, 6, 3), (2, 1, 6)]$	2	36.69
4	4	$\mathbf{g}^1 = [(0, 4, 4), (4, 2, 2)]$ $\mathbf{g}^2 = [(1, 0, 1), (2, 1, 0), (0, 1, 1)]$ $\mathbf{g}^3 = [(3, 1, 2), (6, 4, 4)]$	2	52.97
4	5	$\mathbf{g}^1 = [(3, 4, 4), (0, 4, 2)]$ $\mathbf{g}^2 = [(1, 0, 6), (0, 1, 2), (6, 0, 1)]$ $\mathbf{g}^3 = [(1, 1, 5), (3, 1, 0), (1, 1, 3)]$	3	68.48

Table 4.15 Optimal 8-PSK STTC codes with four transmit antennas for fast fading channels

n_T	ν	generator sequences	δ_H	d_p^2
4	3	$\mathbf{g}^1 = [(2, 1, 2, 4), (2, 4, 2, 1)]$ $\mathbf{g}^2 = [(0, 4, 4, 2), (4, 0, 4, 0)]$ $\mathbf{g}^3 = [(4, 6, 3, 0), (2, 1, 6, 4)]$	2	73.72
4	4	$\mathbf{g}^1 = [(0, 4, 4, 4), (4, 2, 2, 0)]$ $\mathbf{g}^2 = [(1, 0, 1, 1), (2, 1, 0, 5), (0, 1, 1, 5)]$ $\mathbf{g}^3 = [(3, 1, 2, 2), (6, 4, 4, 3)]$	2	96.0
4	5	$\mathbf{g}^1 = [(3, 4, 4, 3), (0, 4, 2, 6)]$ $\mathbf{g}^2 = [(1, 0, 6, 0), (0, 1, 2, 2), (6, 0, 1, 4)]$ $\mathbf{g}^3 = [(1, 1, 5, 2), (3, 1, 0, 1), (1, 1, 3, 0)]$	3	118.63

4.6   Performance Evaluation on Fast Fading Channels

The performance of the optimum codes on fast fading channels is evaluated by simulations. Systems with two transmit and one receive antennas were simulated. Fig. 4.25 shows the FER performance of the optimum QPSK STTC with memory orders of 2 and 4 on a fast fading channel. Their performance is compared with the TSC and the BBH codes of the same memory order. The bandwidth efficiency is 2 bits/s/Hz. In this figure the error rate curves of the codes with the same memory order and number of receive antennas are parallel, as predicted by the same value of δ_H. Different values of d_p^2 yield different coding gains, which are represented by the horizontal shifts of the FER curves. For one receive antenna, the optimum 4-state QPSK STTC is superior to the 4-state TSC and the BBH code by 1.5 and 0.9 dB, respectively, while the optimum 16-state code is better by 1.2 and 0.4 dB, relative to the TSC and the BBH code, respectively.

In addition, it can also be observed from this figure that the error rate curves of all 16-state QPSK STTC have a steeper slope than those of the 4-state ones. This occurs because the

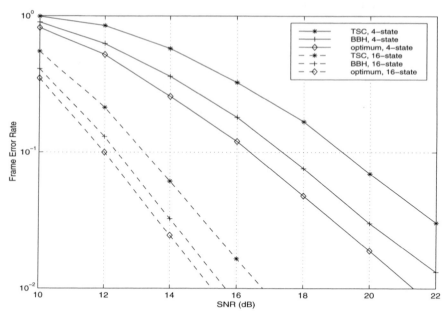

Figure 4.25 Performance comparison of the 4 and 16-state QPSK STTC on fast fading channels

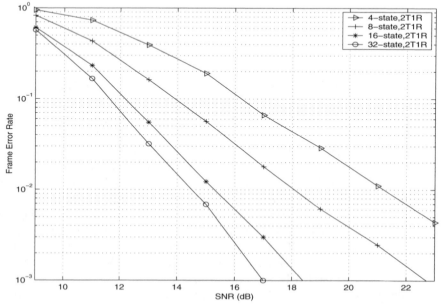

Figure 4.26 Performance of the QPSK STTC on fast fading channels with two transmit and one receive antennas

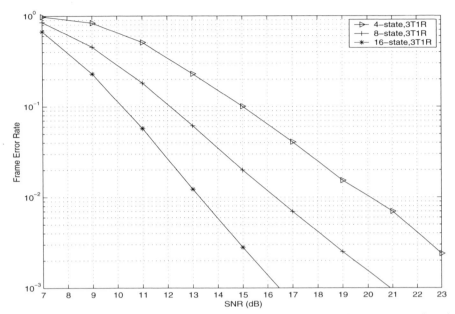

Figure 4.27   Performance of the QPSK STTC on fast fading channels with three transmit and one receive antennas

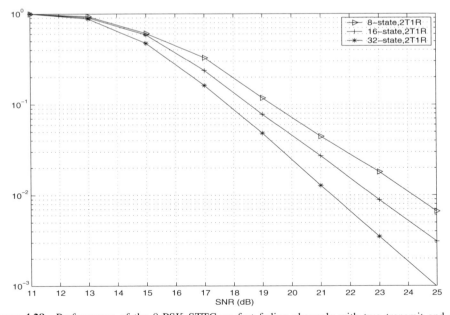

Figure 4.28   Performance of the 8-PSK STTC on fast fading channels with two transmit and one receive antennas

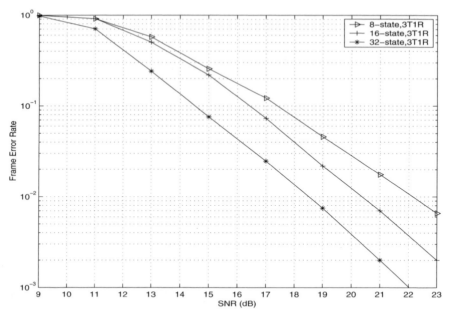

Figure 4.29 Performance of the 8-PSK STTC on fast fading channels with three transmit and one receive antennas

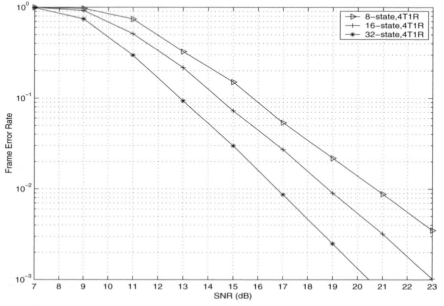

Figure 4.30 Performance of the 8-PSK STTC on fast fading channels with four transmit and one receive antennas

16-state codes have the minimum δ_H of 3, while the 4-state codes have the minimum δ_H of 2.

Furthermore, it is worthwhile to mention that when the number of the receive antennas increases, the performance gain achieved by the optimum QPSK STTC, relative to the TSC and the BBH codes of the same memory order, remains. This is due to the fact that the optimum codes have both a larger minimum product distance and a larger minimum Euclidean distance compared to the known codes.

The performance of the optimum QPSK codes with two and three transmit antennas and various numbers of states on fast fading channels is shown in Figs. 4.26 and 4.27, respectively. The number of the receive antennas was one in the simulations. We can see from the figures that the 16-state QPSK codes are better relative to the 4-state codes by 5.9 dB and 6.8 dB at a FER of 10^{-2} for two and three transmit antennas, respectively.

Figures 4.28, 4.29 and 4.30 illustrate the performance of the optimum 8-PSK codes with various numbers of states on fast Rayleigh fading channels for two, three and four transmit antennas, respectively. In a system with two transmit antennas, a 1.5 dB and 3.0 dB improvement is observed at a FER of 10^{-2} when the number of states increases from 8 to 16 and 32, respectively. As the number of the transmit antennas gets larger, the performance gain achieved from increasing the number of states becomes larger.

Bibliography

[1] G. J. Foschini and M. Gans, "On the limits of wireless communication in a fading environment when using multiple antennas", *Wireless Personal Communication*, vol. 6, pp. 311–335, Mar. 1998.

[2] G. J. Foschini, "Layered space-time architecture for wireless communication in fading environments when using multiple antennas", *Bell Labs Tech. J.*, Autumn 1996.

[3] E. Teletar, "Capacity of multi-antenna Gaussian channels", *Technical Report, AT&T-Bell Labs*, June 1995.

[4] V. Tarokh, N. Seshadri and A. R. Calderbank, "Space-time codes for high data rate wireless communication: performance criterion and code construction", *IEEE Trans. Inform. Theory*, vol. 44, no. 2, pp. 744–765, Mar. 1998.

[5] J.-C. Guey, M. R. Bell, M. P. Fitz and W. Y. Kuo, "Signal design for transmitter diversity wireless communication systems over Rayleigh fading channels", *Proc. IEEE Vehicular Technology Conference*, pp. 136–140, Atlanta, US, 1996; and *IEEE Trans. Commun.* vol. 47, pp. 527–537, Apr. 1998.

[6] A. Naguib, V. Tarokh, N. Seshadri and A. Calderbank, "A space-time coding modem for high-data-rate wireless communications", *IEEE Journal Select. Areas Commun.*, vol. 16, pp. 1459–1478, Oct. 1998.

[7] S. M. Alamouti, "A simple transmit diversity technique for wireless communications", *IEEE Journal Select. Areas Commun.*, Oct. 1998, pp. 1451–1458.

[8] V. Tarokh, H. Jafarkhani and A. R. Calderbank, "Space-time block codes from orthogonal designs", *IEEE Trans. Inform. Theory*, vol. 45, no. 5, July 1999, 1456–1467.

[9] J. Grimm, M. P. Fitz and J. V. Krogmeier, "Further results in space-time coding for Rayleigh fading", *36th Allerton Conference on Communications, Control and Computing Proceedings*, Sept. 1998.

[10] A. R. Hammons and H. E. Gammal, "On the theory of space-time codes for PSK modulation", *IEEE Trans. on. Inform. Theory*, vol. 46, no. 2, Mar. 2000, pp. 524–542.

[11] S. Baro, G. Bauch and A. Hansmann, "Improved codes for space-time trellis coded modulation", *IEEE Commun. Lett.*, vol. 4, no. 1, pp. 20–22, Jan. 2000.

[12] Q. Yan and R. S. Blum, "Optimum space-time convolutional codes", *IEEE WCNC'00*, Chicago, pp. 1351–1355, Sept. 2000.

[13] Z. Chen, J. Yuan and B. Vucetic, "Improved space-time trellis coded modulation scheme on slow Rayleigh fading channels", *IEE Electronics Letters*, vol. 37, no. 7, pp. 440–442, Apr. 2001.

[14] Z. Chen, B. Vucetic, J. Yuan and K. Lo, "Space-time trellis coded modulation with three and four transmit antennas on slow fading channels", *IEEE Commun. Letters*, vol. 6, no. 2, pp. 67–69, Feb. 2002.

[15] J. Yuan, Z. Chen, B. Vucetic and W. Firmanto, "Performance analysis and design of space-time coding on fading channels", submitted to *IEEE Trans. Commun.*, 2000.

[16] W. Firmanto, B. Vucetic and J. Yuan, "Space-time TCM with improved performance on fast fading channels", *IEEE Commun. Letters*, vol. 5, no. 4, pp. 154–156, Apr. 2001.

[17] J. Ventura-Traveset, G. Caire, E. Biglieri and G. Taricco, "Impact of diversity reception on fading channels with coded modulation–Part I: coherent detection", *IEEE Trans. Commun.*, vol. 45, no. 5, pp. 563–572, May 1997.

[18] B. Vucetic and J. Nicolas, "Performance of M-PSK trellis codes over nonlinear fading mobile satellite channels", *IEE Proceedings I*, vol 139, pp. 462–471, Aug. 1992. coding schemes for fading channels", No. 1, pp. 50-61, Jan. 1993.

[19] G. D. Forney, Jr. "Geometrically Uniform Codes", *IEEE Trans. Inform. Theory*, vol. 37, no. 5, pp. 1241–1260, Sept. 1991.

5

Space-Time Turbo Trellis Codes

5.1   Introduction

Turbo codes with iterative decoding are well known for their ability to achieve very low bit error rates [1]. They are constructed by parallel concatenation of two recursive convolutional codes. The code benefits from a suboptimum but very powerful iterative decoding algorithm. Several bandwidth efficient turbo coding techniques have been proposed, combining the principles of turbo codes and trellis codes. A turbo coded modulation scheme with parity check puncturing [4] involves parallel concatenation of two recursive Ungerboeck type trellis codes [2] as component codes. The output symbols from the two component encoders are alternately punctured, to ensure the bandwidth efficiency of k bits/sec/Hz for a signal set of 2^{k+1} points. This is equivalent to alternately puncturing parity symbols from the component codes. The scheme applies symbol interleaving/deinterleaving in the turbo encoder/decoder. In another approach, parallel concatenation of two recursive convolutional codes with puncturing of systematic bits is proposed [5]. The puncturing pattern is selected in such a way that the information bits appear in the output of the concatenated code only once. This scheme uses bit interleaving/deinterleaving of information sequences in the encoder/decoder. In this chapter we consider construction of space-time coding techniques which combine the coding gain benefits of turbo coding with the diversity advantage of space-time coding and the bandwidth efficiency of coded modulation. Bandwidth efficient *space-time turbo trellis code* (ST turbo TC) can be constructed by alternate parity symbol puncturing and applying symbol interleaving [22][7] or by information puncturing and bit interleaving [6]. As in binary turbo codes, in both constructions of bandwidth efficient ST codes, recursive STTC are used as component codes in order to obtain an interleaver gain.

In this chapter we consider the design and performance of various ST turbo TC system structures. We first introduce recursive STTC and show how to convert feedforward STTC designed by applying the criteria developed in Chapter 2 into equivalent recursive codes. This is followed by the encoder structures for ST turbo TC and the discussion of the iterative decoding algorithm. A comparison of various system structures on the basis of performance and implementation complexity is also presented, along with the simulation results.

Space-Time Coding   Branka Vucetic and Jinhong Yuan
© 2003 John Wiley & Sons, Ltd   ISBN: 0-470-84757-3

5.1.1 Construction of Recursive STTC

In this section we will show the construction of systematic and nonsystematic recursive STTC.

Let us consider a feedforward STTC encoder for QPSK and two antennas, as shown in Fig. 5.1 with the memory order of $\nu = \nu_1 + \nu_2$, where $\nu_1 \leq \nu_2$ and $\nu_i = \lfloor \frac{\nu+i-1}{2} \rfloor$, $i = 1, 2$.

The encoded symbol sequence transmitted from antenna i is given by

$$\mathbf{x}(D)^i = \mathbf{c}^1(D)\mathbf{G}_i^1(D) + \mathbf{c}^2(D)\mathbf{G}_i^2(D) \quad \text{mod } 4 \tag{5.1}$$

The relationship in (5.1) can be written in the following form

$$\mathbf{x}^i(D) = \mathbf{c}^1(D)\mathbf{c}^2(D) \begin{bmatrix} \mathbf{G}_i^1(D) \\ \mathbf{G}_i^2(D) \end{bmatrix} \quad \text{mod } 4 \tag{5.2}$$

The feedforward generator matrix from equation (5.2)

$$\mathbf{G}_i(D) = \begin{bmatrix} \mathbf{G}_i^1(D) \\ \mathbf{G}_i^2(D) \end{bmatrix}$$

can be converted into an equivalent recursive matrix by dividing it by a binary polynomial $\mathbf{q}(D)$ of a degree equal or less than ν_1. However, if we choose for $\mathbf{q}(D)$ a primitive polynomial, the resulting recursive code should have a high minimum distance. The generator

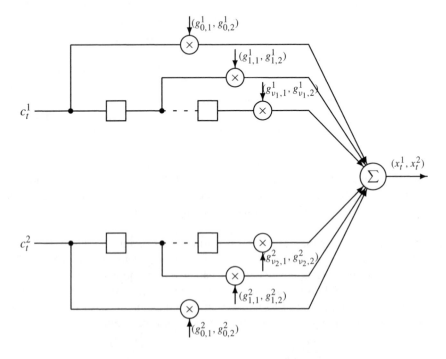

Figure 5.1   A feedforward STTC encoder for QPSK modulation

polynomial for antenna i can be represented as

$$\mathbf{G}_i(D) = \begin{bmatrix} \dfrac{\mathbf{G}_i^1(D)}{\mathbf{q}(D)} \\[2mm] \dfrac{\mathbf{G}_i^2(D)}{\mathbf{q}(D)} \end{bmatrix}$$

where

$$\mathbf{q}(D) = q_0 + q_1 D + q_2 D^2 + \cdots + q_{v_1} D^{v_1}$$

and q_j, $j = 0, 1, 2, \ldots, v_1$, are binary coefficients from $(0, 1)$. A systematic recursive STTC can be obtained by setting

$$\mathbf{G}_1(D) = \begin{bmatrix} 2 \\ 1 \end{bmatrix}$$

which means that the output of the first antenna is obtained by directly mapping the input sequences $\mathbf{c}^1$ and $\mathbf{c}^2$ into a QPSK sequence. A diagram of a recursive QPSK STTC encoder with n_T antennas is shown in Fig. 5.2.

A similar design can be applied to M-PSK modulation. A block diagram of a recursive STTC encoder for M-PSK modulation is shown in Fig. 5.3.

The codes generated by this construction method, with the feedforward coefficients as in the corresponding feedforward STTC, have the same diversity and coding gain as these feedforward STTC.

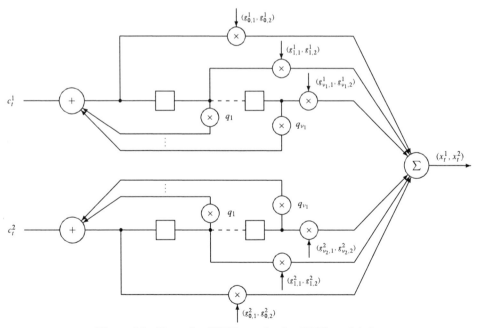

Figure 5.2 Recursive STTC encoder for QPSK modulation

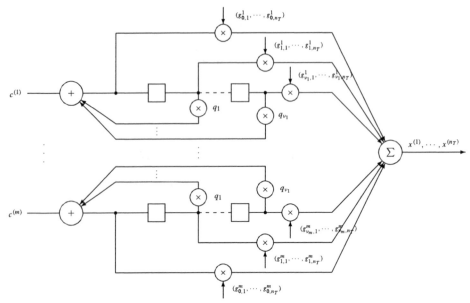

Figure 5.3   Recursive STTC encoder for M-ary modulation

In some communication systems, such as cellular mobile, the channel type might vary from slow to fast, depending on the speed of the mobile terminal. In such cases it is desirable to apply codes which perform well for a whole range of fade rates. Such design criteria are referred to as *hybrid* or *smart and greedy*. It has been shown in Chapter 2 that when $r \cdot n_R \geq 4$ and $\delta_H \cdot n_R \geq 4$, the design criteria for STTC on slow and fast fading channels coincide. Under these conditions, the error probability is minimized when the minimum squared Euclidean distance, d_E^2, of the code is maximized. The feedforward coefficients for a given memory order which maximize d_E^2 are the same as for the corresponding feedforward STTC, designed in the previous chapter. The recursive coefficients are chosen as $q_j = 1$, $j = 1, 2, \ldots, \nu_1$.

Tables 4.5 and 4.8 list the feedforward coefficients for the recursive QPSK and 8-PSK STTC, respectively, with two transmit antennas which best satisfy the design criterion on slow and fast fading channels, provided that $n_R \geq 2$. Each code in both tables has the minimum rank $r = 2$ and the minimum symbol Hamming distance $\delta_H \geq 2$, satisfying the condition on the design criterion. These codes were obtained through an exhaustive computer search [11]. They maintain their squared Euclidean distance, and thus the performance, when converted into a recursive form. For any given memory order, the optimum STTC has the largest d_E^2.

5.2   Performance of Recursive STTC

In this section, we compare the performance of the recursive STTC with their equivalent feedforward codes. The performance is measured in terms of the bit and frame error rate as a function of E_b/N_0, the ratio between the energy per information bit to the noise at each receive antenna. Each frame consists of 130 M-PSK symbol transmissions from each transmit antenna.

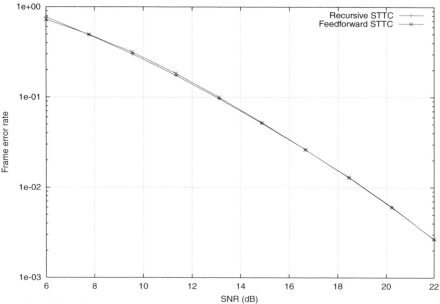

Figure 5.4 FER performance comparison of the 16-state recursive and feedforward STTC on slow fading channels

The recursive STTC has the same frame error rate performance as the corresponding feedforward STTC as demonstrated in an example of the 16-state STTC in Fig. 5.4 on slow fading channels. However, a feedforward STTC has a lower BER than its recursive counterpart, as shown in Fig. 5.5 on the same type of the channel. The same conclusion applies to fast fading channels.

5.3   Space-Time Turbo Trellis Codes

The recursive STTC are used as component codes in a parallel concatenated scheme which benefits from interleaver gain and iterative decoding. Fig. 5.6 shows the encoder structure of a ST turbo TC with n_T transmit antennas, consisting of two recursive STTC encoders, one in the upper and the other in the lower branch, linked by a symbol interleaver [4]. Each encoder operates on a message block of L groups of m information bits, where L is the interleaver size. The message sequence $\mathbf{c}$ is given by $\mathbf{c} = (\mathbf{c}_1, \mathbf{c}_2, \ldots, \mathbf{c}_t, \ldots, \mathbf{c}_L)$, where $\mathbf{c}_t$ is a group of m information bits at time t, given by $\mathbf{c}_t = (c_{t,0}, c_{t,1}, \ldots, c_{t,m-1})$.

The upper recursive STTC encoder in Fig. 5.6 maps the input sequence into n_T streams of L M-PSK symbols, $\mathbf{x}_1^1, \mathbf{x}_1^2, \ldots, \mathbf{x}_1^{n_T}$, where $\mathbf{x}_1^i = (x_{1,1}^i, x_{1,2}^i, \ldots, x_{1,L}^i)$, $i \in \{1, 2, \ldots, n_T\}$ and $M = 2^m$. Prior to encoding by the lower encoder, the information bits are interleaved by a symbol interleaver. The symbol interleaver operates on symbols of m bits instead of on single bits.

The lower encoder also produces n_T streams of L M-PSK symbols. Each stream is deinterleaved before puncturing and multiplexing. The deinterleaved stream can be represented as $\mathbf{x}_2^1, \mathbf{x}_2^2, \ldots, \mathbf{x}_2^{n_T}$, where $\mathbf{x}_2^i = (x_{2,1}^i, x_{2,2}^i, \ldots, x_{2,L}^i)$, $i \in \{1, 2, \ldots, n_T\}$. The streams of symbols generated by the upper and lower encoders, $\mathbf{x}_1^i$ and $\mathbf{x}_2^i$, are alternately punctured,

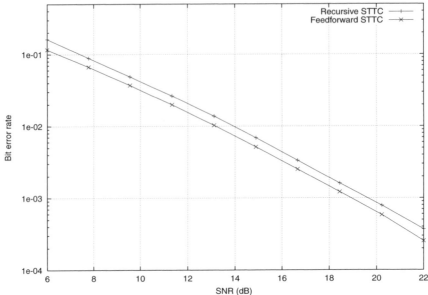

Figure 5.5 BER performance comparison of the 16-state recursive and feedforward STTC on slow fading channels

so that the output from only one encoder is connected to the n_T antennas at a given symbol interval t. For example, in a system with two transmit antennas, if the outputs from the first and second encoder in the first three symbol intervals are $x_{1,1}^1, x_{1,1}^2, x_{1,2}^1, x_{1,2}^2, x_{1,3}^1, x_{1,3}^2$ and $x_{2,1}^1, x_{2,1}^2, x_{2,2}^1, x_{2,2}^2, x_{2,3}^1, x_{2,3}^2$, respectively, the punctured transmitted sequence is $x_{1,1}^1, x_{1,1}^2, x_{2,2}^1, x_{2,2}^2, x_{1,3}^1, x_{1,3}^2$. The spectral efficiency of this scheme is m bits/sec/Hz.

Interleaving can be done on bit rather than symbol streams. If input information sequences are uncoded, they do not need to be interleaved. In general, the encoder output can be only multiplexed, without puncturing, giving the spectral efficiency of $m/2$ bits/sec/Hz. In this case there is no need for symbol interleaving and the deinterleaver in the lower branch.

5.4   Decoding Algorithm

The decoder block diagram for the encoder from Fig. 5.6 is shown in Fig. 5.7.

At time t, the signal received by antenna j, where $j = 1, 2, \ldots, n_R$, can be represented as

$$r_t^j = \sum_{i=1}^{n_T} h_{i,j}^t x_{p,t}^i + n_t^j \tag{5.3}$$

where $x_{p,t}^i$ is the output of the component encoder p at time t, where $p = 1$ for odd time instants t and $p = 2$ for even time instants t.

The received sequence at each antenna j, $j = 1, 2, \ldots, n_R$, is demultiplexed into two vectors, denoted by $\mathbf{r}_1^j$ and $\mathbf{r}_2^j$, contributed by the upper and lower encoder, respectively. These vectors are applied to the first and second decoder, respectively. The punctured

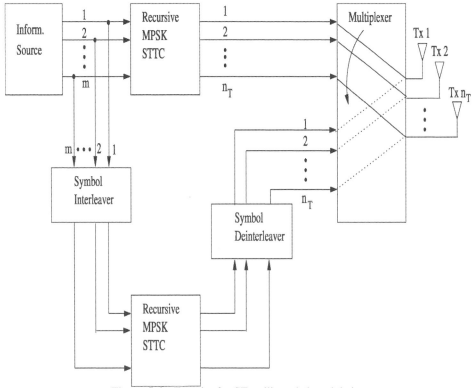

Figure 5.6   Encoder for ST trellis coded modulation

symbols in these decoder input vectors are represented by erasures. They are given by

$$\mathbf{r}_1^j = (r_1^j, 0, r_3^j, 0, r_5^j, \dots,)$$

$$\mathbf{r}_2^j = (0, r_2^j, 0, r_4^j, \dots,)$$

The vector $\mathbf{r}_1^j$ is fed into the first decoder directly, while the vector $\mathbf{r}_2^j$ is fed to the second decoder via the symbol interleaver, identical to the one in the encoder.

The decoding process is very similar to the binary turbo code except that the symbol probability is used as the extrinsic information rather than the bit probability. The MAP decoding algorithm for nonbinary trellises is called symbol-by-symbol MAP algorithm.

The MAP decoder computes the LLR log-likelihood ratio of each group of information bits $\mathbf{c}_t = i$. The soft output $\Lambda(\mathbf{c}_t = i)$ is given by [14]

$$\Lambda(\mathbf{c}_t = i) = \log \frac{Pr\{\mathbf{c}_t = i | \mathbf{r}\}}{Pr\{\mathbf{c}_t = 0 | \mathbf{r}\}}$$

$$= \log \frac{\displaystyle\sum_{(l',l) \in B_t^i} \alpha_{t-1}(l') \gamma_t^i(l', l) \beta_t(l)}{\displaystyle\sum_{(l',l) \in B_t^0} \alpha_{t-1}(l') \gamma_t^0(l', l) \beta_t(l)} \qquad (5.4)$$

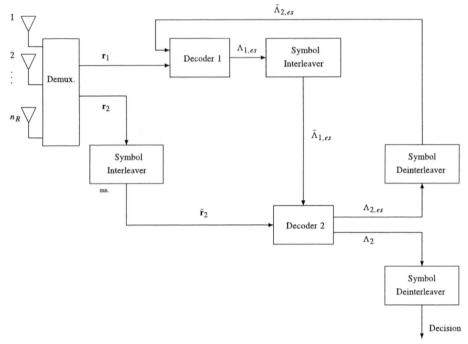

Figure 5.7 Turbo TC decoder with parity symbol puncturing

where i denotes an information group from the set, $\{0, 1, 2, \ldots, 2^m - 1\}$, $\mathbf{r}$ is the received sequence, B_t^i is the set of transitions defined by $S_{t-1} = l' \to S_t = l$, that are caused by the input symbol i, where S_t is a trellis state at time t, and the probabilities $\alpha_t(l)$, $\beta_t(l)$ and $\gamma_t(l', l)$ can be computed recursively [14] (Appendix 5.1). The symbol i with the largest log-likelihood ratio in Eq. (5.4), $i \in \{0, 1, 2, \ldots, 2^m - 1\}$, is chosen as the hard decision output.

The decoder operates on a trellis with M_s states. The forward recursive variables can be computed as follows

$$\alpha_t(l) = \sum_{l'=0}^{M_s-1} \alpha_{t-1}(l') \sum_{i=0}^{2^m-1} \gamma_t^i(l', l) \quad l = 0, 1, \ldots, M_s - 1 \tag{5.5}$$

with the initial condition

$$\alpha_0(0) = 1$$

$$\alpha_0(l) = 0, \quad l \neq 0$$

and the backward recursive variables can be computed as

$$\beta_t(l) = \sum_{l'=0}^{M_s-1} \beta_{t+1}(l') \sum_{i=0}^{2^m-1} \gamma_{t+1}^i(l, l') \quad l = 0, 1, \ldots, M_s - 1 \tag{5.6}$$

with the initial condition

$$\beta_\tau(0) = 1$$

$$\beta_\tau(l) = 0, \quad l \neq 0$$

The branch transition probability at time t, denoted by $\gamma_t^i(l', l)$, is calculated as

$$\gamma_t^i(l', l) = \begin{cases} \dfrac{p_t(i)}{p_t(0)} \exp\left(-\dfrac{\displaystyle\sum_{j=1}^{n_R} |r_t^j - \sum_{n=1}^{n_T} h_{j,n} x_t^n|^2}{2\sigma^2} \right), & \text{for } (l', l) \in B_t^i \\ \\ 0, & \text{otherwise} \end{cases}$$

where r_t^j is the received signal by antenna j at time t, $h_{j,n}$ is the channel attenuation between transmit antenna n and receive antenna j, x_t^n is the modulated symbol at time t, transmitted from antenna n and associated with the transition $S_{t-1} = l'$ to $S_t = l$, and $p_t(i)$ is the a priori probability of $\mathbf{c}_t = i$.

The iterative process of the symbol-by-symbol MAP algorithm for space-time turbo trellis codes is similar to that of binary turbo decoders. However, for binary turbo decoders, a soft output can be split into three terms. They are the a priori information generated by the other decoder, the systematic information generated by the code information symbol and the extrinsic information generated by the code parity symbols. The extrinsic information is independent of the a priori and systematic information. The extrinsic information is exchanged between the two component decoders. For space-time turbo trellis codes, regardless whether component codes are systematic or nonsystematic, it is not possible to separate the influence of the information and the parity-check components within one received symbol, as the symbols transmitted from various antennas interfere with each other. The systematic information and the extrinsic information are not independent. Thus both systematic and extrinsic information will be exchanged between the two component decoders. The joint extrinsic and systematic information of the first MAP decoder, denoted by $\Lambda_{1,es}(\mathbf{c}_t = i)$, can be obtained as

$$\Lambda_{1,es}(\mathbf{c}_t = i) = \Lambda_1(\mathbf{c}_t = i) - \log\frac{p_t(i)}{p_t(0)} \tag{5.7}$$

The joint extrinsic and systematic information $\Lambda_{1,es}(\mathbf{c}_t = i)$ is used as the estimate of the a priori probability ratio at the next decoding stage. After interleaving, it is denoted by $\tilde{\Lambda}_{1,es}(\mathbf{c}_t = i)$. The joint extrinsic and systematic information of the second decoder is given by

$$\Lambda_{2,es}(\mathbf{c}_t = i) = \Lambda_2(\mathbf{c}_t = i) - \tilde{\Lambda}_{1,es}(\mathbf{c}_t = i) \tag{5.8}$$

In the next iteration the a priori probability ratio in Eq. (5.7) is replaced by the deinterleaved joint extrinsic and systematic information from the second decoding stage, denoted by $\tilde{\Lambda}_{2,es}(\mathbf{c}_t = i)$.

Note that each decoder alternately receives the noisy output of its own encoder and that of the other encoder. That is, the parity symbols in every second received signal belong to the other encoder and need to be treated as punctured.

For example, we consider the first decoder. For every odd received signal, the decoding operation proceeds as for the systematic binary turbo codes when the decoder receives the symbol generated by its own encoder, except that the extrinsic information is replaced by the joint extrinsic and systematic information. However, for every even received signal, the decoder receives the punctured symbol in which the parity components are generated by the other encoder. The decoder in this case ignores this symbol by setting the branch transition metric to zero. The only input at this step in the trellis is the a priori component obtained from the other decoder. This component contains the systematic information.

For bit interleaving, decoding can be carried out by converting the joint systematic and extrinsic information computed for a symbol to a bit level, since the exchange of the information between the decoders is on a bit level. After interleaving/deinterleaving operations, the a priori probabilities need to be converted to a symbol level since they will be used in the branch transition probability calculations.

Let us consider a symbol of a group of m information bits given by

$$\mathbf{c} = (c_0, c_1, \ldots, c_{m-1}), \tag{5.9}$$

where $c_j = 0, 1$, $j = 0, 1, 2, \ldots, m - 1$. If we denote the extrinsic information of the symbol $\mathbf{c}$ by $\Lambda_e(\mathbf{c})$, the extrinsic information of extrinsic information the jth bit can be represented by [14]

$$\Lambda_e(c_j) = \log \frac{\displaystyle\sum_{\mathbf{c}:c_j=1} e^{\Lambda_e(\mathbf{c})}}{\displaystyle\sum_{\mathbf{c}:c_j=0} e^{\Lambda_e(\mathbf{c})}} \tag{5.10}$$

After the interleaving/deinterleaving operations, the a priori probability of any symbol can be given by [14]

$$P(\mathbf{c} = (c_0, c_1, \ldots, c_{m-1})) = \prod_{j=0}^{m-1} \frac{e^{c_j \cdot \tilde{\Lambda}_e(c_j)}}{1 + e^{\tilde{\Lambda}_e(c_j)}} \tag{5.11}$$

5.4.1 Decoder Convergence

Consider a turbo code with two component codes. The decoder is based on two component modules, as shown in Fig. 5.8.

The iterative decoder can be viewed as a nonlinear dynamic feedback system [20][21]. Each component decoder can be described by a nonlinear characteristic representing the output versus the input SNR associated with the extrinsic information, denoted by λ. These characteristics are referred to as extrinsic information transfer (EXIT) charts [20]. The pdf of λ can be approximated by a Gaussian distribution. These characteristics are denoted by G_1 and G_2, for the first and second decoder, respectively. For a given channel E_b/N_o, the output SNR of each encoder is a nonlinear function of its input. Thus we have

$$\mathrm{SNR1_{out}} = G_1(\mathrm{SNR1_{in}}, E_b/N_o) \tag{5.12}$$

and

$$\mathrm{SNR2_{out}} = G_2(\mathrm{SNR2_{in}}, E_b/N_o) \tag{5.13}$$

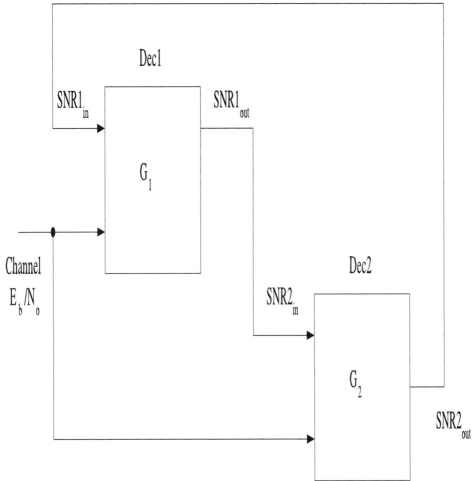

Figure 5.8 Block diagram of an iterative decoder

Also we have

$$\text{SNR2}_{\text{in}} = \text{SNR1}_{\text{out}} \qquad (5.14)$$

and

$$\text{SNR2}_{\text{out}} = G_2(G_1(\text{SNR1}_{\text{in}}, E_b/N_o), E_b/N_o) \qquad (5.15)$$

G_1 and G_2^{-1} at each iteration are shown in Fig. 5.9, for a rate 1/3 CCSDS (Consultative Committee for Space Data Systems) turbo code [14] at Eb/No of 0.8 dB. The encoder for this code is shown in Fig. 5.10.

Note that a nonzero E_b/N_o from the channel enables the first encoder to produce a nonzero output SNR, denoted by SNR1_{out}, though it starts with a zero input SNR, denoted by SNR1_{in}. The second decoder gives a zero output SNR for a zero input SNR, since its information bit is punctured. Figure 5.9 shows the decoder convergence as a function of

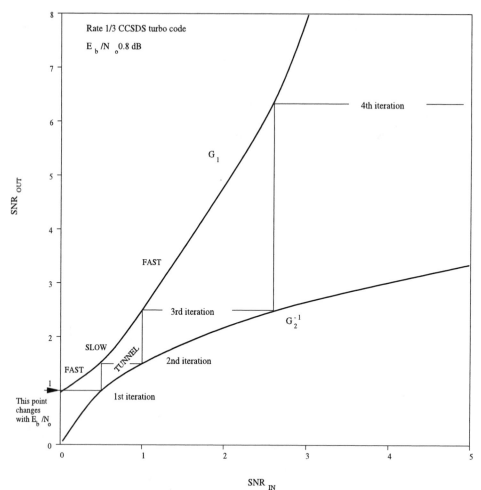

Figure 5.9   EXIT chart for the iterative decoder of the rate 1/3 CCSDS turbo code

iterations. The SNR values of the extrinsic information follow a staircase path between the curves corresponding to G_1 and G_2^{-1}. The steps are large when the curves are far apart and small when they are close. That is, the convergence and consequently the bit error rate improvement, is large when the curves are far apart. The convergence rate is most critical in the narrow passage called the **decoding tunnel**. If the decoder passes the tunnel the convergence becomes fast. If E_b/N_o is reduced from the value in Fig. 5.9, at some point curves G_1 and G_2^{-1} will touch each other. That value represents the iterative **decoding threshold**. The decoder converges for E_b/N_o only above this threshold value.

5.5   ST Turbo TC Performance

This section evaluates the performance of ST turbo TC schemes on fast and slow flat Rayleigh fading channels for various code and system parameters. In each case, it is assumed

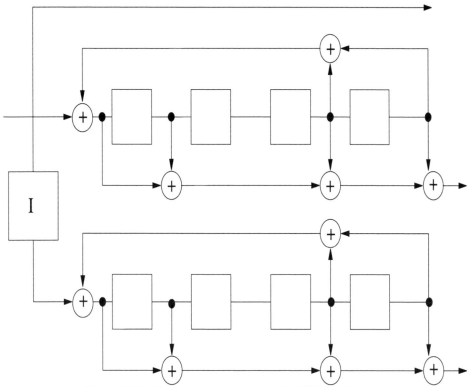

Figure 5.10   The encoder for the rate $1/3$ CCSDS turbo code

that the modulation scheme is either QPSK or 8-PSK, the receiver has two antennas and the interleaver type is S-random [14]. The interleaver size is the same as the block length and ten iterations are performed in decoding ST turbo TC. The bandwidth efficiency is 2 bits/sec/Hz for QPSK and 3 bits/sec/Hz for 8-PSK.

5.5.1   Comparison of ST Turbo TC and STTC

Figure 5.15 shows the frame error rate of a ST turbo TC with 4-state component codes and two transmit antennas on a slow fading channel with a block size of 130 symbols, along with the frame error rate of a 4-state STTC. The feedforward coefficients of the codes are from Table 4.5 and the feedback coefficients $q_j = 1$, $j = 1, 2, \ldots, \nu_1$. The ST turbo TC outperforms the STTC by more than 2 dB at a FER of 10^{-3}. A similar gain is obtained with an 8-state QPSK ST turbo TC relative to an 8-state QPSK STTC, as shown in Fig. 5.16.

5.5.2   Effect of Memory Order and Interleaver Size

Increasing the component code memory order does not improve the FER performance on slow fading channels for the interleaver size of 130 symbols as correlated fading on slow channels makes it more difficult for a higher memory order iterative decoder to converge. This is shown in Fig. 5.11, for a ST turbo TC with two transmit antennas. For the interleaver size of 1024 symbols, the ST turbo TC with an 8-state component codes performs slightly

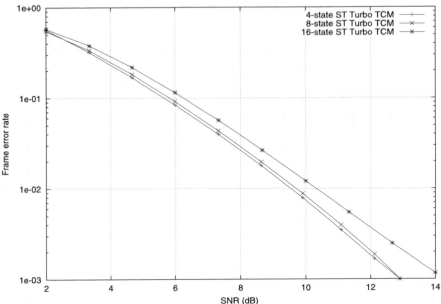

Figure 5.11 FER performance of QPSK ST turbo TC with variable memory order of component codes, two transmit and receive antennas and the interleaver size of 130 symbols on slow fading channels

better than with the 4-state code as illustrated in Fig. 5.12. However, the ST turbo scheme with the 16-state component code is even worse than the one with the 4-state code.

By increasing the number of antennas, the inputs to the decoders become uncorrelated and higher memory codes perform better. This is demonstrated in Fig. 5.13, which compares the FER of a ST turbo TC with 4-state and 16-state component codes and four transmit antennas on a slow fading channel with the interleaver size of 130 symbols. Clearly, the ST turbo TC with the 16-state code has a lower FER than the one with the 4-state component code.

5.5.3 Effect of Number of Iterations

The effect of the number of iterations on the code performance is illustrated in Fig. 5.14. The error rate goes down considerably with increasing the number of iterations from one to six, and at a slower rate between six and ten. It is almost insensitive to an increase in the number of iterations above ten, for the interleaver size of 130. This number of iterations at which the error rate saturates is larger for higher interleaver sizes.

5.5.4 Effect of Component Code Design

The ST turbo TC with the 4-state component code from Table 4.5 outperforms by about 3 dB at FER of $5 \cdot 10^{-3}$ the ST turbo TC in [15] on a slow fading channel, where the component code is a recursive counterpart of the 4-state TSC code [8], as shown in Fig. 5.17. On the other hand, the ST turbo TC with the 4-state component code from Table 4.5 has

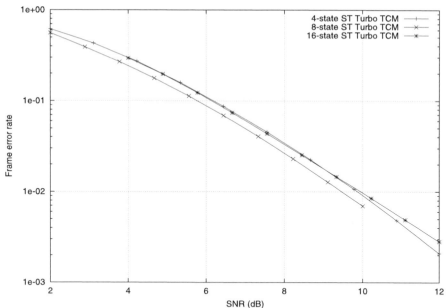

Figure 5.12 FER performance of QPSK ST turbo TC with variable memory order of component codes, two transmit and two receive antennas and the interleaver size of 1024 symbols on slow fading channels

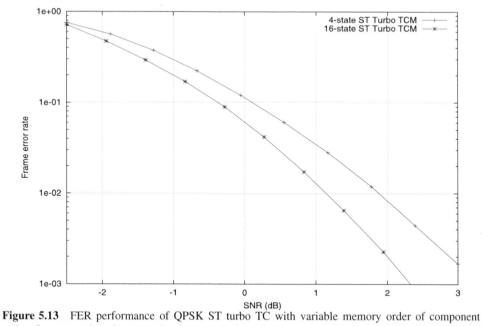

Figure 5.13 FER performance of QPSK ST turbo TC with variable memory order of component codes, four transmit and two receive antennas and the interleaver size of 130 symbols on slow fading channels

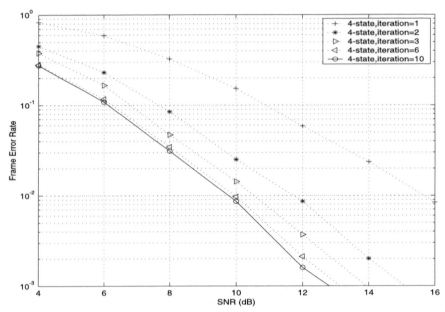

Figure 5.14   FER performance of a 4-state QPSK ST turbo TC with variable number of iterations, two transmit and two receive antennas and the interleaver size of 130 symbols on slow fading channels

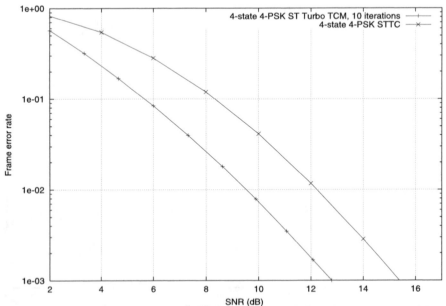

Figure 5.15   FER performance comparison between a 4-state QPSK STTC and a 4-state QPSK ST turbo TC with two transmit and two receive antennas and the interleaver size of 130 on slow fading channels

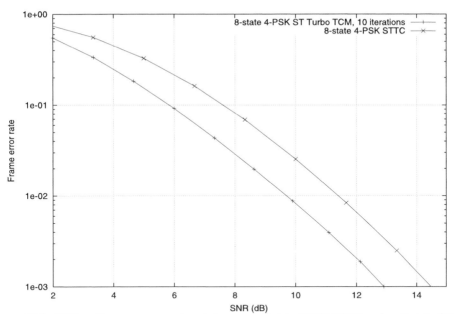

Figure 5.16 FER performance comparison between an 8-state QPSK STTC and an 8-state QPSK ST turbo TC with two transmit and two receive antennas and the interleaver size of 130 on slow fading channels

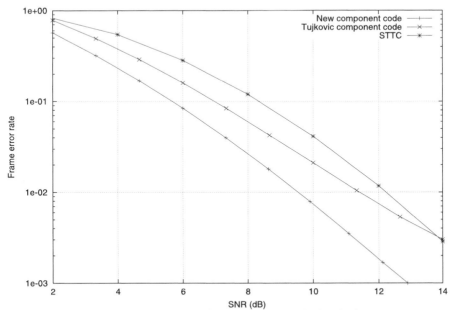

Figure 5.17 FER performance comparison of QPSK ST turbo TC with the 4-state component codes from Table 4.5, from [15] in a system with two transmit and two receive antennas and the interleaver size of 130 symbols on slow fading channels

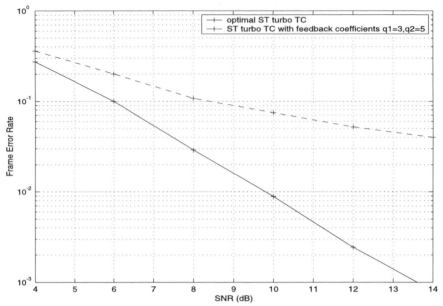

Figure 5.18 FER performance of 8-state QPSK ST turbo TC with variable feedback polynomials of the component codes, two transmit and two receive antennas and the interleaver size of 130 symbols on slow Rayleigh fading channels

approximately the same performance as the ST turbo TC with a 4-state component code presented in [7], which is about 2 dB away from the outage probability.

Variations of the feedback polynomials do not affect the performance of ST turbo TC considerably and for all the presented results the feedback polynomials in all encoder branches have all coefficients equal to one. The only exception is an 8-state ST turbo TC for which the optimum feedback polynomials are $\mathbf{q}^1(D) = 1 + D$, $\mathbf{q}^2(D) = 1 + D + D^2$. However, if the feedback polynomials are $\mathbf{q}^1(D) = 1 + D$ and $\mathbf{q}^2(D) = 1 + D^2$, the performance of the resulting ST turbo TC deteriorates a great deal, as Fig. 5.18 indicates.

5.5.5 Decoder EXIT Charts

The EXIT charts for the 8-state QPSK ST turbo codes with the optimum and non-optimum feedback polynomials are shown in Fig. 5.19. The left EXIT chart for the ST turbo TC with the optimum polynomial in Fig. 5.19 shows that the decoder will converge. The SNR curves for the non-optimum feedback polynomial codes almost touch each other, as illustrated in the right EXIT chart in Fig. 5.19. This means that there is little improvement in SNR in the course of the iteration process and the performance almost does not improve with an increasing number of iterations.

5.5.6 Effect of Interleaver Type

There is a possible choice of bit or symbol interleavers in this scheme. Bit interleavers have an advantage over symbol interleavers only on fast fading channels, by providing higher diversity. However, as Fig. 5.20 depicts, this advantage is only 0.3 dB at the BER of

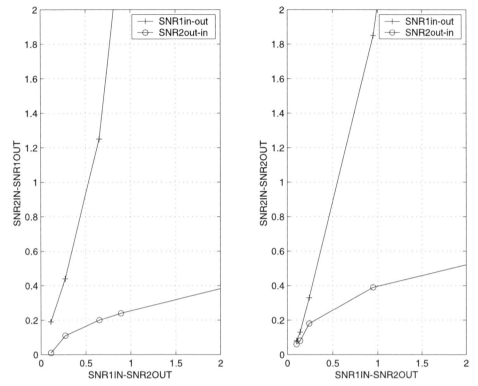

Figure 5.19 EXIT chart for the 8-state QPSK ST turbo TC with the optimum and non-optimum feedback polynomials, two transmit and two receive antennas and the interleaver size of 130 on slow Rayleigh fading channels for Eb/No of 1 dB

10^{-3}, but the decoding complexity with bit interleaving is much higher than with symbol interleaving.

5.5.7 Effect of Number of Transmit and Receive Antennas

The effect of a variable number of transmit antennas on ST turbo TC for a fixed number of receive antennas is shown in Figs. 5.21 and 5.22, for QPSK and 8-PSK signal sets, respectively.

As Fig. 5.21 shows, increasing the number of transmit antennas from two to three, while the number of receive antennas is two, brings a gain of about 3 dB at the FER of 10^{-3}. A further increase in the number of transmit antennas from three to four results in a smaller gain of about 2 dB. A similar behavior can be observed in the performance of 8-PSK ST turbo TC, as the number of transmit antennas varies, as depicted in Fig. 5.22.

The effect of varying the number of receive antennas, given a constant number of transmit antennas, is illustrated in Fig. 5.23. As the number of receive antennas grows from one to two, there is a big gain of about 9 dB, at the FER of 10^{-2}, for two transmit antennas. A further increase in the number of receive antennas to three, yields a gain of about 4 dB.

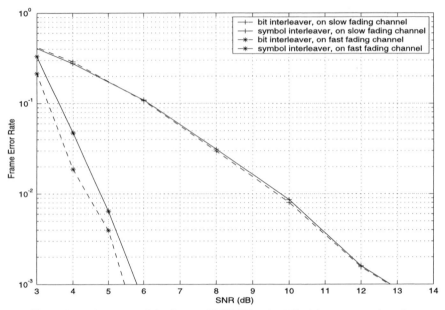

Figure 5.20 FER performance of the 4-state QPSK ST turbo TC with two transmit and two receive antennas, bit and symbol interleavers and the interleaver size of 130 for both interleavers, on slow Rayleigh fading channels

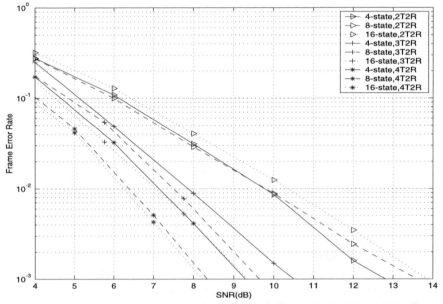

Figure 5.21 FER performance of 4-state QPSK ST turbo TC and STTC with a variable number of transmit and two receive antennas, S-random symbol interleavers of size 130, ten iterations, on slow Rayleigh fading channels

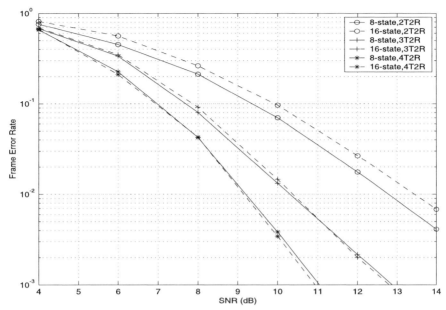

Figure 5.22 FER performance of 8 and 16-state 8-PSK ST turbo TC with a variable number of transmit and two receive antennas, S-random symbol interleavers of memory 130, ten iterations, on slow Rayleigh fading channels

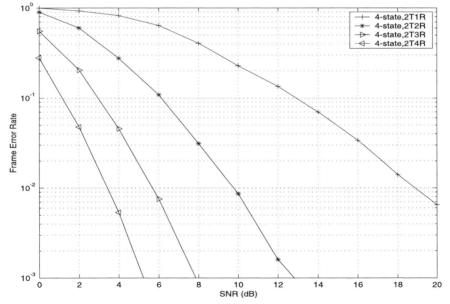

Figure 5.23 FER performance of 4-state 8-PSK ST turbo TC with a variable number of receive and two transmit antennas, S-random symbol interleavers of size 130, on slow Rayleigh fading channels

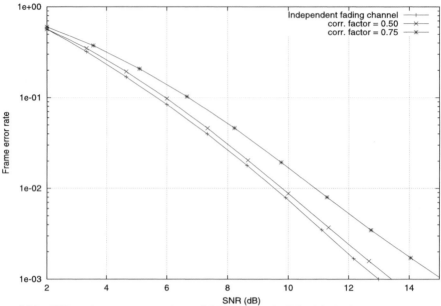

Figure 5.24 FER performance comparison of QPSK ST turbo TC with the 4-state component code from Table 4.5, with uncorrelated and correlated receive antennas in a system with two transmit and two receive antennas and the interleaver size of 130 symbols on slow fading channels

5.5.8   Effect of Antenna Correlation

The ST turbo TC scheme is relatively insensitive to correlation between receive antennas as illustrated in Fig. 5.24. The performance of ST turbo TC with a 4-state component code and two transmit antennas is only slightly degraded on slow fading channels for a correlation factor of 0.5. For the correlation factor of 0.75, the FER deteriorates by about 2 dB relative to the case of uncorrelated antennas.

5.5.9   Effect of Imperfect Channel Estimation

The impact of imperfect channel estimation on the performance of ST turbo TC is depicted in Fig. 5.25. Channel estimation in the simulations was carried out by transmitting orthogonal preambles. The loss of Eb/No at the FER of 10^{-3} is about 1.5 dB relative to the ideal channel estimation.

5.5.10   Performance on Fast Fading Channels

Figure 5.26 shows the FER performance comparison between the 16-state QPSK STTC shown in Table 4.5 and a 16-state QPSK ST turbo TC on a fast fading channel. The 16-state recursive QPSK STTC from Table 4.5 is the constituent code in the ST turbo TC

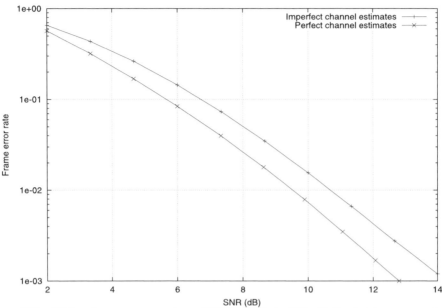

Figure 5.25   FER performance comparison of QPSK ST turbo TC with the 4-state component code from Table 4.5, with ideal and imperfect channel estimation in a system with two transmit and two receive antennas and the interleaver size of 130 symbols on slow fading channels

configuration. The performance curves show that the ST turbo TC configuration offers a considerable improvement. At a frame error rate of 10^{-3}, with ten iterations and an interleaver size of 1024, it achieves a gain of more than 7 dB relative to the STTC. At the same frame error rate, it gains more than 0.2 dB on fast fading channels compared to the ST turbo TC with the constituent code of the same memory order, proposed in [15]. On fast fading channels the performance of ST turbo TC improves as the memory order increases. This is depicted in Figs. 5.27 and 5.28, for systems with two and four transmit antennas, respectively, and the interleaver size of 130 symbols. Comparing Figs. 5.11 and 5.27 it is obvious that the 8-state component code is the best choice if a system is required to operate over both fast and slow fading channels. The ST turbo TC with the 8-state component code is only slightly worse than with the 4-state component code on slow fading channels and better by almost 1.5 dB at FER of 10^{-3} than the system with the 4-state component code on fast fading channels. The effect on variable interleaver size on the performance of ST turbo TC with four transmit and four receive antennas on fast fading channels is illustrated in Figs. 5.29 and 5.30. At the bit error rate of 10^{-5} there is a gain of about 1.5 dB obtained by increasing the interleaver size from 130 to 1024 symbols.

Correlation among antenna elements does not have a significant effect on the performance of ST turbo TC on fast fading channels, as shown in Fig. 5.31. For a correlation factor of 0.75 dB between receive antennas, the loss relative to the uncorrelated antennas is less than 0.5 dB.

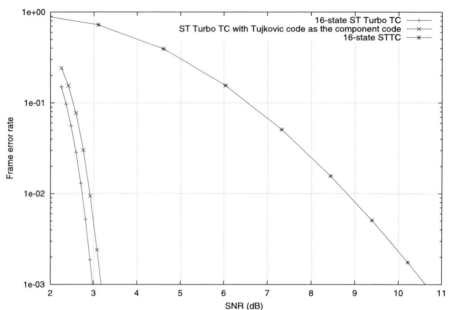

Figure 5.26 FER performance comparison between a 16-state QPSK STTC and a 16-state QPSK ST turbo TC with interleaver size of 1024 on fast fading channels

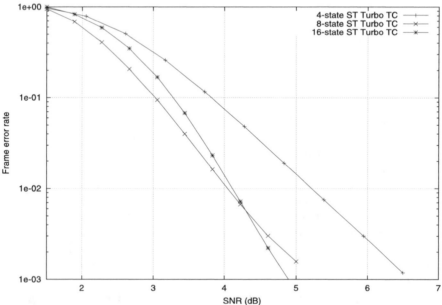

Figure 5.27 FER performance of QPSK ST turbo TC with variable memory component codes from Table 4.5, in a system with two transmit and two receive antennas and the interleaver size of 130 symbols on fast fading channels

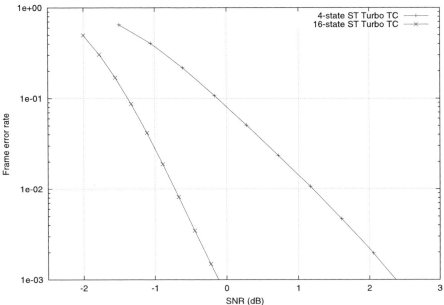

Figure 5.28 FER performance of QPSK ST turbo TC with variable memory component codes from Table 4.7, in a system with four transmit and two receive antennas and the interleaver size of 130 symbols on fast fading channels

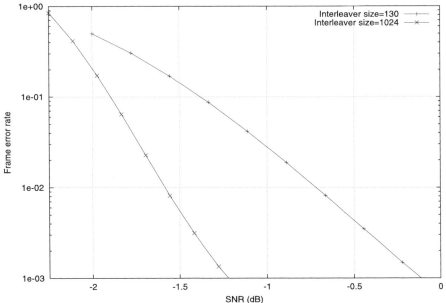

Figure 5.29 FER performance of QPSK ST turbo TC with the 4-state component code from Table 4.7, in a system with four transmit and four receive antennas and a variable interleaver size on fast fading channels

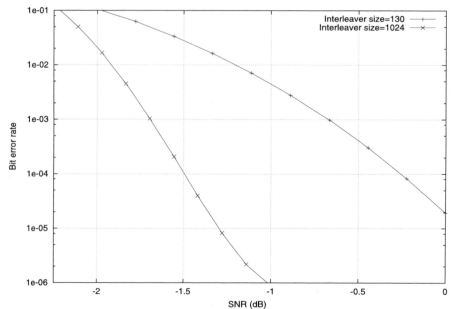

Figure 5.30 BER performance of QPSK ST turbo TC with the 4-state component code from Table 4.7, in a system with four transmit and four receive antennas and a variable interleaver size on fast fading channels

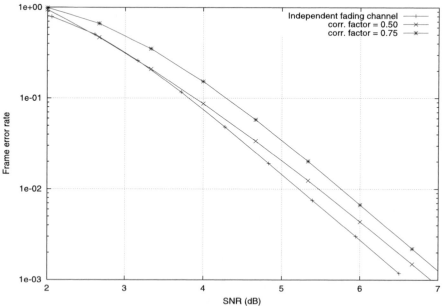

Figure 5.31 FER performance of QPSK ST turbo TC with the 4-state component code from Table 4.5, in a system with two transmit and two receive antennas and an interleaver size of 130 symbols on correlated fast fading channels

Appendix 5.1 MAP Algorithm

We present a MAP decoding algorithm for a system depicted in Fig. 5.32.

In order to simplify the analysis, the following description of the MAP algorithm is specific to $(n, 1, m)$ binary convolutional codes, though it could easily be generalized to include k/n rate convolutional codes, as well as decoding of block codes. A binary message sequence, denoted by $\mathbf{c}$ and given by

$$\mathbf{c} = (c_1, c_2, \dots, c_t, \dots, c_N), \tag{5.16}$$

where c_t is the message symbol at time t and N is the sequence length, is encoded by a linear code. In general the message symbols c_t can be nonbinary but for simplicity we assume that they are independently generated binary symbols and have equal a priori probabilities. The encoding operation is modeled as a discrete time finite-state Markov process. This process can be graphically represented by state and trellis diagrams. In respect to the input c_t, the finite-state Markov process generates an output $\mathbf{v}_t$ and changes its state from S_t to S_{t+1}, where $t + 1$ is the next time instant. The process can be completely specified by the following two relationships

$$\mathbf{v}_t = f(S_t, c_t, t)$$
$$S_{t+1} = g(S_t, c_t, t) \tag{5.17}$$

The functions $f(\cdot)$ and $g(\cdot)$ are generally time varying.

The state sequence from time 0 to t is denoted by $S_0^{t'}$ and is written as

$$\mathbf{S}_0^t = (S_0, S_1, \dots, S_t) \tag{5.18}$$

The state sequence is a Markov process, so that the probability $P(S_{t+1} \mid S_0, S_1, \dots, S_t)$ of being in state S_{t+1}, at time $(t + 1)$, given all states up to time t, depends only on the state S_t, at time t,

$$P(S_{t+1} \mid S_0, S_1, \dots, S_t) = P(S_{t+1} \mid S_t) \tag{5.19}$$

The encoder output sequence from time 0 to t is represented as

$$\mathbf{v}_0^t = (\mathbf{v}_0, \mathbf{v}_1, \dots, \mathbf{v}_t) \tag{5.20}$$

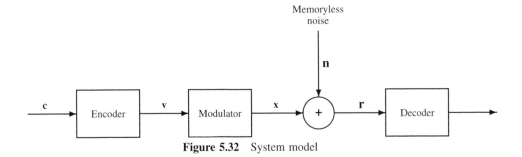

Figure 5.32 System model

where

$$\mathbf{v}_t = (v_{t,0}, v_{t,1}, \ldots, v_{t,n-1}) \tag{5.21}$$

is the code block of length n.

The code sequence $\mathbf{v}_t^{t'}$ is modulated by a BPSK modulator. The modulated sequence is denoted by $\mathbf{x}_t^{t'}$ and is given by

$$\mathbf{x}_0^t = (\mathbf{x}_0, \mathbf{x}_1, \ldots, \mathbf{x}_t) \tag{5.22}$$

where

$$\mathbf{x}_t = (x_{t,0}, x_{t,1}, \ldots, x_{t,n-1}) \tag{5.23}$$

and

$$x_{t,i} = 2v_{t,i} - 1, \quad i = 0, 1, \ldots, n - 1 \tag{5.24}$$

As there is a one-to-one correspondence between the code and modulated sequence, the encoder/modulator pair can be represented by a discrete-time finite-state Markov process and can be graphically described by state or trellis diagrams.

The modulated sequence $\mathbf{x}_t^{t'}$ is corrupted by additive white Gaussian noise, resulting in the received sequence

$$\mathbf{r}_t^{t'} = (\mathbf{r}_t, \mathbf{r}_{t+1}, \ldots, \mathbf{r}_{t'}) \tag{5.25}$$

where

$$\mathbf{r}_t = (r_{t,0}, r_{t,1}, \ldots, r_{t,n-1}) \tag{5.26}$$

and

$$r_{t,i} = x_{t,i} + n_{t,i} \quad i = 0, 1, \ldots, n - 1 \tag{5.27}$$

where $n_{t,i}$ is a zero-mean Gaussian noise random variable with variance σ^2. Each noise sample is assumed to be independent from each other.

The decoder gives an estimate of the input to the discrete finite-state Markov source, by examining the received sequence $\mathbf{r}_t^{t'}$. The decoding problem can be alternatively formulated as finding the modulated sequence $\mathbf{x}_t^{t'}$ or the coded sequence $\mathbf{v}_t^{t'}$. As there is one-to-one correspondence between the sequences $\mathbf{v}_t^{t'}$ and $\mathbf{x}_t^{t'}$, if one of them has been estimated, the other can be obtained by simple mapping.

The discrete-time finite-state Markov source model is applicable to a number of systems in communications, such as linear convolutional and block coding, continuous phase modulation and channels with intersymbol interference.

The MAP algorithm minimizes the symbol (or bit) error probability. For each transmitted symbol it generates its hard estimate and soft output in the form of the a posteriori probability on the basis of the received sequence $\mathbf{r}$. It computes the log-likelihood ratio

$$\Lambda(c_t) = \log \frac{P_r\{c_t = 1 \mid \mathbf{r}\}}{P_r\{c_t = 0 \mid \mathbf{r}\}} \tag{5.28}$$

for $1 \le t \le \tau$, where τ is the received sequence length, and compares this value to a zero threshold to determine the hard estimate c_t as

$$c_t = \begin{cases} 1 & \text{if } \Lambda(c_t) > 0 \\ 0 & \text{otherwise} \end{cases} \tag{5.29}$$

The value $\Lambda(c_t)$ represents the soft information associated with the hard estimate c_t. It might be used in a next decoding stage.

We assume that a binary sequence $\mathbf{c}$ of length N is encoded by a systematic convolutional code of rate $1/n$. The encoding process is modeled by a discrete-time finite-state Markov process described by a state and a trellis diagram with the number of states M_s. We assume that the initial state $S_0 = 0$ and the final state $S_\tau = 0$. The received sequence $\mathbf{r}$ is corrupted by a zero-mean Gaussian noise with variance σ^2.

As an example a rate $1/2$ memory order 2 RSC encoder is shown in Fig. 5.33, and its state and trellis diagrams are illustrated in Figs. 5.34 and 5.35, respectively.

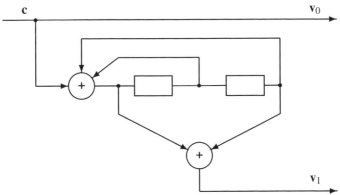

Figure 5.33   A rate 1/2 memory order 2 RSC encoder

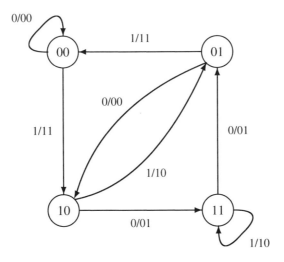

Figure 5.34   State transition diagram for the (2,1,2) RSC code

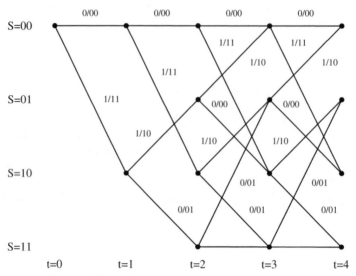

Figure 5.35   Trellis diagram for the (2,1,2) RSC code

The content of the shift register in the encoder at time t represents S_t and it transits into S_{t+1} in response to the input c_{t+1} giving as output the coded block $\mathbf{v}_{t+1}$. The state transition of the encoder is shown in the state diagram.

The state transitions of the encoder are governed by the transition probabilities

$$p_t(l \mid l') = Pr\{S_t = l \mid S_{t-1} = l'\}; \quad 0 \le l, l' \le M_s - 1 \tag{5.30}$$

The encoder output is determined by the probabilities

$$q_t(\mathbf{x}_t \mid l', l) = Pr\{\mathbf{x}_t \mid S_{t-1} = l', S_t = l\}; \quad 0 \le l, l' \le M_s - 1 \tag{5.31}$$

Because of one-to-one correspondence between $\mathbf{x}_t$ and $\mathbf{v}_t$ we have

$$q_t(\mathbf{x}_t \mid l', l) = Pr\{\mathbf{v}_t, \mathbf{v} \mid S_{t-1} = l', S_t = l\}; \quad 0 \le l, l' \le M_s - 1 \tag{5.32}$$

For the encoder in Fig. 5.33, $p_t(l|l')$ is either 0.5, when there is a connection from $S_{t-1} = l'$ to $S_t = l$ or 0 when there is no connection. $q_t(\mathbf{x}|l', l)$ is either 1 or 0. For example, from Figs. 5.34 and 5.35 we have

$$\begin{aligned} p_t(2|0) &= p_t(1) = 0.5; & p_t(1|2) &= p_t(1) = 0.5 \\ p_t(3|0) &= 0; & p_t(1|3) &= p_t(0) = 0.5 \end{aligned}$$

and

$$\begin{aligned} q_t(-1, -1|0, 0) &= 1 & q_t(-1, +1|0, 0) &= 0 \\ q_t(+1, -1|0, 1) &= 0 & q_t(+1, +1|0, 2) &= 1 \end{aligned} \tag{5.33}$$

For a given input sequence

$$\mathbf{c} = (c_1, c_2, \ldots, c_N)$$

the encoding process starts at the initial state $S_0 = 0$ and produces an output sequence $\mathbf{x}_1^\tau$ ending in the terminal state $S_\tau = 0$, where $\tau = N + m$. The input to the channel is $\mathbf{x}_1^\tau$ and the output is $\mathbf{r}_1^\tau = (\mathbf{r}_1, \mathbf{r}_2, \ldots, \mathbf{r}_\tau)$.

The transition probabilities of the Gaussian channel are defined by

$$Pr\{\mathbf{r}_1^\tau | \mathbf{x}_1^\tau\} = \prod_{j=1}^{\tau} R(\mathbf{r}_j | \mathbf{x}_j) \tag{5.34}$$

where

$$R(\mathbf{r}_j | \mathbf{x}_j) = \prod_{i=0}^{n-1} Pr(r_{j,i} | x_{j,i}) \tag{5.35}$$

and

$$Pr\{r_{j,i} | x_{j,i} = -1\} = \frac{1}{\sqrt{2\pi}\,\sigma} e^{-\frac{(r_{j,i}+1)^2}{2\sigma^2}} \tag{5.36}$$

$$Pr\{r_{j,i} | x_{j,i} = 1\} = \frac{1}{\sqrt{2\pi}\,\sigma} e^{-\frac{(r_{j,i}-1)^2}{2\sigma^2}} \tag{5.37}$$

where σ^2 is the noise variance.

Let c_t be the information bit associated with the transition S_{t-1} to S_t, producing as output $\mathbf{v}_t$. The decoder gives an estimate of the input to the Markov source, by examining $\mathbf{r}_1^\tau$. The MAP algorithm provides the log likelihood ratio, denoted by $\Lambda(c_t)$, given the received sequence $\mathbf{r}_1^\tau$, as indicated in Eq. (5.28) where $Pr\{c_t = i | \mathbf{r}_1^\tau\}$, $i = 0, 1$, is the APP of the data bit c_t.

The decoder makes a decision by comparing $\Lambda(c_t)$ to a threshold equal to zero.

We can compute the APPs in (5.28) as

$$Pr\{c_t = 0 | \mathbf{r}_1^\tau\} = \sum_{(l',l) \in B_t^0} Pr\{S_{t-1} = l', S_t = l | \mathbf{r}_1^\tau\} \tag{5.38}$$

where B_t^0 is the set of transitions $S_{t-1} = l' \to S_t = l$ that are caused by the input bit $c_t = 0$. For example, B_t^0 for the diagram in Fig. 5.35 are (3,1), (0,0), (1,2) and (2,3).

Also

$$Pr\{c_t = 1 | \mathbf{r}_1^\tau\} = \sum_{(l',l) \in B_t^1} Pr\{S_{t-1} = l', S_t = l | \mathbf{r}_1^\tau\} \tag{5.39}$$

where B_t^1 is the set of transitions $S_{t-1} = l' \to S_t = l$ that are caused by the input bit $c_t = 1$. For the diagram in Fig. 5.35, B_t^1 consists of (0,2), (2,1), (3,3) and (1,0).

Equation (5.38) can be written as

$$Pr\{c_t = 0 | \mathbf{r}_1^\tau\} = \sum_{(l',l) \in B_t^0} \frac{Pr\{S_{t-1} = l', S_t = l, \mathbf{r}_1^\tau\}}{Pr\{\mathbf{r}_1^\tau\}} \tag{5.40}$$

The APP of the decoded data bit c_t can be derived from the joint probability defined as

$$\sigma_t(l', l) = Pr\{S_{t-1} = l', S_t = l, \mathbf{r}_1^\tau\}, \quad l = 0, 1, \ldots, M_s - 1 \tag{5.41}$$

Equation (5.40) can be written as

$$Pr\{c_t = 0 | \mathbf{r}_1^\tau\} = \sum_{(l', l) \in B_t^0} \frac{\sigma_t(l', l)}{Pr\{\mathbf{r}_1^\tau\}} \tag{5.42}$$

Similarly the APP for $c_t = 1$ is given by

$$Pr\{c_t = 1 | \mathbf{r}_1^\tau\} = \sum_{(l', l) \in B_t^1} \frac{\sigma_t(l', l)}{Pr\{\mathbf{r}_1^\tau\}} \tag{5.43}$$

The log-likelihood ratio $\Lambda(c_t)$ is then

$$\Lambda(c_t) = \log \frac{\displaystyle\sum_{(l', l) \in B_t^1} \sigma_t(l', l)}{\displaystyle\sum_{(l', l) \in B_t^0} \sigma_t(l', l)} \tag{5.44}$$

The log-likelihood $\Lambda(c_t)$ represents the soft output of the MAP decoder. It can be used as an input to another decoder in a concatenated scheme or in the next iteration in an iterative decoder. In the final operation, the decoder makes a hard decision by comparing $\Lambda(c_t)$ to a threshold equal to zero.

In order to compute the joint probability $\sigma_t(l', l)$ required for calculation of $\Lambda(c_t)$ in (5.44), we define the following probabilities

$$\alpha_t(l) = Pr\{S_t = l, \mathbf{r}_1^t\} \tag{5.45}$$

$$\beta_t(l) = Pr\{\mathbf{r}_{t+1}^\tau | S_t = l\} \tag{5.46}$$

$$\gamma_t^i(l', l) = Pr\{c_t = i, S_t = l, \mathbf{r}_t | S_{t-1} = l'\}; \quad i = 0, 1 \tag{5.47}$$

Now we can express $\sigma_t(l', l)$ as

$$\sigma_t(l', l) = \alpha_{t-1}(l') \cdot \beta_t(l) \cdot \sum_{i \in (0,1)} \gamma_t^i(l', l) \tag{5.48}$$

The log-likelihood ratio $\Lambda(c_t)$ can be written as

$$\Lambda(c_t) = \log \frac{\displaystyle\sum_{(l', l) \in B_t^1} \alpha_{t-1}(l') \gamma_t^1(l', l) \beta_t(l)}{\displaystyle\sum_{(l', l) \in B_t^0} \alpha_{t-1}(l') \gamma_t^0(l', l) \beta_t(l)} \tag{5.49}$$

We can obtain α defined in (5.45) as

$$\alpha_t(l) = \sum_{l'=0}^{M_s-1} \alpha_{t-1}(l') \cdot \sum_{i \in (0,1)} \gamma_t^i, (l', l) \tag{5.50}$$

for $t = 1, 2, \ldots \tau$.

For $t = 0$ we have the boundary conditions

$$\alpha_0(0) = 1 \quad \text{and} \quad \alpha_0(l) = 0 \quad \text{for } l \neq 0$$

We can express $\beta_t(l)$ defined in (5.46) as

$$\beta_t(l) = \sum_{l'=0}^{M_s-1} \beta_{t+1}(l') \sum_{i \in (0,1)} \gamma_{t+1}^i(l', l) \tag{5.51}$$

for $t = \tau - 1, \ldots, 1, 0$.

The boundary conditions are $\beta_\tau(0) = 1$ and $\beta_\tau(l) = 0$ for $l \neq 0$.
We can write for $\gamma_t^i(l', l)$ defined in (5.47)

$$\gamma_t^i(l', l) = p_t(l|l') \cdot q_t(\mathbf{x}|l', l) \cdot R(\mathbf{r}_t|\mathbf{x}_t)$$

We can further express $\gamma_t^i(l', l)$ as

$$\gamma_t^i(l', l) = \begin{cases} p_t(i) \exp\left(-\dfrac{\sum\limits_{j=0}^{n-1}(r_{t,j}^i - x_{t,j}^i(l))^2}{2\sigma^2}\right) & \text{for } (l, l') \in B_t^i \\[20pt] 0 & \text{otherwise} \end{cases}$$

where $p_t(i)$ is the a priori probability of $c_t = i$ and $x_{t,j}^i(l)$ is the encoder output associated with the transition $S_{t-1} = l'$ to $S_t = l$ and input $c_t = i$. Note that the expression for $R(\mathbf{r}_t|\mathbf{x}_t)$ is normalized by multiplying (5.35) $(\sqrt{2\pi}\sigma)^n$.

Summary of the MAP Algorithm

1. Forward recursion

    - Initialize $\alpha_0(l)$, $l = 0, 1, \ldots, M_s - 1$
      $\alpha_0(0) = 1$ and $\alpha_0(l) = 0$ for $l \neq 0$
    - For $t = 1, 2, \ldots \tau$, $l = 0, 1, \ldots M_s - 1$ and all branches in the trellis calculate

    $$\gamma_t^i(l', l) = p_t(i) \exp\left(\frac{-d^2(\mathbf{r}_t, \mathbf{x}_t)}{2\sigma^2}\right) \quad \text{for } i = 0, 1 \tag{5.52}$$

    where $p_t(i)$ is the a priori probability of each information bit, $d^2(\mathbf{r}_t, \mathbf{x}_t)$ is the squared Euclidean distance between $\mathbf{r}_t$ and the modulated symbol in the trellis $\mathbf{x}_t$.
    - For $i = 0, 1$ store $\gamma_t^i(l', l)$.
    - For $t = 1, 2, \ldots, \tau$, and $l = 0, 1, \ldots, M_s - 1$ calculate and store $\alpha_t(l)$

    $$\alpha_t(l) = \sum_{l'=0}^{M_s-1} \sum_{i \in (0,1)} \alpha_{t-1}(l')\gamma_t^i(l', l) \tag{5.53}$$

    The graphical representation of the forward recursion is given in Fig. 5.36.

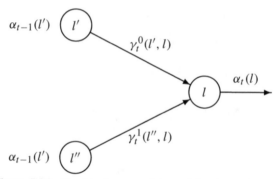

Figure 5.36   Graphical representation of the forward recursion

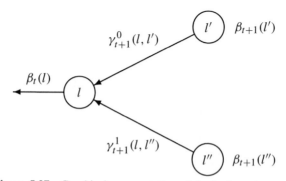

Figure 5.37   Graphical representation of the backward recursion

2.   Backward recursion

- Initialize $\beta_\tau(l)$, $l = 0, 1, \ldots, M_s - 1$
  $\beta_\tau(0) = 1$ and $\beta_\tau(l) = 0$ for $l \neq 0$
- For $t = \tau - 1, \ldots 1, 0$ and $l = 0, 1, \ldots M_s - 1$ calculate $\beta_t(l)$ as

$$\beta_t(l) = \sum_{l'=0}^{M_s-1} \sum_{i \in (0,1)} \beta_{t+1}(l')\gamma_{t+1}^i(l, l') \tag{5.54}$$

  where $\gamma_{t+1}^i(l, l')$ was computed in the forward recursion.
- For $t < \tau$ calculate the log-likelihood $\Lambda(c_t)$ as

$$\Lambda(c_t) = \log \frac{\displaystyle\sum_{l=0}^{M_s-1} \alpha_{t-1}(l')\gamma_t^1(l', l)\beta_t(l)}{\displaystyle\sum_{l=0}^{M_s-1} \alpha_{t-1}(l')\gamma_t^0(l', l)\beta_t(l)} \tag{5.55}$$

The graphical representation of the backward recursion is shown in Fig. 5.37.

Note that because Eq. (5.55) is a ratio, the values for $\alpha_t(l')$ and $\beta_t(l)$ can be normalized at any node which keeps them from overflowing.

If the final state of the trellis is not known, the probability $\beta_\tau(l)$, can be initialized as

$$\beta_\tau(l) = \frac{1}{Ms}, \quad \forall l \tag{5.56}$$

Bibliography

[1] C. Berrou, A. Glavieux and P. Thitimajshima, "Near Shannon limit error-correcting coding and decoding: turbo codes", *Proc. Inter. Conf. Commun.*, 1993, pp. 1064–1070.

[2] G. Ungerboeck, "Channel coding with multilevel phase signals", *IEEE Trans. Inform. Theory*, vol. 28, Jan. 1982, pp. 55–67.

[3] P. Robertson and T. Worz, "Coded modulation scheme employing turbo codes", *IEE Electronics Letters*, vol. 31, no. 18, Aug. 1995, pp. 1546–1547.

[4] P. Robertson and T. Worz, "Bandwidth-efficient turbo trellis coded modulation using punctured component codes", *IEEE Journal on Selec. Areas in Communications*, vol. 16, no. 2, pp. 206–218, Feb. 1998.

[5] S. Benedetto, D. Divsalar, G. Montorsi and F. Pollara, "Parallel concatenated trellis coded modulation", *Proc. IEEE ICC'96*, pp. 974–978.

[6] Y. Liu and M. Fitz, "Space-time turbo codes", *13th Annual Allerton Conf. on Commun. Control and Computing*, Sept. 1999.

[7] Dongzhe Cui and A. Haimovich, "Performance of parallel concatenated space-time codes", *IEEE Commun. Letters*, vol. 5, June 2001, pp. 236–238.

[8] V. Tarokh, N. Seshadri and A. Calderbank, "Space-Time Codes for High Data Rate Wireless Communication: Performance Criterion and Code Construction", *IEEE Trans. Inform. Theory*, vol. 44, no. 2, March 1998, pp. 744–765.

[9] S. Baro, G. Bauch and A. Hansmann, "Improved codes for space-time trellis-coded modulation", *IEEE Trans. Commun. Letters*, vol. 4, Jan. 2000, pp. 20–22.

[10] J. C. Guey, M. Fitz, M. R. Bell and W. Y. Kuo, "Signal design for transmitter diversity wireless communication systems over Rayleigh fading channels", *Proc. of IEEE VTC'96*, pp. 136–140.

[11] Z. Chen, J. Yuan and B. Vucetic, "Improved space-time trellis coded modulation scheme on slow Rayleigh fading channels", *IEE Electronics Letters*, vol. 37, no. 7, March 2001, pp. 440–441.

[12] J. Yuan, B. Vucetic, Z. Chen and W. Firmanto, "Performance of space-time coding on fading channels", *Proc. of Intl. Symposium on Inform. Theory (ISIT) 2001*, Washington D.C, June 2001.

[13] L. R. Bahl, J. Cocke, F. Jelinek and J. Raviv, "Optimal decoding of linear codes for minimizing symbol error rate", *IEEE Trans. Inform. Theory*, vol. IT-20, pp. 284–287, Mar. 1974.

[14] B. Vucetic and J. Yuan, *Turbo Codes Principles and Applications*, Kluwer Publishers, 2000.

[15] D. Tujkovic, "Recursive space-time trellis codes for turbo coded modulation", *Proc. of GlobeCom 2000*, San Francisco.

[16] E. Telatar, "Capacity of multi-antenna Gaussian channels", *The European Transactions on Telecommunications*, vol. 10, no. 6, Nov./Dec. 1999, pp. 585–595.

[17] D. Divsalar, S. Dolinar and F. Pollara, "Low complexity turbo-like codes", *Proc. of 2nd Int'l. Symp. on Turbo Codes and Related Topics*, Brest, 2000, pp. 73–80.

[18] S. Y. Chung, T. Richardson and R. Urbanke, "Analysis of sum-product decoding of low-density-parity-check codes using Gaussian approximation", submitted to *IEEE Trans. Inform. Theory*.

[19] I. S. Gradshteyn and I. M. Ryzhik, *Table of Integrals, Series, and Products*, Fifth edition, Academic Press.

[20] S. ten Brink, "Convergence of iterative decoding", *Electron. Lett.*, vol. 35, no. 13, pp. 806–808, May 24th, 1999.

[21] D. Divsalar, S. Dolinar and F. Pollara, "Iterative turbo decoder analysis based on density evolution", *IEEE Journal on Selected Areas in Communications*, vol. 9, pp. 891–907, May 2001.

[22] W. Firmanto, B. Vucetic, J. Yuan and Z. Chen, "Space-time Turbo Trellis Coded Modulation for Wireless Data Communications", *Eurasip Journal on Applied Signal Processing*, vol. 2002, no. 5, May 2002, pp. 459–470.

[23] W. Firmanto, J. Yuan and B. Vucetic, "Turbo Codes with Transmit Diversity: Performance Analysis and Evaluation", *IEICE Trans. Commun.*, vol. E85-B, no. 5, May 2002.

6

Layered Space-Time Codes

6.1 Introduction

Space-time trellis codes have a potential drawback that the maximum likelihood decoder complexity grows exponentially with the number of bits per symbol, thus limiting achievable data rates. Foschini [35] proposed a layered space-time (LST) architecture that can attain a tight lower bound on the MIMO channel capacity. The distinguishing feature of this architecture is that it allows processing of multidimensional signals in the space domain by 1-D processing steps, where 1-D refers to one dimension in space. The method relies on powerful signal processing techniques at the receiver and conventional 1-D channel codes. In the originally proposed architecture, n_T information streams are transmitted simultaneously, in the same frequency band, using n_T transmit antennas. The receiver uses $n_R = n_T$ antennas to separate and detect the n_T transmitted signals. The separation process involves a combination of interference suppression and interference cancellation. The separated signals are then decoded by using conventional decoding algorithms developed for (1-D)-component codes, leading to much lower complexity compared to maximum likelihood decoding. The complexity of the LST receivers grows linearly with the data rate. Though in the original proposal the number of receive antennas, denoted by n_R, is required to be equal or greater than the number of transmit antennas, the use of more advanced detection/decoding techniques enables this requirement to be relaxed to $n_R \geq 1$.

In this chapter we present the principles of LST codes and discuss transmitter architectures. This is followed by the exposition of the signal processing techniques used to decouple and detect the LST signals. Zero forcing (ZF) and minimum mean square error (MMSE) interference suppression methods are considered, as well as iterative interference cancellation schemes. In these schemes, parallel interference cancellers (PIC) and MMSE nonlinear architectures are used for detection while maximum a posteriori probability (MAP) methods are applied for decoding. A method which can significantly improve the performance of PIC detectors, called decision statistics combining is also presented. The performance of various receiver structures is discussed and illustrated by simulation results.

Space-Time Coding   Branka Vucetic and Jinhong Yuan
© 2003 John Wiley & Sons, Ltd   ISBN: 0-470-84757-3

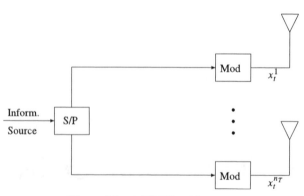

Figure 6.1   A VLST architecture

6.2   LST Transmitters

There is a number of various LST architectures, depending on whether error control coding is used or not and on the way the modulated symbols are assigned to transmit antennas. An uncoded LST structure, known as *vertical layered space-time* (VLST) or *vertical Bell Laboratories layered space-time* (VBLAST) scheme [43], is illustrated in Fig. 6.1. The input information sequence, denoted by $\mathbf{c}$, is first demultiplexed into n_T sub-streams and each of them is subsequently modulated by an M-level modulation scheme and transmitted from a transmit antenna. The signal processing chain related to an individual sub-stream is referred to as a *layer*. The modulated symbols are arranged into a transmission matrix, denoted by $\mathbf{X}$, which consists of n_T rows and L columns, where L is the transmission block length. The tth column of the transmission matrix, denoted by $\mathbf{x}_t$, consists of the modulated symbols $x_t^1, x_t^2, \ldots, x_t^{n_T}$, where $t = 1, 2, \ldots, L$. At a given time t, the transmitter sends the tth column from the transmission matrix, one symbol from each antenna. That is, a transmission matrix entry x_t^i is transmitted from antenna i at time t. Vertical structuring refers to transmitting a sequence of matrix columns in the space-time domain. This simple transmission process can be combined with conventional block or convolutional one-dimensional codes, to improve the performance of the system. This term "one-dimensional" refers to the space domain, while these codes can be multidimensional in the time domain. The block diagrams of various LST architectures with error control coding are shown in Fig. 6.2(a)–(c).

In the *horizontal layered space-time* (HLST) architecture, shown in Fig. 6.2(a), the information sequence is first encoded by a channel code and subsequently demultiplexed into n_T sub-streams. Each sub-stream is modulated, interleaved and assigned to a transmit antenna. If the modulator output symbols are denoted by x_t^i, where i represents the layer number and t is the time interval, the transmission matrix, formed from the modulator outputs, denoted by $\mathbf{X}$, is given by

$$\mathbf{X} = \left[x_t^i \right] \tag{6.1}$$

For example, in a system with three transmit antennas, the transmission matrix $\mathbf{X}$ is given by

$$\mathbf{X} = \begin{bmatrix} x_1^1 & x_2^1 & x_3^1 & x_4^1 & \cdots \\ x_1^2 & x_2^2 & x_3^2 & x_4^2 & \cdots \\ x_1^3 & x_2^3 & x_3^3 & x_4^3 & \cdots \end{bmatrix} \tag{6.2}$$

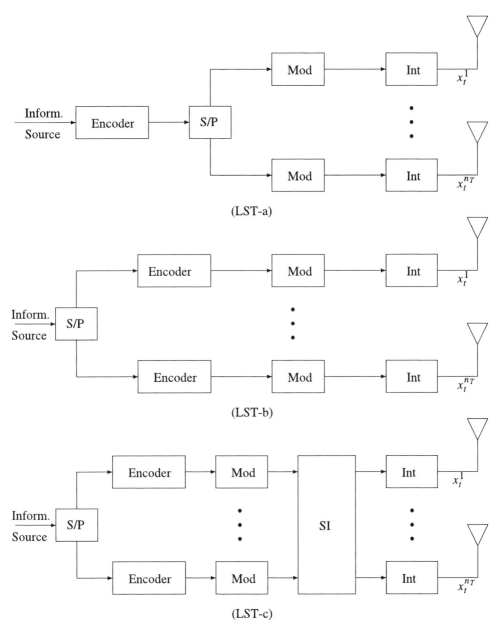

Figure 6.2 LST transmitter architectures with error control coding; (a) an HLST architecture with a single code; (b) an HLST architecture with separate codes in each layer; (c) DLST and TLST architectures

The sequence $x_1^1, x_2^1, x_3^1, x_4^1, \dots$ is transmitted from antenna 1, the sequence $x_1^2, x_2^2, x_3^2, x_4^2, \dots$ is transmitted from antenna 2 and the sequence $x_1^3, x_2^3, x_3^3, x_4^3, \dots$ is transmitted from antenna 3.

An HLST architecture can also be implemented by splitting the information sequence into n_T sub-streams, as shown in Fig. 6.2(b). Each sub-stream is encoded independently by a channel encoder, interleaved, modulated and then transmitted by a particular transmit antenna. It is assumed that channel encoders for various layers are identical. However, different coding in each sub-stream can be used.

A better performance is achieved by a *diagonal layered space-time* (DLST) architecture [35], in which a modulated codeword of each encoder is distributed among the n_T antennas along the diagonal of the transmission array. For example, the DLST transmission matrix, for a system with three antennas, is formed from matrix $\mathbf{X}$ in (6.2), by delaying the ith row entries by $(i-1)$ time units, so that the first nonzero entries lie on a diagonal in $\mathbf{X}$. The entries below the diagonal are padded by zeros. Then the first diagonal is transmitted from the first antenna, the second diagonal from the second antenna, the third diagonal from the third antenna and then the fourth diagonal from the first antenna etc. Hence the codeword symbols of each encoder are transmitted over different antennas. This can be represented by introducing a spatial interleaver SI after the modulators, as shown in Fig. 6.2(c). The spatial interleaving operation for the DLST scheme can be represented as

$$
\begin{bmatrix}
x_1^1 & x_2^1 & x_3^1 & x_4^1 & x_5^1 & x_6^1 & \cdots \\
0 & x_1^2 & x_2^2 & x_3^2 & x_4^2 & x_5^2 & \cdots \\
0 & 0 & x_1^3 & x_2^3 & x_3^3 & x_4^3 & \cdots
\end{bmatrix}
\longrightarrow
\begin{bmatrix}
x_1^1 & x_1^2 & x_1^3 & x_4^1 & x_4^2 & x_4^3 & \cdots \\
0 & x_2^1 & x_2^2 & x_2^3 & x_5^1 & x_5^2 & \cdots \\
0 & 0 & x_3^1 & x_3^2 & x_3^3 & x_6^1 & \cdots
\end{bmatrix}
$$

$$(6.3)$$

The rows of the matrix on the right-hand side of (6.3) are obtained by concatenating the corresponding diagonals of the matrix on the left-hand side. The first row of this matrix is transmitted from the first antenna, the second row from the second antenna and the third row from the third antenna.

The diagonal layering introduces space diversity and thus achieves a better performance than the horizontal one.

It is important to note that there is a spectral efficiency loss in DLST, since a portion of the transmission matrix on the left-hand side of (6.3) is padded with zeros.

A *threaded layered space-time* (TLST) structure [36] is obtained from the HLST by introducing a spatial interleaver SI prior to the time interleavers, as shown in Fig. 6.2(c). In a system with $n_T = 3$, the operation of SI can be expressed as

$$
\begin{bmatrix}
x_1^1 & x_2^1 & x_3^1 & x_4^1 & \cdots \\
x_1^2 & x_2^2 & x_3^2 & x_4^2 & \cdots \\
x_1^3 & x_2^3 & x_3^3 & x_4^3 & \cdots
\end{bmatrix}
\longrightarrow
\begin{bmatrix}
x_1^1 & x_2^3 & x_3^2 & x_4^1 & \ddots \\
x_1^2 & x_2^1 & x_3^3 & x_4^2 & \ddots \\
x_1^3 & x_2^2 & x_3^1 & x_4^3 & \ddots
\end{bmatrix}
\qquad (6.4)
$$

in which an element of the modulation matrix, shown on the left-hand side of (6.4) denoted by x_t^i, represents the modulated symbol of layer i at time t. The matrix on the right-hand side of (6.4), denoted by $\mathbf{X}'$, is the TLST transmission matrix. That is, the modulated symbols $x_1^1, x_2^3, x_3^2, x_4^1, \ldots$, generated by modulators in layers 1, 3, 2 and 1, respectively, are transmitted from antenna 1.

The spatial interleaver of the TLST can be represented by a cyclic-shift interleaver as follows. If we denote the left-hand side matrix in (6.4) by $\mathbf{X}$, the first column of the transmission matrix $\mathbf{X}'$ is identical to the first column of the modulated matrix $\mathbf{X}$. The second column of $\mathbf{X}'$ is obtained by a cyclic shift of the second column of $\mathbf{X}$ by one position from the top to the bottom. The third column of $\mathbf{X}'$ is obtained by a cyclic shift of the third column of $\mathbf{X}$ by two positions, while the fourth column of $\mathbf{X}'$ is identical to the fourth column of $\mathbf{X}$ etc. In general, if we denote the entries of $\mathbf{X}'$ by $x_t^{i'}$, the mapping of x_t^i to $x_t^{i'}$ can be expressed as

$$x_t^{i'} = x_t^i, \quad i' = [(i + t - 2) \bmod n_T] + 1 \tag{6.5}$$

The spectral efficiency of the HLST and TLST schemes is Rmn_T, where R is the code rate and m is the number of bits in a modulated symbol, while the spectral efficiency of the DLST is slightly reduced due to zero padding in the transmission matrix.

6.3 LST Receivers

In this section we consider receiver structures for layered space-time architectures. In order to simplify the analysis, horizontal layering with binary channel codes and BPSK modulation are assumed. Extension to nonbinary codes and to multilevel modulation schemes is straightforward.

The transmit diversity introduces spatial interference. The signals transmitted from various antennas propagate over independently scattered paths and interfere with each other upon reception at the receiver. This interference can be represented by the following matrix operation

$$\mathbf{r}_t = \mathbf{H}\mathbf{x}_t + \mathbf{n}_t \tag{6.6}$$

where $\mathbf{r}_t$ is an n_R-component column matrix of the received signals across the n_R receive antennas, $\mathbf{x}_t$ is the tth column in the transmission matrix $\mathbf{X}$ and $\mathbf{n}_t$ is an n_R-component column matrix of the AWGN noise signals from the receive antennas, where the noise variance per receive antenna is denoted by σ^2. In a structure with spatial interleaving, vector $\mathbf{x}_t$ is the tth column of the matrix at the output of the spatial interleaver, denoted by $\mathbf{X}'$. In order to simplify the notation, we omit the subscripts in vectors $\mathbf{r}_t$, $\mathbf{x}_t$ and $\mathbf{n}_t$ and refer to them as $\mathbf{r}$, $\mathbf{x}$, and $\mathbf{n}$, respectively.

An LST structure can be viewed as a synchronous code division multiple access (CDMA) in which the number of transmit antennas is equal to the number of users. Similarly, the interference between transmit antennas is equivalent to multiple access interference (MAI) in CDMA systems, while the complex fading coefficients correspond to the spreading sequences. This analogy can be further extended to receiver strategies, so that multiuser receiver structures derived for CDMA can be directly applied to LST systems. Under this scenario, the optimum receiver for an uncoded LST system is a maximum likelihood (ML) multiuser detector [8] operating on a trellis. It computes ML statistics as in the Viterbi algorithm. The complexity of this detection algorithm is exponential in the number of the transmit antennas.

For coded LST schemes, the optimum receiver performs joint detection and decoding on an overall trellis obtained by combining the trellises of the layered space-time coded and the channel code. The complexity of the receiver is an exponential function of the product of the number of the transmit antennas and the code memory order. For many systems, the exponential increase in implementation complexity may make the optimal receiver impractical even for a small number of transmit antennas. Thus, in this chapter we will examine a number of less complex receiver structures which have good performance/complexity trade-offs.

The original VLST receiver [43] is based on a combination of interference suppression and cancellation. Conceptually, each transmitted sub-stream is considered in turn to be the desired symbol and the remainder are treated as interferers. These interferers are suppressed by a zero forcing (ZF) approach [43]. This detection algorithm produces a ZF based decision statistics for a desired sub-stream from the received signal vector $\mathbf{r}$, which contains a residual interference from other transmitted sub-streams. Subsequently, a decision on the desired sub-stream is made from the decision statistics and its interference contribution is regenerated and subtracted out from the received vector $\mathbf{r}$. Thus $\mathbf{r}$ contains a lower level of interference and this will increase the probability of correct detection of other sub-streams. This operation is illustrated in Fig. 6.3. In this figure, the first detected sub-stream is n_T. The detected symbol is subtracted from all other layers. These operations are repeated for the lower layers, finishing with layer 1, which, assuming that all symbols at previous layers have been detected correctly, will be free from interference. The soft decision statistics from the detector at each layer is passed to a decision making device in a VBLAST system. In coded LST schemes, the decision statistics is passed to the channel decoder, which makes the hard decision on the transmitted symbol in this sub-stream. The hard symbol estimate is used to reconstruct the interference from this sub-stream, which is then fed back to cancel its contribution while decoding the next sub-stream.

The ZF strategy is only possible if the number of receive antennas is at least as large as the number of transmit antennas. Another drawback of this approach is that achievable diversity depends on a particular layer. If the ZF strategy is used in removing interference

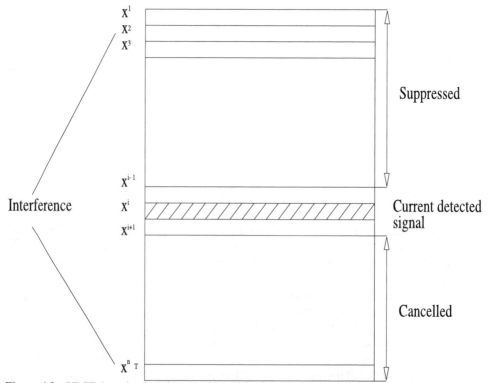

Figure 6.3   VLST detection based on combined interference suppression and successive cancellation

and if n_R receive antennas are available, it is possible to remove

$$n_i = n_R - d_o \tag{6.7}$$

interferers with diversity order of d_o [9]. The diversity order can be expressed as

$$d_o = n_R - n_i \tag{6.8}$$

If the interference suppression starts at layer n_T, then at this layer $(n_T - 1)$ interferers need to be suppressed. Assuming that $n_R = n_T$, the diversity order in this layer, according to (6.7) is 1. In the 1st layer, there are no interferers to be suppressed, so the diversity order is $n_R = n_T$. As different layers have different diversity orders, the diagonal layering is required to achieve equal performance of various encoded streams.

Apart from the original BLAST receivers we will consider minimum mean square error (MMSE) detectors and iterative receivers. The iterative receiver, [20][21] based on the turbo processing principle, can be singled out as the architecture with the best complexity/performance trade-off. Its complexity grows linearly with the number of transmit antennas and transmission rate.

6.3.1 QR Decomposition Interference Suppression Combined with Interference Cancellation

Any $n_R \times n_T$ matrix $\mathbf{H}$, where $n_R \geq n_T$, can be decomposed as

$$\mathbf{H} = \mathbf{U}_R \mathbf{R}, \tag{6.9}$$

where $\mathbf{U}_R$ is an $n_R \times n_T$ unitary matrix and $\mathbf{R}$ is an $n_T \times n_T$ upper triangular matrix, with entries $(R_{i,j})_t = 0$, for $i > j$, $i, j = 1, 2, \ldots n_T$, represented as

$$\mathbf{R} = \begin{bmatrix} (R_{1,1})_t & (R_{1,2})_t & \cdots & (R_{1,n_T})_t \\ 0 & (R_{2,2})_t & \cdots & (R_{2,n_T})_t \\ 0 & 0 & \cdots & (R_{3,n_T})_t \\ \vdots & \vdots & \vdots & \vdots \\ 0 & 0 & \cdots & (R_{n_T,n_T})_t \end{bmatrix} \tag{6.10}$$

The decomposition of the matrix $\mathbf{H}$, as in (6.9), is called *QR factorization*. Let us introduce an n_T-component column matrix $\mathbf{y}$ obtained by multiplying from the left the receive vector $\mathbf{r}$, given by Eq. (6.6), by $\mathbf{U}_R^T$

$$\mathbf{y} = \mathbf{U}_R^T \mathbf{r} \tag{6.11}$$

or

$$\mathbf{y} = \mathbf{U}_R^T \mathbf{H} \mathbf{x} + \mathbf{U}_R^T \mathbf{n} \tag{6.12}$$

Substituting the QR decomposition of $\mathbf{H}$ from (6.9) into (6.12), we get for $\mathbf{y}$

$$\mathbf{y} = \mathbf{R} \mathbf{x} + \mathbf{n}' \tag{6.13}$$

where $\mathbf{n'} = U_R^T n$ is an n_T-component column matrix of i.i.d AWGN noise signals. As $\mathbf{R}$ is upper-triangular, the ith component in $\mathbf{y}$ depends only on the ith and higher layer transmitted symbols at time t, as follows

$$y_t^i = (R_{i,i})_t x_t^i + n_t^{'i} + \sum_{j=i+1}^{n_T} (R_{i,j})_t x_t^j \qquad (6.14)$$

Consider x_t^i as the current desired detected signal. Eq. (6.14) shows that y_t^i contains a lower level of interference than in the received signal $\mathbf{r}_t$, as the interference from x_t^l, for $l < i$, are suppressed. The third term in (6.14) represents contributions from other interferers, $x_t^{i+1}, x_t^{i+2}, \ldots, x_t^{n_T}$, which can be cancelled by using the available decisions $\hat{x}_t^{i+1}, \hat{x}_t^{i+2}, \ldots, \hat{x}_t^{n_T}$, assuming that they have been detected. The decision statistics on x_t^i, denoted by y_t^i, can be rewritten as

$$y_t^i = \sum_{j=i}^{n_T} (R_{i,j})_t x_t^j + n_t^{'i} \qquad i = 1, 2, \ldots, n_T \qquad (6.15)$$

The estimate on the transmitted symbol x_t^i is given by

$$\hat{x}_t^i = q \left(\frac{y_t^i - \displaystyle\sum_{j=i+1}^{n_T} (R_{i,j})_t \hat{x}_t^j}{(R_{i,i})_t} \right) \qquad i = 1, 2, \ldots, n_T \qquad (6.16)$$

where $q(x)$ denotes the hard decision on x.

A QR factorization algorithm [7] is presented in Appendix 6.1.

Example 6.1

For a system with three transmit antennas, the decision statistics for various layers can be expressed as

$$y_t^1 = (R_{1,1})_t x_t^1 + (R_{1,2})_t x_t^2 + (R_{1,3})_t x_t^3 + n^{'1} \qquad (6.17)$$

$$y_t^2 = (R_{2,2})_t x_t^2 + (R_{2,3})_t x_t^3 + n^{'2} \qquad (6.18)$$

$$y_t^3 = (R_{3,3})_t x_t^3 + n^{'3} \qquad (6.19)$$

The estimate on the transmitted symbol x_t^3, denoted by $\hat{x}_t^3$, can be obtained from Eq. (6.19) as

$$\hat{x}_t^3 = q \left(\frac{y_t^3}{(R_{3,3})_t} \right) \qquad (6.20)$$

The contribution of $\hat{x}_t^3$ is cancelled from Eq. (6.18) and the estimate on x_t^2 is obtained as

$$\hat{x}_t^2 = q \left(\frac{y_t^2 - (R_{2,3})_t \hat{x}_t^3}{(R_{2,2})_t} \right) \qquad (6.21)$$

Finally, after cancelling out $\hat{x}_t^3$ and $\hat{x}_t^2$, we obtain for $\hat{x}_t^1$

$$\hat{x}_t^1 = q\left(\frac{y_t^1 - (R_{1,3})_t \hat{x}_t^3 - (R_{1,2})_t \hat{x}_t^2}{(R_{1,1})_t}\right) \tag{6.22}$$

The described algorithm applies to VBLAST. In coded LST schemes, the soft decision statistics on x_t^i, given by the arguments in the $q(\cdot)$ expressions on the right-hand side in Eqs. (6.20), (6.21) and (6.22), are passed to the channel decoder, which estimates $\hat{x}_t^i$.

In the above example the decision statistics $y_t^{n_T}$ is computed first, then $y_t^{n_T-1}$, and so on. The performance can be improved if the layer with the maximum SNR is detected first, followed by the one with the next largest SNR and so on [49].

6.3.2 Interference Minimum Mean Square Error (MMSE) Suppression Combined with Interference Cancellation

In the MMSE detection algorithm, the expected value of the mean square error between the transmitted vector $\mathbf{x}$ and a linear combination of the received vector $\mathbf{w}^H \mathbf{r}$ is minimized

$$\min E\{(\mathbf{x} - \mathbf{w}^H \mathbf{r})^2\} \tag{6.23}$$

where $\mathbf{w}$ is an $n_R \times n_T$ matrix of linear combination coefficients given by [8]

$$\mathbf{w}^H = \left[\mathbf{H}^H \mathbf{H} + \sigma^2 \mathbf{I}_{n_T}\right]^{-1} \mathbf{H}^H \tag{6.24}$$

σ^2 is the noise variance and $\mathbf{I}_{n_T}$ is an $n_T \times n_T$ identity matrix. The decision statistics for the symbol sent from antenna i at time t is obtained as

$$y_t^i = \mathbf{w}_i^H \mathbf{r} \tag{6.25}$$

where $\mathbf{w}_i^H$ is the ith row of $\mathbf{w}^H$ consisting of n_R components. The estimate of the symbol sent by antenna i, denoted by $\hat{x}_t^i$, is obtained by making a hard decision on y_t^i

$$\hat{x}_i^t = q(y_i^t) \tag{6.26}$$

In an algorithm with interference suppression only, the detector calculates the hard decisions estimates by using (6.25) and (6.26) for all transmit antennas.

In a combined interference suppression and interference cancellation, the receiver starts from antenna n_T and computes its signal estimate by using (6.25) and (6.26). The received signal $\mathbf{r}$ in this level is denoted by $\mathbf{r}^{n_T}$. For calculation of the next antenna signal ($n_T - 1$), the interference contribution of the hard estimate $\hat{x}_t^{n_T}$ is subtracted from the received signal $\mathbf{r}^{n_T}$ and this modified received signal denoted by $\mathbf{r}^{n_T-1}$ is used in computing the decision statistics for antenna ($n_T - 1$) in Eq. (6.25) and its hard estimate from (6.26). In the next level, corresponding to antenna ($n_T - 2$), the interference from $n_T - 1$ is subtracted from the received signal $\mathbf{r}^{n_T-1}$ and this signal is used to calculate the decision statistics in (6.25) for antenna ($n_T - 2$). This process continues for all other levels up to the first antenna.

After detection of level i, the hard estimate $\hat{x}_t^i$ is subtracted from the received signal to remove its interference contribution, giving the received signal for level $i - 1$

$$\mathbf{r}^{i-1} = \mathbf{r}^i - \hat{x}_t^i \mathbf{h}_i \tag{6.27}$$

where $\mathbf{h}_i$ is the ith column in the channel matrix $\mathbf{H}$, corresponding to the path attenuations from antenna i. The operation $\hat{x}_t^i \mathbf{h}_i$ in (6.27) replicates the interference contribution caused by $\hat{x}_t^i$ in the received vector. $\mathbf{r}^{i-1}$ is the received vector free from interference coming from $\hat{x}_t^{n_T}, \hat{x}_t^{n_T-1}, \ldots, \hat{x}_t^i$. For estimation of the next antenna signal x_t^{i-1}, this signal $\mathbf{r}^{i-1}$ is used in (6.25) instead of $\mathbf{r}$. Finally, a deflated version of the channel matrix is calculated, denoted by $\mathbf{H}_d^{i-1}$, by deleting column i from $\mathbf{H}_d^i$. The deflated matrix $\mathbf{H}_d^{i-1}$ at the $(n_T - i + 1)$th cancellation step is given by

$$
\mathbf{H}_d^{i-1} = \begin{bmatrix} h_{1,1} & h_{1,2} & \cdots & h_{1,i-1} \\ h_{2,1} & h_{2,2} & \cdots & h_{2,i-1} \\ \vdots & \vdots & \vdots & \vdots \\ h_{n_R,1} & h_{n_R,2} & \cdots & h_{n_R,i-1} \end{bmatrix} \tag{6.28}
$$

This deflation is needed as the interference associated with the current symbol has been removed. This deflated matrix $\mathbf{H}_d^{i-1}$ is used in (6.24) or computing the MMSE coefficients and the signal estimate from antenna $i - 1$. Once the symbols from each antenna have been estimated, the receiver repeats the process on the vector $\mathbf{r}_{t+1}$ received at time $(t + 1)$. The summary of this algorithm is given below.

Summary of Linear MMSE Suppression and Successive Cancellation

```
Set  i = n_T
and  r^n_T = r.
while i ≥ 1
{
```

$$
\begin{aligned}
\mathbf{w}^H &= [\mathbf{H}^H \mathbf{H} + \sigma^2 \mathbf{I}_{n_T}]^{-1} \mathbf{H}^H \\
y_t^i &= \mathbf{w}_i^H \mathbf{r}^i \\
\hat{x}_i^t &= q(y_i^t) \\
\mathbf{r}^{i-1} &= \mathbf{r}^i - \hat{x}_t^i \mathbf{h}_i
\end{aligned}
$$

```
Compute H_d^{i-1} by deleting column i from H_d^i.
H = H_d^{i-1}
i = i - 1

}
```

The receiver can be implemented without the interference cancellation step (6.27). This will reduce system performance but some computational cost can be saved. Using cancellation requires that MMSE coefficients be recalculated at each iteration, as $\mathbf{H}$ is deflated. With no cancellation, the MMSE coefficients are only computed once, as $\mathbf{H}$ remains unchanged. The most computationally intensive operation in the detection algorithm is the computation of the MMSE coefficients. A direct calculation of the MMSE coefficients based on (6.24), has a complexity polynomial in the number of transmit antennas. However, on slow fading channels, it is possible to implement adaptive MMSE receivers with the complexity being linear in the number of transmit antennas.

The described algorithm is for uncoded LST systems. The same detector can be applied to coded systems. The receiver consists of the described MMSE interference suppressor/

canceller followed by the decoder. The decision statistics, y_t^i, from (6.25), is passed to the decoder which makes the decision on the symbol estimate $\hat{x}_t^i$.

The performance of a QR decomposition receiver (QR), the linear MMSE (LMMSE) detector (LMMSE) and the performance of the last detected layer in an MMSE detector with successive interference cancellation (MMSE-IC) are shown for a VBLAST structure with $n_T = 4$, $n_R = 4$ and BPSK modulation on a slow Rayleigh fading channel in Fig. 6.4. Figure 6.4 also shows the interference free (single layer) BER which is given by [3]

$$P_b = \left[\frac{1}{2}(1 - \mu)\right]^{n_R} \sum_{k=0}^{k=n_R-1} \left[\frac{1}{2}(1 + \mu)\right]^k \tag{6.29}$$

where $\mu = \sqrt{\dfrac{\frac{\gamma_b}{n_R}}{1+\frac{\gamma_b}{n_R}}}$ and $\gamma_b = \dfrac{E_b}{N_o}$.

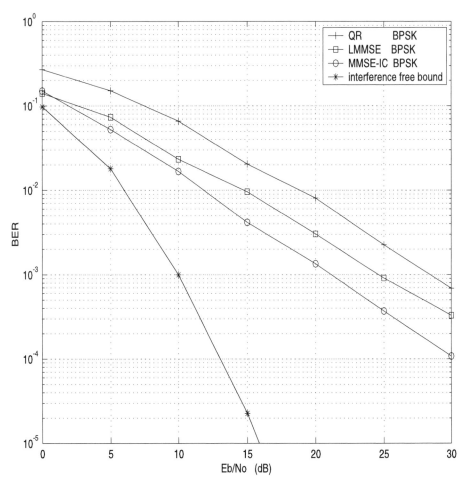

Figure 6.4 V-BLAST example, $n_T = 4$, $n_R = 4$, with QR decomposition, MMSE interference suppression and MMSE interference suppression/successive cancellation

One of the disadvantages of the MMSE scheme with successive interference cancellation is that the first desired detected signal to be processed sees all the interference from the remaining $(n_T - 1)$ signals, whereas each antenna signal to be processed later sees less and less interference as the cancellation progresses. This problem can be alleviated either by ordering the layers to be processed in the decreasing signal power or by assigning power to the transmitted signals according to the processing order. Another disadvantage of the successive scheme is that a delay of n_T computation stages is required to carry out the cancellation process.

The complexity of the LST receiver can be further reduced by replacing the MMSE interference suppressor by a matched filter, resulting in interference cancellation only.

A laboratory prototype of a VLST system was constructed in Bell Laboratories [43]. The prototype operates at a carrier frequency of 1.9 GHz, uncoded 16-QAM modulation and a symbol rate of 24.3 k symbols/sec, in a bandwidth of 30 kHz with 8 transmit and 12 receive antennas. The system achieves a frame error rate of 10^{-2} at an SNR of 25 dB. The frame length is 100 symbols, 20 of which are used to estimate the channel in each frame, so that the efficiency within a frame is 80%. The ideal spectral efficiency is 25.9 bits/s/Hz, but if the bandwidth loss due to transmission of training sequences is included, the reduced spectral efficiency is 20.7 bits/s/Hz. This is much higher than the achievable spectral efficiency in the second generation of cellular mobile systems with a single element transmit/receive antenna.

6.3.3   Iterative LST Receivers

The challenge in the detection of space-time signals is to design a low-complexity detector, which can efficiently remove multilayer interference and approach the interference free bound. The iterative processing principle, as applied in turbo coding [10], has been successfully extended to joint detection and decoding [11]–[21]. This receiver can be applied only in coded LST systems. Block diagrams of the iterative receivers for LST (a)–(c) architectures are shown in Fig. 6.5. In all three receivers, the detector provides joint soft-decision estimates of the n_T transmitted symbol sequences. In LST (a) the detected sequence is decoded by a single decoder with soft inputs/soft outputs, while in LST (b) each of the detected sequences is decoded by a separate channel decoder with soft inputs/soft outputs. At each iteration, the decoder soft outputs are used to update the a priori probabilities of the transmitted signals. These updated probabilities are then used to calculate the symbol estimate in the detector. Note that each of the coded streams is independently interleaved to enable the receiver convergence. In LST (c), apart from time interleaving/deinterleaving, there is space interleaving/deinterleaving across transmit antennas.

The decoder can apply a number of the soft output decoding algorithms. The maximum a posteriori (MAP) approach [32] is optimum in the sense that it minimizes the bit error probability at the decoder output. The log-MAP decoding [1] is an additive version of the MAP algorithm, that operates in the log-domain and thus has a lower complexity. The soft output Viterbi algorithm (SOVA) [1] is a modified Viterbi algorithm generating soft outputs. It has a lowest complexity, and somewhat degraded performance compared to the MAP decoder. As the overall receiver complexity is mainly dominated by the decoder complexity, the choice of the decoding algorithm depends on the available processing power at the receiver.

COLLEGE LANE LRC
ITEMS CHARGED - SELF ISSUE

12/08/05
03:55 pm

Ahmad El Tom

DUE DATE:
2005-09-30 23:59:00

TITLE:Space-time coding / Branka Vucetic,
Jinhong Yuan.

ITEM:60007372224

Please note that the due date is in
the American format: year-month-day.

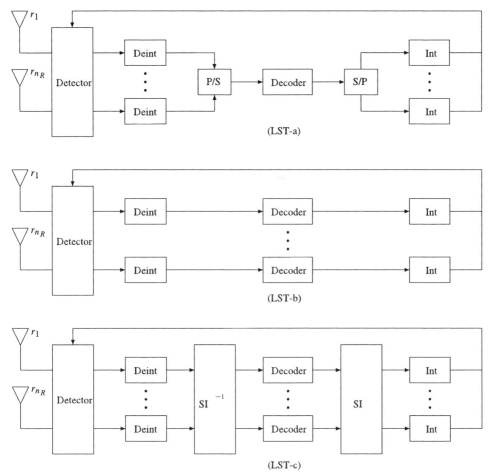

Figure 6.5 Block diagrams of iterative LSTC receivers; (a) HLST with a single decoder; (b) HLST with separate decoders; (c) DLST and TLST receivers

Another important algorithm that is essential for the receiver complexity is multilayer interference suppression. We consider low complexity detector architectures, including a parallel interference canceller (PIC) and a nonlinear MMSE detector.

6.3.4 An Iterative Receiver with PIC

A block diagram of the standard iterative receiver with a parallel interference canceller (PIC-STD) is shown in Fig. 6.6. In order to simplify the presentation we assume that an HLST architecture with separate error control coding in each layer is used. In addition, the same convolutional codes with BPSK modulation are selected in each layer.

In the first iteration, the PIC detectors are equivalent to a bank of matched filters. The detectors provide decision statistics of the n_T transmitted symbol sequences. The decision statistics in the first iteration, for antenna i and time t, denoted by $y_t^{i,1}$, is determined as

$$y_t^{i,1} = \mathbf{h}_i^H \mathbf{r} \tag{6.30}$$

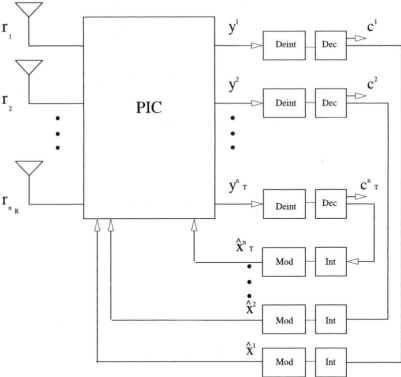

Figure 6.6   Block diagram of an iterative receiver with PIC-STD

where $\mathbf{h}_i^H$ is the ith row of matrix $\mathbf{H}^H$. These decision statistics are passed to the respective decoders, which generate soft estimates on the transmitted symbols.

In the second and later iterations, the decoder soft output is used to update the PIC detector decision statistics.

The decision statistics in the kth iteration at time t, for transmit antenna i, denoted by $y_t^{i,k}$, is given by

$$y_t^{i,k} = \mathbf{h}_i^H (\mathbf{r} - \mathbf{H}\underline{\hat{\mathbf{X}}}_i^{k-1}) \tag{6.31}$$

where $\underline{\hat{\mathbf{X}}}_i^{k-1}$ is an $n_T \times 1$ column matrix with the symbol estimates from the $(k-1)$th iteration as elements, except for the ith element which is set to zero. It can be written as

$$\underline{\hat{\mathbf{x}}}_i^{k-1} = (\hat{x}_t^{1,k-1}, \dots, \hat{x}_t^{i-1,k-1}, 0, \hat{x}_t^{i+1,k-1}, \dots, \hat{x}_t^{n_T,k-1})^T \tag{6.32}$$

The detection outputs for layer i for a whole block of transmitted symbols form a vector, $\mathbf{y}^{i,k}$, which is interleaved and then passed to the i-the decoder.

The decoder in the kth iteration calculates the log-likelihood ratios (LLR) for antenna i at time t, denoted by $\lambda_t^{i,k}$ and given by

$$\lambda_t^{i,k} = \log \frac{P(x_t^{i,k} = 1 | \mathbf{y}^{i,k})}{P(x_t^{i,k} = -1 | \mathbf{y}^{i,k})} \tag{6.33}$$

where $P(x_t^{i,k} = j|\mathbf{y}^{i,k})$, $j = 1, -1$, are the symbol a posteriori probabilities (APP). The LLR can be calculated by the iterative MAP algorithm (Appendix 5.1).

The symbol a posteriori probabilities $P(x_t^{i,k} = j|\mathbf{y}^{i,k})$, $j = 1, -1$, can be expressed as

$$P(x_t^{i,k} = 1|\mathbf{y}^{i,k}) = \frac{e^{\lambda_t^{i,k}}}{1 + e^{\lambda_t^{i,k}}} \tag{6.34}$$

$$P(x_t^{i,k} = -1|\mathbf{y}^{i,k}) = \frac{1}{1 + e^{\lambda_t^{i,k}}} \tag{6.35}$$

The estimates of the transmitted symbols in (6.32) are calculated by finding their mean

$$\hat{x}_t^{i,k} = 1 \cdot P(x_t^{i,k} = 1|\mathbf{y}^{i,k}) + (-1) \cdot P(x_t^{i,k} = -1|\mathbf{y}^{i,k}) \tag{6.36}$$

By combining Eqs. (6.36), (6.34) and (6.35), we express the symbol estimates as functions of the LLR

$$\hat{x}_t^{i,k} = \frac{e^{\lambda_t^{i,k}} - 1}{e^{\lambda_t^{i,k}} + 1} \tag{6.37}$$

When the LLR is calculated on the basis of the a posteriori probabilities, it is obtained as

$$\lambda_t^{i,k} = \log \frac{\displaystyle\sum_{m,m'=0,x_t^i=1}^{m,m'=M_s-1} \alpha_{j-1}(m') p_t(x_t^i = 1) \exp\left(-\frac{\displaystyle\sum_{l=(j-1)n}^{jn} (y_l^{i,k} - x_l^i)^2}{2(\sigma^{i,k})^2}\right) \beta_j(m)}{\displaystyle\sum_{m,m'=0,x_t^i=-1}^{m,m'=M_s-1} \alpha_{j-1}(m') p_t(x_t^i = -1) \exp\left(-\frac{\displaystyle\sum_{l=(j-1)n}^{jn} (y_l^{i,k} - x_l^i)^2}{2(\sigma^{i,k})^2}\right) \beta_j(m)} \tag{6.38}$$

where $\lambda_t^{i,k}$ denotes the LLR ratio for the pth symbol within the jth codeword transmitted at time $t = (j - 1)n + p$ and n is the code symbol length. m' and m are the pair of states connected in the trellis, x_t^i is the tth BPSK modulated symbol in a code symbol connecting the states m' and m, $y_t^{i,k}$ is the detector output in iteration k, for antenna i, at time t, $(\sigma^{i,k})^2$ is the noise variance for layer i and iteration k, M_s is the number of states in the trellis and $\alpha(m')$ and $\beta(m)$ are the feed-forward and the feedback recursive variables, defined as for the LLR (Appendix 5.1).

In computing the LLR value in (6.38) the decoder uses two inputs. The first input is the decision statistics, $y_t^{i,k}$, which depends on the transmitted signal x_t^i. The second input is the a priori probability on the transmitted signal x_t^i, computed as

$$p_t(x_t^i = l) = \frac{1}{\sqrt{2\pi}\sigma} e^{-\frac{(y_t^{i,k} - l\mu_t^i)^2}{2\sigma^2}}, \quad l = 1, -1 \tag{6.39}$$

where μ_t^i is the mean of the received amplitude after matched filtering, given by $\mu_t^i = \mathbf{h}_i^H \mathbf{h}_i$. As $p_t(x_t^i = l)$ in (6.39) depends also on x_t^i, the inputs to the decoder in iteration k, where $k > 1$, are correlated. This causes the decision statistics mean value, conditional on x_t^i, to be biased [20][27]. The bias always has a sign opposite of x_t^i. That is, the bias reduces the useful signal term and degrades the system performance. This bias is particularly significant for a large number of interferers.

The bias effect can be eliminated by estimating the mean of the transmitted symbols based on the a posteriori extrinsic information ratio instead of the LLR [16][20]. The extrinsic information represents the information on the coded bit of interest calculated from the a priori information on the other coded bits and the code constraints. The EIR does not include the metric for the symbol x_t^i that is being estimated. That is

$$
\lambda_{t,e}^{i,k} = \log \frac{\displaystyle\sum_{m,m'=0,x_t^i=1}^{m,m'=M_s-1} \alpha_{j-1}(m') p_t(x_t^i = 1) \exp\left(-\frac{\displaystyle\sum_{l=(j-1)n,l\neq t}^{jn} (y_l^{i,k} - x_l^i)^2}{2(\sigma^{i,k})^2}\right)\beta_j(m)}{\displaystyle\sum_{m,m'=0,x_t^i=-1}^{m,m'=M_s-1} \alpha_{j-1}(m') p_t(x_t^i = -1) \exp\left(-\frac{\displaystyle\sum_{l=(j-1)n,l\neq t}^{jn} (y_l^{i,k} - x_l^i)^2}{2(\sigma^{i,k})^2}\right)\beta_j(m)}
$$

(6.40)

where $\lambda_{t,e}^{i,k}$ denotes the EIR for the pth symbol within the jth codeword transmitted at time $t = (j-1)n + p$, $y_t^{i,k}$ is the detector output in iteration k, for antenna i, $\alpha(m')$ and $\beta(m')$ are defined as for the LLR (Appendix 5.1). However, excluding the contribution of the bit of interest reduces the extrinsic information SNR, which leads to a degraded system performance.

A decision statistics combining (DSC) method is effective in minimizing these effects. In the iterative parallel interference canceller with decision statistics combining (PIC-DSC) [20], shown in Fig. 6.7, a DSC module is added to the PIC-DSC structure. The decision statistics of the PIC-DSC is generated as a weighted sum of the current PIC output and the DSC output from the previous operation. In each stage, except in the first one, the PIC output is passed to the DSC module. The DSC module performs recursive linear combining of the detector output in iteration k for layer i, denoted by $y^{i,k}$, with the DSC output from the previous iteration for the same layer, denoted by $y_c^{i,k-1}$. The output of the decision statistics combiner, in iteration k and for layer i, denoted by $y_c^{i,k}$, is given by

$$
y_c^{i,k} = p_1^{i,k} y^{i,k} + p_2^{i,k} y_c^{i,k-1}
$$

(6.41)

where $p_1^{i,k}$ and $p_2^{i,k}$ are the DSC weighting coefficients in stage k, respectively. They are estimated by maximizing the signal-to-noise plus interference ratio (SINR) at the output of DSC in iteration k under the assumption that $y^{i,k}$ and $y_c^{i,k-1}$ are Gaussian random variables with the conditional means $\mu^{i,k}$ and $\mu_c^{i,k-1}$, given that x^i is the transmitted symbol for antenna i, and variances $(\sigma^{i,k})^2$ and $(\sigma_c^{i,k-1})^2$, respectively.

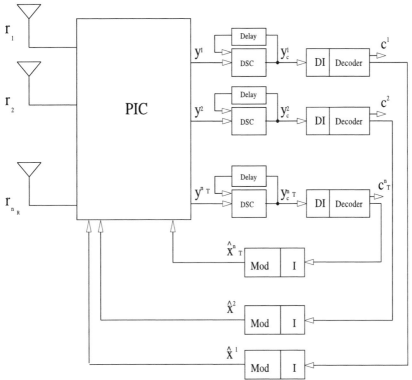

Figure 6.7 Block diagram of an iterative receiver with PIC-DSC

Coefficients $p_1^{i,k}$ and $p_2^{i,k}$ can be normalized in the following way

$$E\{y_c^{i,k}\} = p_1^{i,k}\mu^{i,k} + p_2^{i,k}\mu_c^{i,k-1} = 1 \tag{6.42}$$

The SINR at the output of the DSC for layer i and in iteration k is then given by

$$\text{SINR}^{i,k} = \cfrac{1}{(p_1^{i,k})^2(\sigma^{i,k})^2 + 2p_1^{i,k}\left(\dfrac{1-p_1^{i,k}\mu^{i,k}}{\mu_c^{i,k-1}}\right)\rho_{k,k-1}^i\sigma^{i,k}\sigma_c^{i,k-1} + \left(\dfrac{1-p_1^{i,k}\mu^{i,k}}{\mu_c^{i,k-1}}\right)^2(\sigma_c^{i,k-1})^2} \tag{6.43}$$

where $\rho_{k,k-1}^i$ is the correlation coefficient for layer i, between the detector output in the kth and $(k-1)$th iterations defined as

$$\rho_{k,k-1}^i = \frac{E\{(y^{i,k} - \mu^{i,k}x^i)(y_c^{i,k-1} - \mu_c^{i,k-1}x^i)|x^i\}}{\sigma^{i,k}\sigma_c^{i,k-1}} \tag{6.44}$$

The optimal combining coefficient is given by

$$p_1^{i,k}{}_{\text{opt}} = \cfrac{\dfrac{\mu^{i,k}}{(\mu_c^{i,k-1})^2}(\sigma_c^{i,k-1})^2 - \dfrac{1}{\mu_c^{i,k-1}}\rho_{k,k-1}^i\sigma^{i,k}\sigma_c^{i,k-1}}{(\sigma^{i,k})^2 - 2\dfrac{\mu^{i,k}}{\mu_c^{i,k-1}}\rho_{k,k-1}^i\sigma^{i,k}\sigma_c^{i,k-1} + \left(\dfrac{\mu^{i,k}}{\mu_c^{i,k-1}}\right)^2(\sigma_c^{i,k-1})^2} \tag{6.45}$$

In the derivation of the optimal coefficients we assume that $\mu^{k,i}$, $\mu_c^{k,i-1}$, $(\sigma^{k,i})^2$ and $(\sigma_c^{k,i-1})^2$ are the true conditional means and the true variances of the detector outputs.

The parameters required for the calculation of the optimal combining coefficients in Eq. (6.45) are difficult to estimate, apart from the signal variances.

However, in a system with a large number of interferers, which happens when the number of transmit antennas is large relative to the number of receive antennas, and for the APP based symbol estimates, the DSC inputs in the first few iterations are low correlated. Thus, it is possible to combine them, in a way similar to receive diversity maximum ratio combining.

Under these conditions, the weighting coefficient in this receiver can be obtained from Eq. (6.45) by assuming that the correlation coefficient is zero and neglecting the reduction of the received signal conditional mean caused by interference. The DSC coefficients are then given by

$$p_1^{i,k} = \frac{(\sigma_c^{i,k-1})^2}{(\sigma_c^{i,k-1})^2 + (\sigma^{i,k})^2} \qquad (6.46)$$

The DSC output, in the second and higher iterations, with coefficients from (6.46) can be expressed as

$$y_c^{i,k} = \frac{(\sigma_c^{i,k-1})^2}{(\sigma_c^{i,k-1})^2 + (\sigma^{i,k})^2} y^{i,k} + \frac{(\sigma^{i,k})^2}{(\sigma^{i,k})^2 + (\sigma_c^{i,k-1})^2} y_c^{i,k-1} \qquad i > 1 \qquad (6.47)$$

The complexity of both PIC-STD and PIC-DSC is linear in the number of transmit antennas.

We demonstrate the performance of an HLST scheme with separate $R = 1/2$, 4-state convolutional component encoders, the frame size of $L = 206$ symbols and BPSK modulation. In simulations decoding is performed by a MAP algorithm. The HLSTC with n_T transmit and n_R receive antennas is denoted as an (n_T, n_R) HLSTC. The channel is modelled as a frequency flat slow Rayleigh fading channel. The results are shown in the form of the frame error rate (FER) versus E_b/N_0. The SNR is related to E_b/N_0 as

$$SNR = \eta E_b/N_0 \qquad (6.48)$$

where $\eta = R m n_T$ is the spectral efficiency and m is the number of bits per modulation symbol. Figure 6.8 compares the performance of the PIC-STD with EIR and LLR based symbol estimates and the PIC-DSC for a (6,2) HLSTC. The spectral efficiency of the HLSTC is $\eta = 3$ bits/s/Hz. The results show that for the PIC-STD with LLR based symbol estimates the error floor is higher than for the other two schemes. With 8 iterations the error floor for the PIC-STD(LLR) appears at FER of 0.1, while for the PIC-STD (EIR) the error floor is about 0.04. However, the PIC-DSC receiver has an error floor below 0.007.

Figure 6.9 shows the performance for the HLSTC code with $n_T = 4$ transmit and $n_R = 2$ receive antennas. The spectral efficiency of the HLSTC is $\eta = 2$ bits/s/Hz. The relative performance of the three receivers, PIC-STD(LLR), PIC-STD(EIR) and PIC-DSC, is the same as in the previous figure. Note that in both $(4, 2)$ HLSTC and $(6, 2)$ HLSTC the FER in second iteration for the PIC-STD (LLR) is better than FER for the PIC-STD (EIR). Generally, if the number of interferers is low, the receiver with LLR symbol estimates converges faster than the receiver with EIR symbol estimates. This can be explained by the fact that under low interference the bias effect is negligible and the LLR estimates have a lower variance relative to the EIR estimates resulting in a faster convergence.

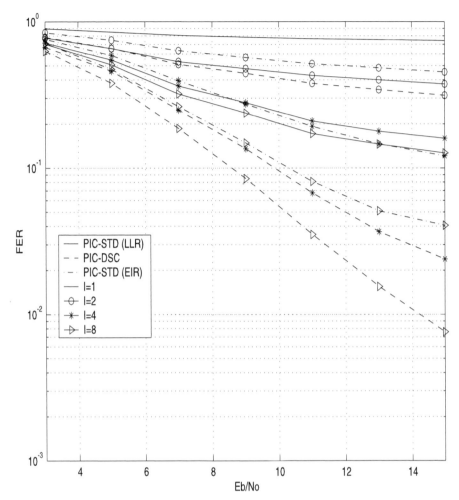

Figure 6.8 FER performance of HLSTC with $n_T = 6$, $n_R = 2$, $R = 1/2$, BPSK, a PIC-STD and PIC-DSC detection on a slow Rayleigh fading channel

Figures 6.8 and 6.9 show the performance results for various PIC receivers when the PIC output variance in iteration i is estimated assuming that the receiver ideally recovers the transmitted symbols, as

$$(\sigma_{mv}^{i,k})^2 = \frac{1}{L} \sum_{t=1}^{L} (y_t^{i,k} - \mu_t^i x_t^i)^2 \tag{6.49}$$

where x_t^i is the transmitted symbol, $\mu_t^i = \mathbf{h}_i{}^H \mathbf{h}_i$ is the nominal mean of the received amplitudes after the maximum-ratio-combining (MRC) and $y_t^{i,k}$ is PIC output. The variance of the DSC output is estimated in the same way. The variance in (6.49) is called a measured variance.

In a real system the transmitted symbols are not known at the receiver. The variance can be calculated by using the symbol estimate of the transmitted symbol from the previous

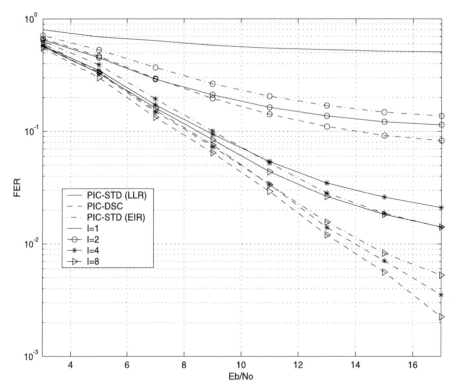

Figure 6.9 FER performance of HLSTC with $n_T = 4$, $n_R = 2$, $R = 1/2$, BPSK, the PIC-STD and the PIC-DSC detection on a slow Rayleigh fading channel

decoder output as

$$(\sigma_{ev}^{i,k})^2 = \frac{1}{L} \sum_{t=1}^{L} (y_t^{i,k} - \mu_t^i \hat{x}_t^{i,k-1})^2 \tag{6.50}$$

where $\hat{x}_t^{i,k-1}$ is a symbol estimate in iteration $k - 1$. The variance in (6.50) is called an estimated variance.

Figures 6.10 and 6.11 compare the performance of the PIC-DSC with a measured variance as in Eq. (6.49) (PIC-DSC mv) and estimated variance as in Eq. (6.50) (PIC-DSC ev) for a $(6, 2)$ and a $(8, 2)$ HLSTC. Clearly, until the number of interfering layers relative to the number of receive antennas becomes very high as in the example of the $(8, 2)$ HLSTC, the differences between the performance of the PIC-DSC mv and PIC-DSC ev is not large.

Figure 6.12 compares the performance of the iterative PIC-STD and iterative PIC-DSC decoder for a (4,4) HLSTC code with a rate $R = 1/2$, 4 state convolutional component code, BPSK modulation on a slow Rayleigh fading channel.

Figure 6.13 illustrates the performance of an HLSTC (4,4) system on a two-path Rayleigh fading channel with PIC-STD detection. As the results show, the error rate is very close to the one achieved in an interference free system. This proves that the PIC-STD receiver is also able to remove the interference coming from frequency selective fading. The overall performance is better than on a single path Rayleigh fading channel due to a diversity gain.

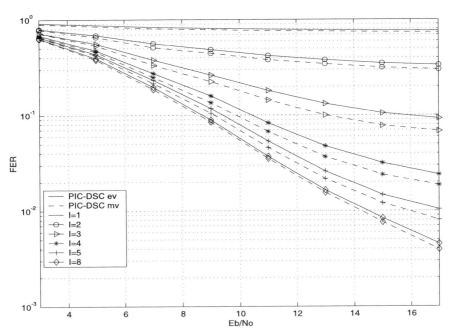

Figure 6.10   Effect of variance estimation for an HLSTC with $n_T = 6$, $n_R = 2$, $R = 1/2$, BPSK and a PIC-DSC receiver on a slow Rayleigh fading channel

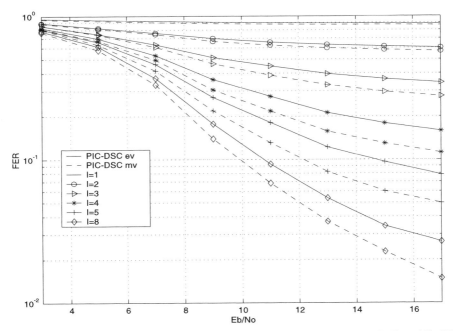

Figure 6.11   Effect of variance estimation on an HLSTC with $n_T = 8$, $n_R = 2$, $R = 1/2$, BPSK and PIC-DSC detection on a slow Rayleigh fading channel

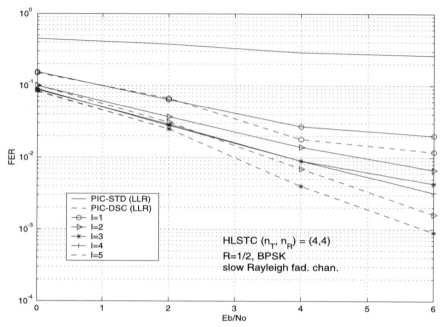

Figure 6.12 FER performance of HLSTC with $n_T = 4$, $n_R = 4$, $R = 1/2$, BPSK, PIC-STD and PIC-DSC detection on a slow Rayleigh fading channel

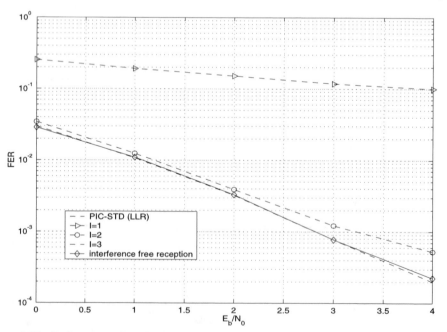

Figure 6.13 Performance of an HLSTC $(4, 4)$, $R = 1/2$ with BPSK modulation on a two path slow Rayleigh fading channel with PIC-STD detection

6.3.5 An Iterative MMSE Receiver

We consider an iterative receiver with a multiuser detector consisting of a feed-forward module which performs interference suppression followed by a feedback module which performs parallel interference cancellation, as proposed in [36]. We refer to this receiver structure as an iterative MMSE receiver. A block diagram of the iterative MMSE receiver is shown in Fig. 6.14.

The decision statistics vector obtained at the output of the feedback module in the kth iteration at time t, for layer i, denoted by $y_t^{i,k}$, is given by

$$y_t^{i,k} = (\mathbf{w_f}^{i,k})^H \mathbf{r} + w_b^{i,k} \tag{6.51}$$

where $\mathbf{w_f}^{i,k}$ is an $n_R \times 1$ optimized feed-forward coefficients column matrix and $w_b^{i,k}$ is a single coefficient which represents the cancellation term. The coefficients $\mathbf{w_f}^{i,k}$ and $w_b^{i,k}$ are calculated by minimizing the mean square error between the transmitted symbol and its estimate, given by

$$e = E\{|\ (\mathbf{w_f}^{i,k})^H \mathbf{r} + w_b^{i,k} - x_t^i\ |^2\} \tag{6.52}$$

Let us denote by $\underline{\mathbf{x}}^i$ a column matrix with $(n_T - 1)$ components, consisting of the transmitted symbols from all antennas except antenna i.

$$(\underline{\mathbf{x}}^i)^T = (x_t^1, x_t^2, \ldots, x_t^{i-1}, x_t^{i+1}, \ldots, x_t^{n_T}) \tag{6.53}$$

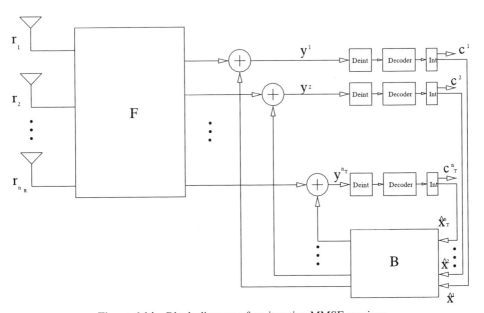

Figure 6.14 Block diagram of an iterative MMSE receiver

Similarly, we define a vector $\hat{\underline{\mathbf{x}}}^{i,k}$ of the symbol estimates from other antennas in the kth iteration

$$(\hat{\underline{\mathbf{x}}}^{i,k})^T = (\hat{x}_t^{1,k}, \hat{x}_t^{2,k}, \ldots, \hat{x}_t^{i-1,k}, \hat{x}_t^{i+1,k}, \ldots, \hat{x}_t^{n_T,k}) \tag{6.54}$$

The decoder calculates the LLRs for the transmitted symbols at a particular time instant for each transmit antenna, by using (6.33). These LLR values are used to calculate the transmitted symbol estimates $\hat{x}_t^{l,k}$, $l = 1, 2, \ldots, i-1, i+1, \ldots, n_T$ in $(\hat{\underline{\mathbf{x}}}^k)^T$ as in (6.36).

Let us denote by $\mathbf{h}_i$ the ith column of the channel matrix $\mathbf{H}$, representing a column matrix with n_R complex channel gains for the ith transmit antenna and by $\underline{\mathbf{H}}^i$ an $n_R \times (n_T - 1)$ matrix composed of the complex channel gains for the other $(n_T - 1)$ transmit antennas. To simplify the notation, we define the following matrices

$$\mathbf{A} = \mathbf{h}_i \mathbf{h}_i^H \tag{6.55}$$

$$\mathbf{B} = \underline{\mathbf{H}}^i \left[\mathbf{I}_{n_T-1} - \text{diag}((\hat{\underline{\mathbf{x}}}^{i,k})^H \hat{\underline{\mathbf{x}}}^{i,k}) + \hat{\underline{\mathbf{x}}}^{i,k} (\hat{\underline{\mathbf{x}}}^{i,k})^H \right] (\underline{\mathbf{H}}^i)^H \tag{6.56}$$

where $\mathbf{I}_{n_T-1}$ is an $(n_T - 1) \times (n_T - 1)$ identity matrix.

$$\mathbf{D} = \underline{\mathbf{H}}^i \hat{\underline{\mathbf{x}}}^{i,k} \tag{6.57}$$

$$\mathbf{R}_n = \sigma^2 \mathbf{I}_{n_R} \tag{6.58}$$

where σ^2 is the noise variance. The optimum feed-forward and feedback coefficients are given by

$$(\mathbf{w_f}^{i,k})^H = \mathbf{h}_i^H (\mathbf{A} + \mathbf{B} + \mathbf{R_n} - \mathbf{D}\mathbf{D}^H) \tag{6.59}$$

$$w_b^{i,k} = -(\mathbf{w_f}^{i,k})^H \mathbf{D} \tag{6.60}$$

The MMSE coefficients were derived assuming perfect interleaving and feedback symbol estimates based on the extrinsic information ratio (EIR) [19]. In the first iteration, since the a priori probabilities of the transmitted symbols are the same, the symbol estimates $\hat{\underline{\mathbf{x}}}^{i,1}$ obtained from (6.36) are zeros. Thus in the first iteration, the feed-forward coefficients $\mathbf{w_f}^{i,1}$ are obtained in a similar way as in Eq. (6.24) and the feedback coefficient $w_b^{i,1} = 0$. In the second and higher iterations, the symbol estimates, computed by the decoder as in (6.37), are used to recalculate the new set of feed-forward and feedback coefficients as described above.

In the case of hard decision decoding $(\hat{x}_t^{i,k})^2 = 1$ for all i and $k > 1$. The iterative MMSE receiver which employs hard decision decoders is equivalent to the receiver which performs linear MMSE suppression in the first iteration and parallel interference cancellation in the following iterations. This filter would be optimal in the MMSE sense if perfect symbol estimates were fed back.

The a priori probability on the transmitted signal x_t^i, used in the decoder, are computed as

$$p_t(x_t^i = l) = \frac{1}{\sqrt{2\pi}\sigma} e^{-\frac{(y_t^i - l)^2}{2\sigma^2}} \quad l = 1, -1 \tag{6.61}$$

It has been observed that the iterative MMSE receiver performs better if LLRs are used for symbol estimation instead of EIRs, though the MMSE filter coefficients were derived assuming EIR symbol estimation and uncorrelated decoder outputs. If LLRs are used there is a bias between symbol estimates. However, the bias effect is less relevant in the iterative MMSE receiver than in the iterative PIC receiver since the MMSE detector performs interference suppression as well as cancellation. Thus the use of DSC in iterative MMSE receivers is less effective than for iterative PIC receivers.

6.3.6   Comparison of the Iterative MMSE and the Iterative PIC-DSC Receiver

In this section we compare the performance of the iterative MMSE receiver and the iterative PIC-DSC receiver. It is demonstrated that the PIC-DSC receiver is able to achieve similar performance as the MMSE-STD and even outperform MMSE-STD in a high interference environment.

The direct implementation of the iterative MMSE receiver based on matrix inversion has complexity which is polynomial in the number of transmit antennas [36]. Furthermore, the iterative MMSE filter coefficients need to be recalculated for each symbol in iterations $k > 1$, as well as from iteration to iteration. The complexity/performance trade-off of the iterative PIC-DSC is therefore significantly better than that of the iterative MMSE receiver. However, for slow fading channels, it is possible to implement adaptive MMSE receivers with the complexity being linear in the number of transmit antennas.

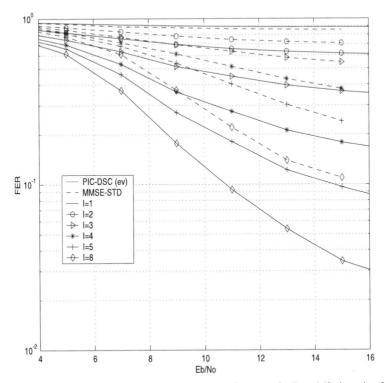

Figure 6.15   FER performance of a HLSTC with $n_T = 8$, $n_R = 2$, $R = 1/2$, iterative MMSE and iterative PIC-DSC receivers, BPSK modulation on a slow Rayleigh fading channel

We demonstrate the performance of an HLST scheme with separate $R = 1/2$, 4-state convolutional component encoders, the frame size of $L = 206$ symbols, BPSK modulation and MAP decoding. The channel is modelled as a frequency flat slow Rayleigh fading channel. The results are shown in the form of the frame error rate (FER) versus E_b/N_0. Figure 6.15 compares the iterative MMSE and iterative PIC-DSC performance for an (8,2) HLSTC. The results show that the iterative PIC-DSC outperforms the iterative MMSE in terms of the achieved FER after 2 iterations. The error floor in the FER performance of the MMSE-STD appears at $E_b/N_0 = 13$ dB and for FER $= 0.1$, while for the PIC-DSC receiver the error floor appears at $E_b/N_0 = 15$ dB and for FER $= 0.03$.

Figure 6.16 shows the performance of MMSE-STD and PIC-DSC for a (4,4) HLSTC. Both receivers achieve the same FER after 4 iterations. The PIC-DSC needs one more iteration than the MMSE-STD to achieve the interference free bound. No error floor has been observed in both receiver structures for simulated FER ≥ 0.0025.

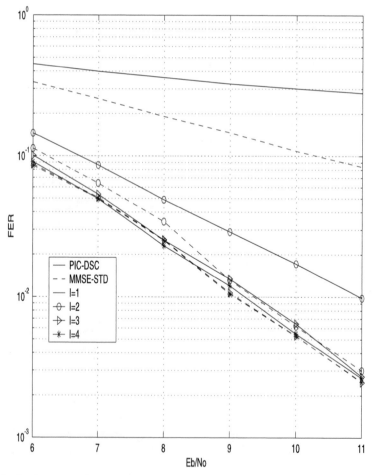

Figure 6.16 Performance of a HLSTC with $n_T = 4$, $n_R = 4$, $R = 1/2$, iterative MMSE and iterative PIC receivers, BPSK modulation on a slow Rayleigh fading channel

6.4 Comparison of Various LST Architectures

We compare the three LST structures performance with convolutional component codes. Two rate 1/2 convolutional codes with memory order $\nu = 2$ and $\nu = 5$ are considered. We denote by (n, k, ν) a rate k/n convolutional code with memory ν. The generator polynomials in octal form of these codes are (5,7) and (53,75), and the free Hamming distances d_{free} are 5 and 8, respectively. The channel is a flat slow Rayleigh fading channel. The modulation format is QPSK and the number of symbols per frame is 252. The MAP algorithm is employed to decode convolutional codes and the iterative PIC-DSC is applied in detection with five iterations between the decoder and the detector. Figs. 6.17 and 6.18 show the performance of three LST structures with $(n_T, n_R) = (2, 2)$ and $\nu = 2$ and $\nu = 5$, respectively. The performance results of these two codes in LST structures with $(n_T, n_R) = (4, 4)$ are shown in Figs. 6.19 and 6.20. For a given memory order, LST-c outperforms LST-b considerably and LST-a slightly. The LST-a has a lower error rate than the LST-b architecture on slow fading channels, as in LST-a a codeword from one encoder is distributed to various antennas resulting in a higher diversity order. However, LST-a is more sensitive to interference and when the number of interferers increases, or when a weaker interference canceller is used, its performance deteriorates. The convolutional code with $\nu = 5$ achieves about 1 and 2 dB gain compared to the code with $\nu = 2$ in LST-c and LST-b, respectively.

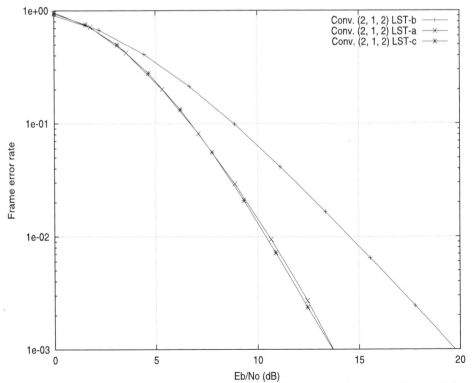

Figure 6.17 Performance comparison of three different LST structures with the (2,1,2) convolutional code as a constituent code for $(n_T, n_R) = (2, 2)$

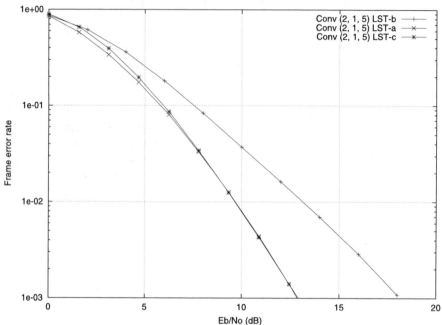

Figure 6.18   Performance comparison of three different LST structures with the (2,1,5) convolutional code as a constituent code for $(n_T, n_R) = (2, 2)$

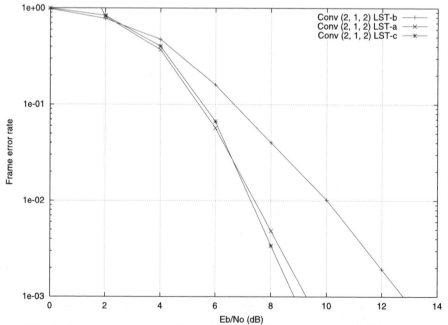

Figure 6.19   Performance comparison of three different LST structures with the (2,1,2) convolutional code as a constituent code for $(n_T, n_R) = (4, 4)$

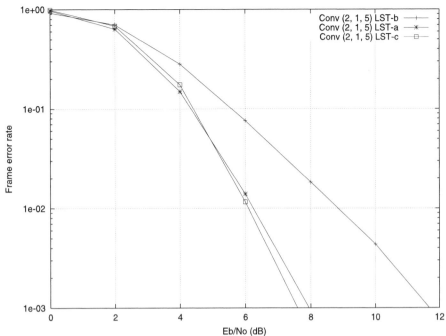

Figure 6.20 Performance comparison of three different LST structures with the (2,1,5) convolutional code as a constituent code for $(n_T, n_R) = (4, 4)$

6.4.1 Comparison of HLST Architectures with Various Component Codes

We compare the performance and decoding complexity of convolutional and low density parity check (LDPC) codes. The convolutional codes are the same as in the previous figures. The LDPC code is a regular rate 1/2 Gallager LDPC (500,250) code. Its parity check matrix has a fixed column weight of $\gamma = 3$ and a fixed row weight of $\rho = 6$. The minimum Hamming distance $d_{\min}$ of this LDPC code is 11. The $d_{\min}$ and the squared Euclidean distance d_E^2 of these three codes are given in Table 6.1.

The MAP and sum-product algorithms are employed to decode convolutional and LDPC codes, respectively. Other system parameters are the same as in the previous figures with convolutional component codes. An LDPC code is represented by a factor graph. The sum-product algorithm is a probabilistic suboptimal method for decoding graph based codes. This is a syndrome decoding method which finds the most probable vector to satisfy all syndrome constraints. The decoding complexity of the MAP algorithm increases exponentially with

Table 6.1 Comparison of convolutional and LDPC code distances

	Conv. $\nu = 2$	Conv. $\nu = 5$	LDPC
$d_{\min}$	5	8	11
d_E^2	20	32	44

Table 6.2 Performance comparison of convolutional and the LDPC codes

	Conv. $\nu = 2$	Conv. $\nu = 5$	LDPC
LST-a	9.2	8.0	9.2
LST-b	12.7	11.6	11.0
LST-c	8.8	7.6	8.8
LST-c (perfect decoding feedback)	7.2	8.2	4.9

the memory order ν. On the other hand, the complexity of decoding the LDPC code is linearly proportional to the number of entries in the parity check matrix **H**.

Table 6.2 shows the required E_b/N_o (in dB) of the simulated codes to achieve FER of 10^{-3} in three LST structures with $(n_T, n_R) = (4, 4)$, five iterations between the decoder and the detector and ten iterations in the sum-product algorithm.

In LST-b, the LDPC outperforms both convolutional codes. The LDPC code achieves a similar performance as the (2,1,2) convolutional code but has a worse performance compared to (2,1,5) convolutional code in both LST-a and LST-c structures, although the LDPC code has a higher distance than the convolutional codes. In addition, there exist error floors for the LDPC code in LST structures with $n_R = 2$. However no error floor occurs for any of the convolutional codes with $n_R = 2$ in Figs. 6.17 and 6.18. The reason for this is that the sum-product algorithm is more sensitive to error propagation than the MAP decoder used for the convolutional codes.

The last row of Table 6.2 shows the required E_b/N_o (in dB) of three different codes achieving FER of 10^{-3} in the (4,4) LST-c system with perfect decoding feedback. It shows that the performance difference between perfect and non-perfect decoding feedback of convolutional and LDPC codes are about 0.4 and 3.9 dB, respectively. This means that the iterative joint detection and MAP decoding algorithm approaches the performance with no interference. On the other hand, the iterative detection with the sum-product algorithm of LDPC codes is far from the optimum performance.

As the number of receive antennas increases, the detector can provide better estimates of the transmitted symbols to the channel decoder. In this situation, the distance of the code dominates the LST system performance. Figure 6.21 shows that the LDPC code outperforms both convolutional codes in a (4,8) LST-c system. We conclude that the LDPC code has a superior error correction capability, but the performance is limited by error propagation in the LST-a and LST-c structures.

Several rate 1/3 turbo codes with information length 250 were chosen as the constituent codes in LST systems on a MIMO slow Rayleigh fading channel. Gray mapping and QPSK modulation are employed in all simulations. Ten iterations are used between the detector and the decoder; and ten iterations for each turbo channel decoder. A PIC-DSC is used as the detector and a MAP algorithm in the turbo channel decoder. Figure 6.22 shows the performance of LST-b and LST-c structures with a turbo constituent code. The generator polynomials in octal form of the component recursive convolutional code are (13,15). The performance of LST-c structure is better than the LST-b structure due to a higher diversity gain. Figure 6.22 also shows the performance of LST-b and LST-c with perfect decoding feedback. The performance of LST-b is very close to a system performance with no interference. On the other hand, there is about 2 dB difference between non-perfect and perfect decoding feedback in LST-c at FER of 10^{-3}. An error floor is observed in both structures, due to a low minimum free distance of the turbo code.

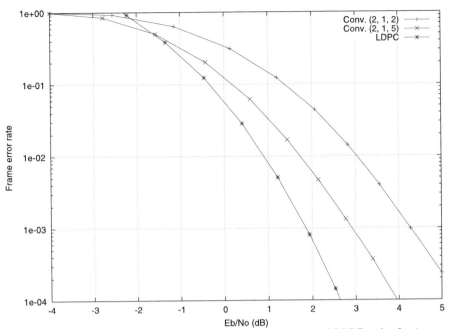

Figure 6.21 Performance comparison of LST-c with convolutional and LDPC codes for $(n_T, n_R) =$ (4, 8)

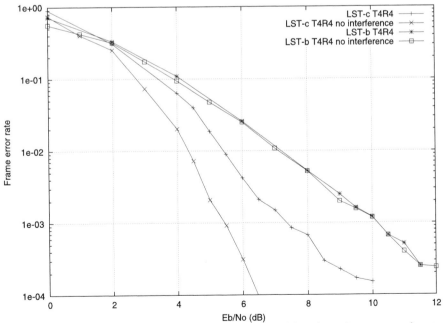

Figure 6.22 Performance comparison of LST-b and LST-c with turbo codes as a constituent code for $(n_T, n_R) = (4, 4)$

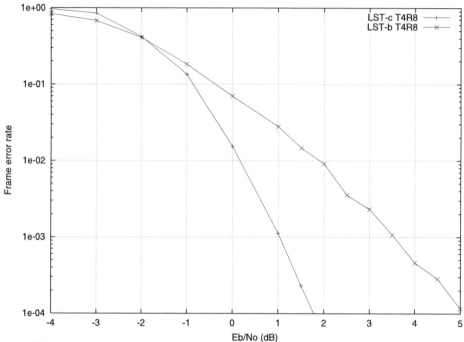

Figure 6.23 Performance comparison of LST-b and LST-c with turbo codes as a constituent code for $(n_T, n_R) = (4, 8)$

Figure 6.23 shows the performance of LST-b and LST-c structures with turbo constituent code for $(n_T, n_R) = (4, 8)$. No error floor exists in this scheme.

Figures 6.24 and 6.25 show the bit error rate performance of LST-a with interleaver sizes 256 and 1024 for a (4,4) and (4,8) systems, respectively. The performance of LST-a with interleaver size 1024 is superior than 252 in both cases. From Fig. 6.24, one can see that the performance of LST-a structure with the turbo code is much worse than in the system with no interference. There is about 2.0 dB and 1.5 dB difference between non-perfect and perfect decoding feedback in LST-a structure with interleaver sizes 252 and 1024 at the BER of 10^{-3}, respectively. Significant error floors are observed in Fig. 6.24. The error floor is due to both low minimum free distance of the turbo code and the decoding feedback error in LST-a structure.

Appendix 6.1 QR Decomposition

Orthogonal matrix triangularization (QR decomposition) reduces a real (m, n) matrix $\mathbf{A}$ with $m \geq n$ and full rank to a much simpler form. A suitably chosen orthogonal matrix Q will triangularize the given matrix:

$$\mathbf{A} = \mathbf{Q} \begin{bmatrix} \mathbf{R} \\ 0 \end{bmatrix}$$

with the (n, n) upper triangular matrix $\mathbf{R}$. One only has then to solve the triangular system $\mathbf{Rx} = \mathbf{Pb}$, where $\mathbf{P}$ consists of the first n rows of $\mathbf{Q}$.

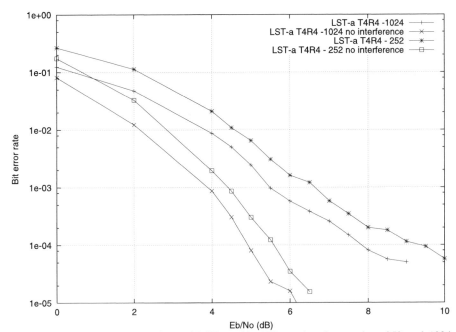

Figure 6.24 Performance comparison of LST-a with different interleaver sizes 252 and 1024 for $(n_T, n_R) = (4, 4)$

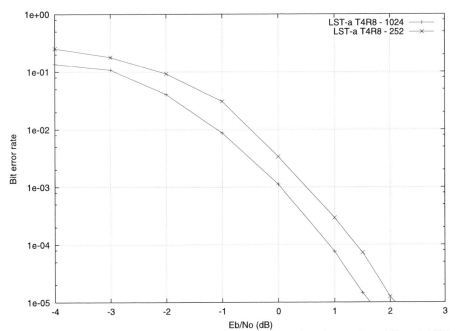

Figure 6.25 Performance comparison of LST-a with different interleaver sizes 252 and 1024 for $(n_T, n_R) = (4, 8)$

The least squares problem $\mathbf{A}\mathbf{x} \approx \mathbf{b}$ is easy to solve with $\mathbf{A} = \mathbf{Q}\mathbf{R}$ and $\mathbf{Q}^T\mathbf{Q} = \mathbf{I}$. The solution

$$\mathbf{x} = (\mathbf{A}^T\mathbf{A})^{-1}\mathbf{A}^T\mathbf{b}$$

becomes

$$\mathbf{x} = (\mathbf{R}^T\mathbf{Q}^T\mathbf{Q}\mathbf{R})^{-1}\mathbf{R}^T\mathbf{Q}^T\mathbf{b} = (\mathbf{R}^T\mathbf{R})^{-1}\mathbf{R}^T\mathbf{Q}^T\mathbf{b} = \mathbf{R}^{-1}\mathbf{Q}^T\mathbf{b}.$$

This is a matrix-vector multiplication $\mathbf{Q}^T\mathbf{b}$, followed by the solution of the triangular system $\mathbf{R}\mathbf{x} = \mathbf{Q}^T b$ by back-substitution.

Many different methods exist for the QR decomposition, e.g. the *Householder transformation*, the *Givens rotation*, or the *Fram-Schmidt decomposition*.

Householder Transformation

The most frequently applied algorithm for QR decomposition uses the Householder transformation $\mathbf{u} = \mathbf{H}\mathbf{v}$, where the Householder matrix H is a symmetric and orthogonal matrix of the form:

$$\mathbf{H} = \mathbf{I} - 2\mathbf{x}\mathbf{x}^T$$

with the identity matrix $\mathbf{I}$ and any normalized vector $\mathbf{x}$ with $\|\mathbf{x}\|_2^2 = \mathbf{x}^T\mathbf{x} = 1$.

Householder transformations zero the $m - 1$ elements of a column vector $\mathbf{v}$ below the first element:

$$\begin{bmatrix} v_1 \\ v_2 \\ \vdots \\ v_n \end{bmatrix} \longrightarrow \begin{bmatrix} c \\ 0 \\ \vdots \\ 0 \end{bmatrix} \quad \text{with } c = \pm\left(\sum_{i=1}^{m} v_i^2\right)^{1/2}$$

One can verify that

$$\mathbf{x} = f \begin{bmatrix} v_1 - c \\ v_2 \\ \vdots \\ v_m \end{bmatrix} \quad \text{with } f = \frac{1}{\sqrt{2c(c - v_1)}}$$

fulfils $\mathbf{x}^T\mathbf{x} = 1$ and that with $\mathbf{H} = \mathbf{I} - 2\mathbf{x}\mathbf{x}^2$ one obtains the vector $[c0 \cdots 0]^T$.

To perform the decomposition of the (m, n) matrix $\mathbf{A} = \mathbf{Q}\mathbf{R}$ (with $m \geq n$) we construct in this way an (m, m) matrix $\mathbf{H}^{(1)}$ to zero the $m - 1$ elements of the first column. An $(m - 1, m - 1)$ matrix $\mathbf{G}^{(2)}$ will zero the $m - 2$ elements of the second column. With $\mathbf{G}^{(2)}$ we produce the (m, m) matrix

$$H^{(2)} = \begin{bmatrix} 1 & 0 & \cdots & 0 \\ 0 & & & \\ \vdots & & G^{(2)} & \\ 0 & & & \end{bmatrix}, \quad \text{etc.}$$

After $n(n-1$ for $m=n)$ such orthogonal transforms $\mathbf{H}^{(i)}$ we obtain:

$$\mathbf{R} = \mathbf{H}^{(n)} \dots \mathbf{H}^{(2)}\mathbf{H}^{(1)}\mathbf{A}.$$

$\mathbf{R}$ is upper triangular and the orthogonal matrix $\mathbf{Q}$ becomes:

$$\mathbf{Q} = \mathbf{H}^{(1)}\mathbf{H}^{(2)} \dots \mathbf{H}^{(n)}.$$

In practice the $\mathbf{H}^{(i)}$ are never explicitly computed.

Bibliography

[1] B. Vucetic and J. Yuan, *Turbo Codes: Principles and Applications*, Kluwer, 2000.

[2] J. C. Liberti and T. S. Rappaport, *Smart Antennas for Wireless Communications*, Prentice Hall PTR, 1999.

[3] J. G. Proakis, *Digital Communications*, McGraw-Hill series in electrical and computer engineering, 1995.

[4] L. H. C. Lee, *Convolutional Coding, Fundamentals and Applications*, Artech House, Inc, 1997.

[5] D. Zwillinger, *Standard Mathematical Tables and Formulae*, CRC Press LLC, 1996.

[6] S. Lin and D. Costello, *Error Control Coding: Fundamentals and Applications*, Prentice-Hall series in computer applications in electrical engineering, 1983.

[7] G. Golub and C. van Loan, *Matrix Computations*, The Johns Hopkins University Press, 1996.

[8] S. Verdu, *Multiuser Detection*, Cambridge University Press, 1998.

[9] J. H. Winters, J. Salz and R. D. Gitlin, "The impact of antenna diversity on the capacity of wireless communication systems", *IEEE Trans. Commun.*, vol. 42, pp. 1740–1751, Apr. 1994.

[10] C. Berrou, A. Glavieux and P. Thitimajshima, "Near Shannon limit error-correcting coding and decoding: turbo codes (1)", *Proc. ICC'93*, Geneva, Switzerland, pp. 1064–1070, May 1993.

[11] P. Alexander, A. J. Grant and M. Reed, "Iterative multiuser detection in code-division multiple-access with error control coding", *European Trans. Telecommun.*, Special Issue on CDMA Techniques in Wireless Commun. Syst., vol. 9, pp. 419–425, Sept./Oct. 1998.

[12] M. Moher, "An iterative multiuser decoder for near-capacity communications", *IEEE Trans. Commun.*, vol. 46, no. 7, pp. 870–880, July 1998.

[13] M. Reed, "Iterative receiver techniques for coded multiple access communication systems", PhD Thesis, The University of South Australia, Australia, 1999.

[14] P. Alexander, M. Reed, J. Asenstorfer and C. Schlegel, "Iterative multiuser detection for CDMA with FEC: near-single user performance", *IEEE Trans. Commun.*, vol. 46, no. 12, pp. 1693–1699, Dec. 1998.

[15] P. Alexander, M. Reed, J. Asenstorfer and C. Schlegel, "Iterative multiuser interference reduction: turbo CDMA", *IEEE Trans. Commun.*, vol. 47, no. 7, pp. 1008–1014, July 1999.

[16] S. Marinkovic, B. Vucetic and J. Evans, "Improved iterative parallel interference cancellation for coded CDMA systems", in *Proc. ISIT 2001*, Washington DC, USA, p.34, June 2001.

[17] X. Wang and H. V. Poor, "Iterative (turbo) soft interference cancellation and decoding for coded CDMA", *IEEE Transactions on Communications*, vol. 47, no. 7, July 1999, pp. 1046–1061.

[18] H. El Gamal and E. Geraniotis, "Iterative multiuser detection for coded CDMA signals in AWGN and fading channels", *IEEE Journal on Selected Areas in Communications*, vol. 18. no. 1, January 2000, pp. 30–41.

[19] H. El Gamal and R. Hammons, "A new approach to layered space-time coding and signal processing", *IEEE Trans. Inform. Theory*, vol. 47, no. 6, Sept. 2001, pp. 2321–2334.

[20] S. Marinkovic, B. Vucetic and Akihisa Ushirokawa, "Space-time iterative and multi-stage receiver structures for CDMA mobile communication systems", *IEEE Journal on Selected Areas in Communications*, vol. 19, no. 8, August 2001, pp. 1594–1604.

[21] S.L. Ariyavisitakul, "Turbo space-time processing to improve wireless channel capacity", *IEEE Trans. Commun.*, vol. 48, pp. 1347–1358, Aug. 2000.

[22] P. Rapajic and B. Vucetic, "Adaptive receiver structures for synchronous CDMA systems", *IEEE Journal on Selected Areas in Commun.*, vol. 12, pp. 685–697, May 1994.

[23] A. Duel-Hallen, "Equalizers for multiple input/multiple output channels and PAM systems with cyclostationary input sequences", *IEEE Journal on Selected Areas in Communications*, vol. 10, no. 3, pp. 630–639, April 1992.

[24] A. Duel-Hallen, "A family of multiuser decision-feedback detectors for asynchronous code-division multiple access channels", *IEEE Transactions on Communications*, vol. 43, no. 2/3/4, pp. 421–434, February/March/April 1995.

[25] D. Divsalar, M. K. Simon and D. Rapheli, "Improved parallel interference cancellation for CDMA", *IEEE Trans. Commun.*, vol. 46, pp. 258–268, February 1998.

[26] P. G. Renucci and B. D. Woerner, "Analysis of soft cancellation to minimize BER in DS-CDMA interference cancellation", *Proc. 1998 Int. Conf. Telecommun.*, 1998.

[27] R. M. Buehrer, S. P. Nicoloso and S. Gollamudi, "Linear versus nonlinear interference cancellation", *J. Commun. and Networks*, vol. 1, no. 2, pp. 118–132, June 1999.

[28] R. M. Buehrer and B. D. Woerner, "Analysis of adaptive multistage interference cancellation for CDMA using an improved Gaussian approximation", *IEEE Transactions on Communications*, vol. 44, no. 10, October 1996.

[29] N. S. Correal, R. M. Buehrer and B. D. Woerner, "A DSP-based DS-CDMA multiuser receiver employing partial parallel interference cancellation", *IEEE J. Select. Areas Commun.*, vol. 17, pp. 613–630, Apr. 1999.

[30] T. R. Giallorenzi and S.G. Wilson, "Multiuser ML sequence estimator for convolutionally coded asynchronous DS-CDMA systems", *IEEE Transactions on Communications*, vol. 44, no. 8, August 1996, pp. 997–1008.

[31] T. R. Giallorenzi and S.G. Wilson, "Suboptimum multiuser receivers for convolutionally coded asynchronous DS-CDMA systems", *IEEE Transactions on Communications*, vol. 44, no. 9, September 1996, pp. 1183–1196.

[32] L. Bahl, J. Cocke, F. Jelinek and J. Raviv, "Optimal decoding of linear codes for minimizing symbol error rate", *IEEE Trans. Inform. Theory*, vol. IT-20, pp. 284–287, March 1979.

[33] G. J. Foschini and M. J. Gans, "On limits of wireless communications in a fading environment when using multiple antennas", *Wireless Personal Communications*, Mar. 1998, pp. 311–335.

[34] E. Telatar, "Capacity of multi-antenna Gaussian channels", *The European Transactions on Telecommunications*, vol. 10, no. 6, Nov./Dec. 1999, pp. 585–595.

[35] G. Foschini, "Layered space-time architecture for wireless communication in a fading environment when using multi-element antennas", *Bell Labs Technical Journal*, Autumn 1996, pp. 41–59.

[36] H. El Gamal and A.R. Hammons, "The layered space-time architecture: a new perspective", *IEEE Trans. Inform. Theory*, vol. 47, pp. 2321–2334, Sept. 2001.

[37] "Physical layer standard for CDMA2000 spread spectrum systems, Version 3.0 Release 0", June 2001, 3rd Generation Partnership Project 2 (3GPP2), www.3gpp2.org/Public_html

[38] "Air interface for fixed broadband wireless access systems, Part A: Systems between 2 and 11 GHz", IEEE 802.16 Task Group 3/4, June 2001, www.ieee802.org/16/tg3_4/index.html.

[39] "Universal mobile telecommunication systems (UMTS): Physical channels and mapping of transport channels onto physical channels (FDD) (3GPP TS 25.211 version 4.1.0 release 4)", ETSI, June 2001, www.etsi.org/key/.

[40] "Universal mobile telecommunication Systems (UMTS): Physical layer procedures (FDD) (3GPP TS 25.214 version 4.1.0 release 4)", ETSI, June 2001, www.etsi.org/key/.

[41] "Universal mobile telecommunication Systems (UMTS): Physical layer procedures (TDD) (3GPP TS 25.224 version 4.1.0 release 4)", June 2001, www.etsi.org/key/

[42] L. Herault, "Synthesis report on current status of standardization of UMTS concerning the use of smart antennas", Advanced Signal Processing Schemes for Link Capacity Increase in UMTS (ASILUM), Feb. 2000, http://www.nari.ee.ethz.ch/ asilum/

[43] G. D. Golden, G. J. Foschini, R. A. Valenzuela and P. W. Wolniansky, "Detection algorithm and initial laboratory results using the V-BLAST space-time communication architecture", *Electronics Letters*, vol. 35, no. 1, Jan. 7, 1999, pp. 14–15.

[44] J. Yuan, B. Vucetic, W. Feng and M. Tan, "Design of cyclic shift interleavers for turbo codes", *Annals of Telecomms.*, vol. 56, no. 7–8, July-Aug. 2001, pp. 384–393.

[45] W. Firmanto, J. Yuan, K. Lo and B. Vucetic, "Layered space-time coding: Performance analysis and design criteria", Globecom 2001, San Antonio, Dec. 2001.

[46] B Lu and X. Wang, "Iterative receivers for multiuser space-time coding systems", *IEEE Journal on Selected Areas in Communications*, vol. 18 no. 11, November 2000, pp. 2322–2336.

[47] D. Shiu and J. M. Kahn, "Layered space-time codes for wireless communications using multiple transmit antennas," ICC'99, Vancouver, Canada, June 1999.

[48] D. Shiu and J. M. Kahn, "Scalable layered space-time codes for wireless communications; Performance analysis and design criteria", WCNC, 1999, *IEEE Wireless Communications and Networking Conference*, part I, vol. 1, 1999, pp. 159–163, Piscataway, NJ, USA.

[49] G. Foschini, G. Golden, R. Valenzuela and P. Wolniansky, "Simplified processing for high spectral efficiency wireless communication employing multi-element arrays", *IEEE Journal on Selected Areas in Communications*, vol. 17, pp. 1841–1852, Nov. 2000.

[50] D. Wubben, R. Bohnke, J. Rinas, V. Kuhn and K. D. Kammeyer, "Efficient algorithm for decoding layered space-time codes", *Electronics Letters*, vol. 37, pp. 1348–1350, March 2001.

[51] K. Lo, S. Marinkovic, Z Chen and B. Vucetic, "Performance comparison of layered space time codes", ICC 2002, May 2002, New York, USA.

[52] K. Lo, Z. Chen, P. Alexander and B. Vucetic, "Layered space time coding with joint iterative detection, channel estimation and decoding", ISSSTA 2002, Sept. 2002, Prague, Czech Republic.

7

Differential Space-Time Block Codes

7.1 Introduction

In the previous chapters, we show that space-time coding with multiple transmit and receive antennas minimizes the effect of multipath fading and improves the performance and capacity of digital transmission over wireless radio channels. Thus far, it has been assumed that perfect channel estimates, e.g., perfect channel state information, are available at the receiver and coherent detection is employed. When the channel changes slowly compared to the symbol rate, the transmitter can send pilot sequences which enable the receiver to estimate the channel accurately. However, in some situations, such as high-mobility environment or channel fading conditions changing rapidly, it may be difficult or costly to estimate the channel accurately. For such situations, it is useful to develop space-time coding techniques that do not require channel estimates either at the receiver or at the transmitter.

For a single transmit antenna, it is well known that differential schemes, such as differential phase-shift keying (DPSK), can be demodulated without the use of channel estimates. Differential schemes have been widely used in practical cellular mobile communication systems. For example, the standard for United State digital cellular systems, IS-54, employs $\frac{\pi}{4}$-DPSK [14].

It is natural to consider extensions of differential schemes to MIMO systems. Various space-time coding schemes have been proposed such that they can be demodulated and decoded without channel estimates at the receiver [7][8][9][10]. In this chapter, we present differential space-time block codes based on the orthogonal designs [7][8]. These schemes provide simple differential encoding and decoding algorithms. The performance of the schemes is worse by 3 dB relative to the codes with ideal channel state information at the receiver. Differential space-time modulation based on group codes [9] and unitary-space-time block codes [10] are also discussed.

Space-Time Coding Branka Vucetic and Jinhong Yuan
© 2003 John Wiley & Sons, Ltd ISBN: 0-470-84757-3

7.2 Differential Coding for a Single Transmit Antenna

First, we consider a DPSK scheme in a single-antenna system, where the channel has a phase response that is approximately constant from one symbol period to the next. Differential schemes encode the transmitted information into phase differences between two consecutive symbols. Information is essentially transmitted by first providing a reference symbol followed by differentially phase-shifted symbols. The receiver decodes the information in the current symbol by comparing its phase to the phase of the previous symbol.

Consider a differential M-PSK modulation with M signal points and the spectral efficiency of $\eta = \log_2 M = m$ bits/s/Hz. The modulation signal constellation can be represented by

$$A = \{e^{2\pi kj/M}; \quad k = 0, 1, 2, \ldots, M - 1\} \tag{7.1}$$

where $j = \sqrt{-1}$. Let us assume that a data sequence

$$c_1, c_2, c_3, \ldots, c_t, \ldots \tag{7.2}$$

is transmitted, where $c_t \in \{0, 1, 2, \ldots, M - 1\}$. The data sequence is mapped into the signal constellation A, to generate a modulated symbol sequence given by

$$s_1, s_2, s_3, \ldots, s_t, \ldots \tag{7.3}$$

where

$$s_t = e^{j\theta_t} = e^{2\pi c_t j/M} \tag{7.4}$$

The transmitter generates the differential modulated sequence

$$x_0, x_1, x_2, \ldots, x_t, \ldots \tag{7.5}$$

where the differential encoded signal x_t is obtained as

$$x_t = x_{t-1} \cdot s_t$$
$$= x_{t-1} \cdot e^{2\pi c_t j/M} \tag{7.6}$$

Thus, the data information is sent in the difference of the phases of two consecutive symbols. The initial symbol $x_0 = 1$ does not carry any data information and can be thought of as a reference.

Let us represent the received data sequence by

$$r_0, r_1, r_2, \ldots, r_t, \ldots \tag{7.7}$$

The received data are processed by computing the differential phases between any two consecutive symbols. The differential phases are given by

$$\hat{\theta}_t = \arg r_{t-1}^* r_t \tag{7.8}$$

Since

$$e^{j\theta_t} = e^{2\pi c_t j/M} \tag{7.9}$$

and

$$c_t = M\theta_t/2\pi \tag{7.10}$$

we can formulate the decision rule as

$$\text{For} \quad i - 1/2 \leq \frac{M\hat{\theta}_t}{2\pi} \leq i + 1/2, \quad \hat{c}_t = i \tag{7.11}$$

where $\hat{c}_t$ is the estimate of the transmitted data symbol c_t and $i \in \{0, 1, 2, \ldots, M-1\}$. The decision rule can also be expressed as

$$\hat{c}_t = \lfloor \hat{\theta}_t \cdot M/(2\pi) + 1/2 \rfloor \bmod M \tag{7.12}$$

From (7.8) and (7.12), it is clear that the decision output does not depend on earlier demodulation decisions and channel state information, but only on the received symbols in every two consecutive symbol periods. If the channel is approximately constant for a time at least two symbol periods, the differential demodulation performs within 3 dB of the coherent demodulation in Gaussian channels.

7.3 Differential STBC for Two Transmit Antennas

7.3.1 Differential Encoding

The block diagram of the differential space-time block encoder based on orthogonal designs is given in Fig. 7.1 [7]. For two transmit antennas, this scheme begins the transmission by sending two reference modulated signals x_1 and x_2. According to the Alamouti encoding operation, the transmitter sends signals x_1 and x_2 at time one from two transmit antennas simultaneously, and signals $-x_2^*$ and x_1^* at time two from the two transmit antennas. These two transmissions do not carry any data information. Then the transmitter encodes the data sequence in a differential manner and sends them subsequently as follows.

Let us assume that x_{2t-1} and x_{2t} are sent from transmit antennas one and two, respectively, at time $2t - 1$, and that signals $-x_{2t}^*$ and x_{2t-1}^* are sent from transmit antennas one and two, respectively, at time $2t$. At time $2t + 1$, a block of $2m$ information bits, denoted by c_{2t+1}, arrives at the encoder. The block of message is used to choose two complex coefficients R_1 and R_2. Then, based on the previous transmitted signals and the complex coefficients, the encoder computes the modulated symbols for the next two transmissions as

$$(x_{2t+1}, x_{2t+2}) = R_1(x_{2t-1}, x_{2t}) + R_2(-x_{2t}^*, x_{2t-1}^*) \tag{7.13}$$

The transmitter therefore sends x_{2t+1} and x_{2t+2} at time $2t + 1$ from antennas one and two, respectively, and sends $-x_{2t+2}^*$ and x_{2t+1}^* at time $2t + 2$ from antennas one and two, respectively. The process is repeated until the end of the data frame. The encoding scheme is called the *differential space-time block code* for two transmit antennas and (7.13) is referred to as the differential encoding rule.

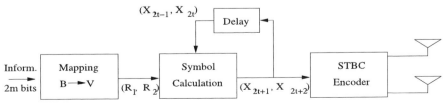

Figure 7.1 A differential STBC encoder

According to the differential encoding rule (7.13), we see that the signals to be transmitted at times $2t + 1$ and $2t + 2$ are represented in terms of a linear combination of those at times $2t - 1$ and $2t$. The coefficient vector (R_1, R_2) of the linear combination is determined by the transmitted data. The key procedure to generate the differential space-time block codes is to compute a set of the coefficient vectors (R_1, R_2) and map a block of $2m$ information bits into the coefficient vector set.

Note that the coefficient vector represents the transmitted data information. Now we show how to compute the elements of the coefficient vector.

We consider an M-PSK modulation with the constellation signal set

$$A = \left\{ \frac{e^{2\pi kj/M}}{\sqrt{2}}; \ k = 0, 1, 2, \ldots, M - 1 \right\} \tag{7.14}$$

where the signal amplitude is divided by $\sqrt{2}$ so that the total transmitted power of the baseband signals from two transmit antennas is one. Note that the complex reference vectors (x_{2t-1}, x_{2t}) and $(-x_{2t}^*, x_{2t-1}^*)$ are orthogonal to each other and each vector has a unit length, because of the power normalization constraint in (7.14). Any complex signal vector (x_{2t+1}, x_{2t+2}) can be uniquely represented in the orthonormal basis of (x_{2t-1}, x_{2t}) and $(-x_{2t}^*, x_{2t-1}^*)$ as shown in (7.13). The coefficients of the representation are given by

$$R_1 = x_{2t+1} x_{2t-1}^* + x_{2t+2} x_{2t}^*$$
$$R_2 = -x_{2t+1} x_{2t} + x_{2t+2} x_{2t-1} \tag{7.15}$$

For a given constellation A and reference vector (x_{2t-1}, x_{2t}), there are $M^2 = 2^{2m}$ distinct coefficient vectors (R_1, R_2), corresponding to M^2 distinct signal vectors (x_{2t+1}, x_{2t+2}). Let us denote by V the set of the coefficient vectors (R_1, R_2). Therefore, there exists a one-to-one correspondence between a block of $2m$ information bits to the coefficient vector of V.

Now we consider some properties of the differential space-time block codes. Note that all the elements of the coefficient vector set have equal lengths of one. Given a vector (x_{2t-1}, x_{2t}), a mapping from (R_1, R_2) to (x_{2t+1}, x_{2t+2}) preserves the distance between the points of the two dimensional complex spaces. In other words, assume that (R_1, R_2) and $(\hat{R}_1, \hat{R}_2)$ are two distinct elements in the set V, and (x_{2t+1}, x_{2t+2}) and $(\hat{x}_{2t+1}, \hat{x}_{2t+2})$ are their corresponding constellation signal vectors generated by using the differential encoding rule. Then, the squared Euclidean distance between the two distinct coefficient vectors (R_1, R_2) and $(\hat{R}_1, \hat{R}_2)$ is equivalent to the squared Euclidean distance between their corresponding constellation signal vectors (x_{2t+1}, x_{2t+2}) and $(\hat{x}_{2t+1}, \hat{x}_{2t+2})$.

$$\|(R_1, R_2) - (\hat{R}_1, \hat{R}_2)\| = \|(x_{2t+1}, x_{2t+2}) - (\hat{x}_{2t+1}, \hat{x}_{2t+2})\| \tag{7.16}$$

Therefore, the mapping can be regarded as changing the orthonormal basis from standard vectors $\{(1, 0), (0, 1)\}$ to $\{(x_{2t-1}, x_{2t}), (-x_{2t}^*, x_{2t-1}^*)\}$. In addition, the minimum distance between any two distinct coefficient vectors of V is equal to the minimum distance of the signal points in the M-PSK constellation A [7].

Now we illustrate the differential space-time block coding scheme by an example. Consider a BPSK constellation of two signal points $-\frac{1}{\sqrt{2}}$ and $+\frac{1}{\sqrt{2}}$. The coefficient vector set computed by using (7.15) is given by

$$V = \{(1, 0), (0, 1), (-1, 0), (0, -1)\} \tag{7.17}$$

At each encoding operation, a block of $2m = 2$ bits arrives at the encoder and is mapped into V. The mapping rule is defined as follows. Let the two reference modulated signals for the differential scheme be

$$x_1 = x_2 = +\frac{1}{\sqrt{2}} \tag{7.18}$$

Given any block of 2 bits at the encoder input, denoted by c_1 and c_2, the first bit c_1 is mapped into the BPSK constellation symbol x_3 and the second bit c_2 is mapped into the BPSK constellation symbol x_4. The mapping is represented as

$$c_1 \rightarrow x_3 = \frac{(-1)^{c_1}}{\sqrt{2}}, \quad c_1 = 0, 1$$

$$c_2 \rightarrow x_4 = \frac{(-1)^{c_2}}{\sqrt{2}}, \quad c_2 = 0, 1 \tag{7.19}$$

The mapping function from each block of information bits to a coefficient vector can be computed by using (7.15) and (7.19) as [7]

$$M(00) = (1, 0)$$
$$M(10) = (0, 1)$$
$$M(01) = (0, -1)$$
$$M(11) = (-1, 0) \tag{7.20}$$

The mapping function is defined based on the reference signals in (7.18). It is used throughout the whole data sequence.

Assume that at time $2t - 1$, $x_{2t-1} = -\frac{1}{\sqrt{2}}$ and $x_{2t} = -\frac{1}{\sqrt{2}}$ are sent from antennas one and two, respectively, and at time $2t$, $-x_{2t}^* = \frac{1}{\sqrt{2}}$ and $x_{2t-1}^* = -\frac{1}{\sqrt{2}}$ are sent from antennas one and two, respectively. If the two information bits at the encoder input at time $2t + 1$ are 11, according to the mapping function $M(11) = (-1, 0)$, the coefficients used to compute the transmitted signals for the next two transmissions are $R_1 = -1$ and $R_2 = 0$. Thus, we have

$$(x_{2t+1}, x_{2t+2}) = -1 \cdot \left(-\frac{1}{\sqrt{2}}, -\frac{1}{\sqrt{2}}\right) + 0 \cdot \left(+\frac{1}{\sqrt{2}}, -\frac{1}{\sqrt{2}}\right)$$

$$= \left(+\frac{1}{\sqrt{2}}, +\frac{1}{\sqrt{2}}\right) \tag{7.21}$$

At time $2t + 1$, $x_{2t+1} = +\frac{1}{\sqrt{2}}$ and $x_{2t+2} = +\frac{1}{\sqrt{2}}$ are sent from antennas one and two, respectively, and at time $2t + 2$, $-x_{2t+2}^* = -\frac{1}{\sqrt{2}}$ and $x_{2t+1}^* = +\frac{1}{\sqrt{2}}$ are sent from antennas one and two, respectively.

This example shows how to compute the transmitted signals for given input bits and previous symbols. For all other possible previous symbols and input bits, the transmitted signals at time $2t + 1$ and $2t + 2$ can be computed in a similar way. We list all the combinations in Table 7.1.

Table 7.1 Transmitted symbols for a differential scheme

(x_{2t-1}, x_{2t})	input bits	(x_{2t+1}, x_{2t+2})
$\left(+\frac{1}{\sqrt{2}}, +\frac{1}{\sqrt{2}}\right)$	00	$\left(+\frac{1}{\sqrt{2}}, +\frac{1}{\sqrt{2}}\right)$
$\left(+\frac{1}{\sqrt{2}}, +\frac{1}{\sqrt{2}}\right)$	01	$\left(+\frac{1}{\sqrt{2}}, -\frac{1}{\sqrt{2}}\right)$
$\left(+\frac{1}{\sqrt{2}}, +\frac{1}{\sqrt{2}}\right)$	10	$\left(-\frac{1}{\sqrt{2}}, +\frac{1}{\sqrt{2}}\right)$
$\left(+\frac{1}{\sqrt{2}}, +\frac{1}{\sqrt{2}}\right)$	11	$\left(-\frac{1}{\sqrt{2}}, -\frac{1}{\sqrt{2}}\right)$
$\left(+\frac{1}{\sqrt{2}}, -\frac{1}{\sqrt{2}}\right)$	00	$\left(+\frac{1}{\sqrt{2}}, -\frac{1}{\sqrt{2}}\right)$
$\left(+\frac{1}{\sqrt{2}}, -\frac{1}{\sqrt{2}}\right)$	01	$\left(-\frac{1}{\sqrt{2}}, -\frac{1}{\sqrt{2}}\right)$
$\left(+\frac{1}{\sqrt{2}}, -\frac{1}{\sqrt{2}}\right)$	10	$\left(+\frac{1}{\sqrt{2}}, +\frac{1}{\sqrt{2}}\right)$
$\left(+\frac{1}{\sqrt{2}}, -\frac{1}{\sqrt{2}}\right)$	11	$\left(-\frac{1}{\sqrt{2}}, +\frac{1}{\sqrt{2}}\right)$
$\left(-\frac{1}{\sqrt{2}}, +\frac{1}{\sqrt{2}}\right)$	00	$\left(-\frac{1}{\sqrt{2}}, +\frac{1}{\sqrt{2}}\right)$
$\left(-\frac{1}{\sqrt{2}}, +\frac{1}{\sqrt{2}}\right)$	01	$\left(+\frac{1}{\sqrt{2}}, +\frac{1}{\sqrt{2}}\right)$
$\left(-\frac{1}{\sqrt{2}}, +\frac{1}{\sqrt{2}}\right)$	10	$\left(-\frac{1}{\sqrt{2}}, -\frac{1}{\sqrt{2}}\right)$
$\left(-\frac{1}{\sqrt{2}}, +\frac{1}{\sqrt{2}}\right)$	11	$\left(+\frac{1}{\sqrt{2}}, -\frac{1}{\sqrt{2}}\right)$
$\left(-\frac{1}{\sqrt{2}}, -\frac{1}{\sqrt{2}}\right)$	00	$\left(-\frac{1}{\sqrt{2}}, -\frac{1}{\sqrt{2}}\right)$
$\left(-\frac{1}{\sqrt{2}}, -\frac{1}{\sqrt{2}}\right)$	01	$\left(-\frac{1}{\sqrt{2}}, +\frac{1}{\sqrt{2}}\right)$
$\left(-\frac{1}{\sqrt{2}}, -\frac{1}{\sqrt{2}}\right)$	10	$\left(+\frac{1}{\sqrt{2}}, -\frac{1}{\sqrt{2}}\right)$
$\left(-\frac{1}{\sqrt{2}}, -\frac{1}{\sqrt{2}}\right)$	11	$\left(+\frac{1}{\sqrt{2}}, +\frac{1}{\sqrt{2}}\right)$

7.3.2 Differential Decoding

Assume that one receive antenna is employed. Let us denote by r_t the received signal at time t, by n_t the noise sample at time t, and by h_1 and h_2 the fading coefficients from transmit antennas one and two to the receive antenna, respectively. The received signals at times $2t - 1$, $2t$, $2t + 1$ and $2t + 2$ can be represented by

$$r_{2t-1} = h_1 \cdot x_{2t-1} + h_2 \cdot x_{2t} + n_{2t-1}$$

$$r_{2t} = -h_1 \cdot x_{2t}^* + h_2 \cdot x_{2t-1}^* + n_{2t}$$

$$r_{2t+1} = h_1 \cdot x_{2t+1} + h_2 \cdot x_{2t+2} + n_{2t+1}$$

$$r_{2t+2} = -h_1 \cdot x_{2t+2}^* + h_2 \cdot x_{2t+1}^* + n_{2t+2} \tag{7.22}$$

Let

$$\mathbf{H} = \begin{pmatrix} h_1 & h_2^* \\ h_2 & -h_1^* \end{pmatrix} \qquad (7.23)$$

and

$$N_{2t-1} = (n_{2t-1}, n_{2t}^*)$$

$$N_{2t} = (n_{2t}, -n_{2t-1}^*)$$

For simplicity, we consider a vector representation of the received signals [7]. The received signals in the vector form are given by

$$(r_{2t-1}, r_{2t}^*) = (x_{2t-1}, x_{2t}) \cdot \mathbf{H} + N_{2t-1} \qquad (7.24)$$

$$(r_{2t+1}, r_{2t+2}^*) = (x_{2t+1}, x_{2t+2}) \cdot \mathbf{H} + N_{2t+1} \qquad (7.25)$$

$$(r_{2t}, -r_{2t-1}^*) = (-x_{2t}^*, x_{2t-1}^*) \cdot \mathbf{H} + N_{2t} \qquad (7.26)$$

Let us define a decision statistics signal at the receiver, denoted by $\tilde{R}_1$, as the inner product of the two received signal vectors (7.24) and (7.25).

$$\tilde{R}_1 = (r_{2t+1}, r_{2t+2}^*) \cdot (r_{2t-1}, r_{2t}^*)$$

$$= r_{2t+1} r_{2t-1}^* + r_{2t+2}^* r_{2t} \qquad (7.27)$$

The inner product can be computed as

$$\tilde{R}_1 = (x_{2t+1}, x_{2t+2}) \mathbf{H} \mathbf{H}^H (x_{2t-1}, x_{2t})^H$$

$$+ (x_{2t+1}, x_{2t+2}) \mathbf{H} N_{2t-1}^H + N_{2t+1} \mathbf{H}^H (x_{2t-1}, x_{2t})^H + N_{2t+1} N_{2t-1}^H \qquad (7.28)$$

Substituting (7.23) into (7.28), we have

$$\tilde{R}_1 = (|h_1|^2 + |h_2|^2) \left(x_{2t+1} x_{2t-1}^* + x_{2t+2} x_{2t}^* \right)$$

$$+ (x_{2t+1}, x_{2t+2}) \mathbf{H} N_{2t-1}^H + N_{2t+1} \mathbf{H}^H (x_{2t-1}, x_{2t})^H + N_{2t+1} N_{2t-1}^H \qquad (7.29)$$

Now, let us define

$$\tilde{N}_1 = (x_{2t+1}, x_{2t+2}) \mathbf{H} N_{2t-1}^H + N_{2t+1} \mathbf{H}^H (x_{2t-1}, x_{2t})^H + N_{2t+1} N_{2t-1}^H \qquad (7.30)$$

Then, the decision statistics (7.29) can be rewritten as

$$\tilde{R}_1 = (|h_1|^2 + |h_2|^2) R_1 + \tilde{N}_1 \qquad (7.31)$$

Next, we construct another decision statistics signal, denoted by $\tilde{R}_2$, as the inner product of the two received signal vectors (7.25) and (7.26).

$$\tilde{R}_2 = (r_{2t+1}, r_{2t+2}^*) \cdot (r_{2t}, -r_{2t-1}^*)$$

$$= r_{2t+1} r_{2t}^* - r_{2t+2}^* r_{2t-1} \qquad (7.32)$$

The inner product can be computed as

$$\tilde{R}_2 = (x_{2t+1}, x_{2t+2}) \mathbf{H} \mathbf{H}^H (-x_{2t}^*, x_{2t-1}^*)^H$$

$$+ (x_{2t+1}, x_{2t+2}) \mathbf{H} N_{2t}^H + N_{2t+1} \mathbf{H}^H (-x_{2t}^*, x_{2t-1}^*)^H + N_{2t+1} N_{2t}^H \qquad (7.33)$$

Substituting (7.23) into (7.33), we have

$$\tilde{R}_2 = (|h_1|^2 + |h_2|^2)(-x_{2t+1}x_{2t} + x_{2t+2}x_{2t-1})$$
$$+(x_{2t+1}, x_{2t+2})\mathbf{H}N_{2t}^H + N_{2t+1}\mathbf{H}^H(-x_{2t}^*, x_{2t-1}^*)^H + N_{2t+1}N_{2t}^H \qquad (7.34)$$

Now, let us define

$$\tilde{N}_2 = (x_{2t+1}, x_{2t+2})\mathbf{H}N_{2t}^H + N_{2t+1}\mathbf{H}^H(-x_{2t}^*, x_{2t-1}^*)^H + N_{2t+1}N_{2t}^H \qquad (7.35)$$

Then, the decision statistics signal (7.34) can be rewritten as

$$\tilde{R}_2 = (|h_1|^2 + |h_2|^2)R_2 + \tilde{N}_2 \qquad (7.36)$$

Now we can write the decision statistics in a vector form

$$(\tilde{R}_1, \tilde{R}_2) = (|h_1|^2 + |h_2|^2)(R_1, R_2) + (\tilde{N}_1, \tilde{N}_2) \qquad (7.37)$$

For a given channel realization h_1 and h_2, the decision statistics signals $\tilde{R}_1$ and $\tilde{R}_2$ are only a function of differential coefficients R_1 and R_2, respectively. Since all the coefficient vectors in the set V have equal lengths, the receiver now chooses the closest coefficient vector from V to the decision statistics signal vector $(\tilde{R}_1, \tilde{R}_2)$ as the detector output. Then the inverse mapping of (7.20) is applied to decode the transmitted block of bits [7]. The overall receiver block diagram is shown in Fig. 7.2.

7.3.3 Performance Simulation

Comparing (7.37) and (3.12), we can see that the decision statistics signals for differential detection have a very similar format to those for coherent detection. For both cases, the decision statistics signals have the same multiplication coefficient $|h_1|^2 + |h_2|^2$. Intuitively, we can say that both the differential and coherent schemes achieve the same diversity gain. However, since the noise terms for the decision statistics signals in (7.37) and (3.12) are different, the differential and coherent schemes have different coding gains.

The performance of the differential scheme with two transmit and one receive antennas on slow Rayleigh fading channels is evaluated by simulations. The results for BPSK, QPSK and 8-PSK constellations are presented in Figs. 7.3, 7.4, and 7.5, respectively. The fading is assumed to be constant over a frame of 130 symbols and independent between the frames. The performance of the corresponding STBC with coherent detection is also shown in the figures. From the figures, we can see that the performance curves of the differential schemes are parallel to those of coherent schemes, which indicates that the differential schemes also achieve full transmit diversity due to the orthogonal designs. However, since neither the

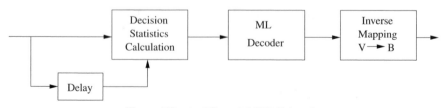

Figure 7.2 A differential STBC decoder

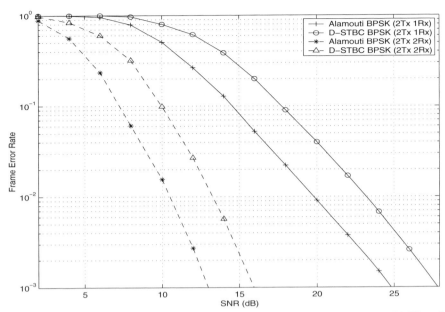

Figure 7.3 Performance comparison of the coherent and differential STBC with BPSK and two transmit antennas on slow fading channels

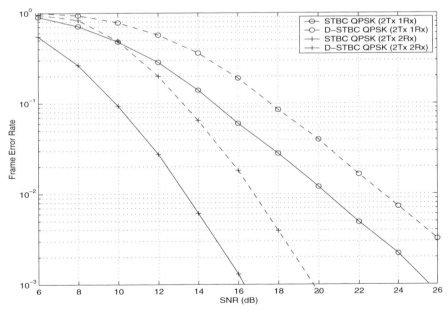

Figure 7.4 Performance comparison of the coherent and differential STBC with QPSK and two transmit antennas on slow fading channels

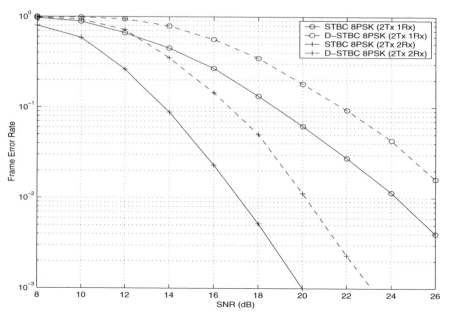

Figure 7.5 Performance comparison of the coherent and differential STBC with 8-PSK and two transmit antennas on slow fading channels

receiver nor the transmitter requires the channel state information, the differential schemes are 3 dB worse than the respective space-time block codes with coherent detection. This is the same as in the single transmit antenna case.

It is worth noting that the same decoding method can be used for more than one receive antennas and similar to the coherent detection, only linear signal processing is employed in the receiver.

7.4 Differential STBC with Real Signal Constellations for Three and Four Transmit Antennas

The differential space-time block codes described in the previous section can be extended to $n_T \geq 2$ transmit antennas. In this section, we present differential schemes with real signal constellations for three and four transmit antennas based on the codes of $\mathbf{X}^3$ and $\mathbf{X}^4$, respectively. The derivation of the schemes can be found in [8] and is not included here.

7.4.1 Differential Encoding

Let us recall the encoding operation for space-time block codes with a real signal constellation of 2^m points. At each encoding operation, km information bits arrive at the encoder and select k constellation signals to be transmitted. The encoder takes k modulated signals and generates codeword sequences based on the transmission matrix of a space-time block code.

For the code with four transmit antennas, the number of the modulated signals that the encoder takes at each block is $k = 4$. Let us denote the modulated signals in the vth encoding operation by $\mathbf{x}_\nu$

$$\mathbf{x}_\nu = (x_\nu^1, x_\nu^2, x_\nu^3, x_\nu^4) \qquad (7.38)$$

where x_ν^i, $i = 1, 2, 3$ and 4, is the ith modulated signal in the νth encoding operation. For the $(\nu + 1)$th encoding operation, a block of $km = 4m$ input bits arrives to the encoder. The message block is used to choose a unit-length real coefficient vector, denoted by $\mathcal{R}$, and given by

$$\mathcal{R} = (R_1, R_2, R_3, R_4) \qquad (7.39)$$

Then, based on the previous modulated signals $\mathbf{x}_\nu$ and the current coefficients $\mathcal{R}$, the encoder computes the modulated signals for the $(\nu + 1)$th block as

$$\mathbf{x}_{\nu+1} = \sum_{i=1}^{4} R_i \cdot V_i(\mathbf{x}_\nu) \qquad (7.40)$$

where $V_i(\mathbf{x}_\nu)$ are vectors defined as

$$V_1(\mathbf{x}_\nu) = (x_\nu^1, x_\nu^2, x_\nu^3, x_\nu^4)$$
$$V_2(\mathbf{x}_\nu) = (x_\nu^2, -x_\nu^1, x_\nu^4, -x_\nu^3)$$
$$V_3(\mathbf{x}_\nu) = (x_\nu^3, -x_\nu^4, -x_\nu^1, x_\nu^2)$$
$$V_4(\mathbf{x}_\nu) = (x_\nu^4, x_\nu^3, -x_\nu^2, -x_\nu^1) \qquad (7.41)$$

The four-dimensional vectors form an orthonormal basis for a four-dimensional real signal space. If the both sides of (7.40) are multiplied by $V_i^T(\mathbf{x}_\nu)$, $i = 1, 2, 3$, and 4, due to orthogonality of the vectors of $V_i(\mathbf{x}_\nu)$, we can represent the coefficients by

$$R_i = \mathbf{x}_{\nu+1} V_i^T(\mathbf{x}_\nu) \qquad (7.42)$$

Let us denote by V the set of the unit-length vectors $\mathcal{R}$. Given the signal vector $\mathbf{x}_\nu$ for the νth block, there exist 2^{4m} coefficient vectors that generate valid signal vectors $\mathbf{x}_{\nu+1}$ for the $(\nu + 1)$th block, corresponding to the all possible $4m$ input bits. Therefore, there is a one-to-one mapping between the unit-length vectors in V and the input message blocks. The key issue in constructing a differential scheme is to compute the set of coefficient vectors V and map each block of $4m$ information bits to the set.

For example, let us consider a BPSK constellation of two signal points $-\frac{1}{2}$ and $+\frac{1}{2}$. The signal amplitude is divided by 2 such that the total transmitted power of the baseband signals in a system with four transmit antennas is one. Let the four reference modulated signals at the initial transmission, denoted by $x_0^1, x_0^2, x_0^3, x_0^4$, for the differential scheme be $1/2$, i.e.

$$\mathbf{x}_0 = (x_0^1, x_0^2, x_0^3, x_0^4) = \left(\tfrac{1}{2}, \tfrac{1}{2}, \tfrac{1}{2}, \tfrac{1}{2}\right) \qquad (7.43)$$

For four input bits at the first encoding block $\mathbf{c}_1 = (c_1^1, c_1^2, c_1^3, c_1^4)$, we choose four BPSK signals, denoted by $x_1^1, x_1^2, x_1^3, x_1^4$, based on the following mapping

$$c_1^i \rightarrow x_1^i = \frac{(-1)^{c_1^i}}{2}, \quad i = 1, 2, 3, 4 \qquad (7.44)$$

The coefficients can be calculated from (7.42) as

$$R_1 = \frac{1}{2}\left(x_1^1 + x_1^2 + x_1^3 + x_1^4\right)$$

$$R_2 = \frac{1}{2}\left(x_1^1 - x_1^2 + x_1^3 - x_1^4\right)$$

$$R_3 = \frac{1}{2}\left(x_1^1 - x_1^2 - x_1^3 + x_1^4\right)$$

$$R_4 = \frac{1}{2}\left(x_1^1 + x_1^2 - x_1^3 - x_1^4\right) \tag{7.45}$$

Therefore, a one-to-one mapping from each block of four input bits to a coefficient vector is defined by (7.44) and (7.45). This mapping is based on the reference signals given by (7.43) and is used throughout the whole data sequence.

7.4.2 Differential Decoding

Now we consider differential decoding for the code $\mathbf{X}^4$. Note that $\mathbf{x}_\nu$ and $\mathbf{x}_{\nu+1}$ are the modulated signal vectors for the νth and $(\nu + 1)$th block message, respectively. According to the transmission matrix $\mathbf{X}^4$, for each block of data, there are four signals received subsequently, assuming that one receive antenna is used. We denote the received signals for the νth block message by $r_1^\nu, r_2^\nu, r_3^\nu, r_4^\nu$, and the received signals for the $(\nu+1)$th block message by $r_1^{\nu+1}, r_2^{\nu+1}, r_3^{\nu+1}, r_4^{\nu+1}$. The received signals can be written in the following vector form

$$\mathbf{r}_\nu^1 = \left(r_1^\nu, r_2^\nu, r_3^\nu, r_4^\nu\right)$$
$$= V_1(\mathbf{x}_\nu)\Lambda + \left(n_1^\nu, n_2^\nu, n_3^\nu, n_4^\nu\right) \tag{7.46}$$

$$\mathbf{r}_\nu^2 = \left(-r_2^\nu, r_1^\nu, r_4^\nu, -r_3^\nu\right)$$
$$= V_2(\mathbf{x}_\nu)\Lambda + \left(-n_2^\nu, n_1^\nu, n_4^\nu, -n_3^\nu\right) \tag{7.47}$$

$$\mathbf{r}_\nu^3 = \left(-r_3^\nu, -r_4^\nu, r_1^\nu, r_2^\nu\right)$$
$$= V_3(\mathbf{x}_\nu)\Lambda + \left(-n_3^\nu, -n_4^\nu, n_1^\nu, n_2^\nu\right) \tag{7.48}$$

$$\mathbf{r}_\nu^4 = \left(-r_4^\nu, r_3^\nu, -r_2^\nu, r_1^\nu\right)$$
$$= V_4(\mathbf{x}_\nu)\Lambda + \left(-n_4^\nu, n_3^\nu, -n_2^\nu, n_1^\nu\right) \tag{7.49}$$

$$\mathbf{r}_{\nu+1} = \left(r_1^{\nu+1}, r_2^{\nu+1}, r_3^{\nu+1}, r_4^{\nu+1}\right)$$
$$= V_1(\mathbf{x}_{\nu+1})\Lambda + \left(n_1^{\nu+1}, n_2^{\nu+1}, n_3^{\nu+1}, n_4^{\nu+1}\right) \tag{7.50}$$

where Λ is a 4×4 matrix defined by

$$\Lambda = \begin{bmatrix} h_1 & h_2 & h_3 & h_4 \\ h_2 & -h_1 & -h_4 & h_3 \\ h_3 & h_4 & -h_1 & -h_2 \\ h_4 & -h_3 & h_2 & -h_1 \end{bmatrix} \tag{7.51}$$

and h_i is the fading coefficient for the channel from transmit antenna i to the receive antenna, n_i^ν is the noise sample at the ith symbol period in the νth block. Let us define the noise vectors in (7.46)–(7.50) as N_ν^1, N_ν^2, N_ν^3, N_ν^4, and $N_{\nu+1}$, respectively. Thus, we can construct the decision statistics signals $\tilde{R}_i$, $i = 1, 2, 3, 4$, as the inner product of vectors $\mathbf{r}_{\nu+1}$ and $\mathbf{r}_\nu^i$. The decision statistics signal can be computed as [8]

$$
\tilde{R}_i = \mathbf{r}_{\nu+1} \cdot \mathbf{r}_\nu^i
$$

$$
= \sum_{j=1}^{4} |h_j|^2 \cdot V_1(\mathbf{x}_{\nu+1}) V_i^T(\mathbf{x}_\nu) + \tilde{N}_i
$$

$$
= \sum_{j=1}^{4} |h_j|^2 \cdot \mathbf{x}_{\nu+1} V_i^T(\mathbf{x}_\nu) + \tilde{N}_i
$$

$$
= \sum_{j=1}^{4} |h_j|^2 \cdot R_i + \tilde{N}_i \tag{7.52}
$$

where $\tilde{N}_i$ is a noise term given by

$$
\tilde{N}_i = \mathbf{x}_{\nu+1} \Lambda (N_\nu^i)^H + N_{\nu+1} \Lambda^H V_i^T(\mathbf{x}_\nu) + N_{\nu+1} (N_\nu^i)^H. \tag{7.53}
$$

Let us write the decision statistics signals in a vector form

$$
(\tilde{R}_1, \tilde{R}_2, \tilde{R}_3, \tilde{R}_4) = \left(\sum_{j=1}^{4} |h_j|^2 \right) \cdot (R_1, R_2, R_3, R_4) + (\tilde{N}_1, \tilde{N}_2, \tilde{N}_3, \tilde{N}_4) \tag{7.54}
$$

Clearly, the decision statistics signal vector is only a function of the differential coefficient vector. Since all the coefficient vectors in the set V have equal length, the receiver now chooses the closest coefficient vector from V to the decision statistics signal vector as the differential detection output. Then the inverse mapping from the coefficient vector to the block of information bits is applied to decode the transmitted signal [8]. From (7.54), we can see that the scheme provides a four-level transmit diversity with four transmit antennas.

Thus far, we have described differential encoding and decoding algorithms for space-time block codes with four transmit antennas $\mathbf{X}^4$. For codes with three transmit antennas $\mathbf{X}^3$, the code rate is the same as for $\mathbf{X}^4$ and the transmission matrix $\mathbf{X}_3$ is identical to the first three rows of $\mathbf{X}_4$. Therefore, the same differential encoding and decoding algorithms for $\mathbf{X}_4$ can be applied directly for the code $\mathbf{X}_3$. In this case, the received signal can also be represented by (7.46)–(7.50), where h_4 in matrix Λ is set to zero.

$$
\Lambda = \begin{bmatrix} h_1 & h_2 & h_3 & 0 \\ h_2 & -h_1 & 0 & h_3 \\ h_3 & 0 & -h_1 & -h_2 \\ 0 & -h_3 & h_2 & -h_1 \end{bmatrix} \tag{7.55}
$$

The decision statistics signal vector at the receiver is given by

$$
(\tilde{R}_1, \tilde{R}_2, \tilde{R}_3, \tilde{R}_4) = \left(\sum_{j=1}^{3} |h_j|^2 \right) \cdot (R_1, R_2, R_3, R_4) + (\tilde{N}_1, \tilde{N}_2, \tilde{N}_3, \tilde{N}_4) \tag{7.56}
$$

Therefore, this scheme can achieve three-level diversity with three transmit antennas and one receive antenna.

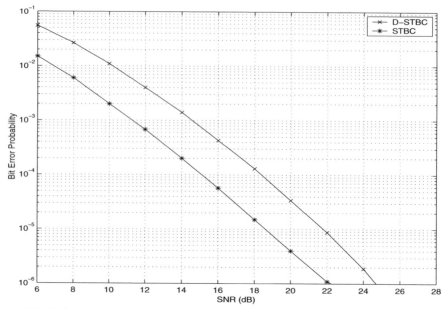

Figure 7.6 Performance comparison of the coherent and differential BPSK STBC with three transmit and one receive antenna on slow Rayleigh fading channels

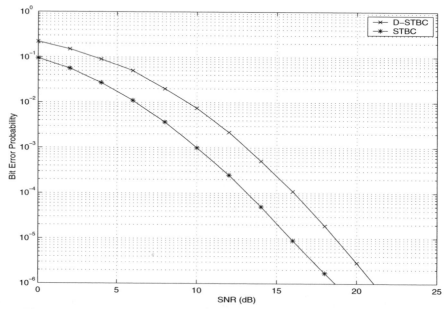

Figure 7.7 Performance comparison of the coherent and differential BPSK STBC with four transmit and one receive antenna on slow Rayleigh fading channels

7.4.3   Performance Simulation

Figures 7.6 and 7.7 depict the FER performance of the BPSK differential STBC with three and four transmit antennas, respectively, on slow Rayleigh fading channels. The frame was assumed to be 130 symbols and one receive antenna was used in the simulations. The performance curves of the corresponding STBC with coherent detection are also plotted in the figures for comparison. For both cases, the differential scheme is worse by about 3 dB relative to the coherent scheme.

7.5   Differential STBC with Complex Signal Constellations for Three and Four Transmit Antennas

7.5.1   Differential Encoding

Differential space-time block coding scheme with four transmit antennas can be defined based on the transmission matrix $\mathbf{X}_4^c$. The differential encoder over complex constellations is similar to that over real signal constellations. At the νth encoding operation, the encoder takes $4m$ input bits and chooses a unit-length real coefficient vector

$$\mathcal{R} = (R_1, R_2, R_3, R_4) \tag{7.57}$$

The differential encoding rule is represented by

$$\mathbf{x}_{\nu+1} = \sum_{i=1}^{4} R_i \cdot V_i(\mathbf{x}_\nu) \tag{7.58}$$

where $\mathbf{x}_\nu$ and $\mathbf{x}_{\nu+1}$ are a modulated signal vector for the νth and $(\nu+)1$th encoding block, respectively, and $V_i(\mathbf{x}_\nu)$ is given by (7.41). Let us define a new set of vectors as

$$V_1^c(\mathbf{x}_\nu) = (x_\nu^1, x_\nu^2, x_\nu^3, x_\nu^4, (x_\nu^1)^*, (x_\nu^2)^*, (x_\nu^3)^*, (x_\nu^4)^*)$$

$$V_2^c(\mathbf{x}_\nu) = (x_\nu^2, -x_\nu^1, x_\nu^4, -x_\nu^3, (x_\nu^2)^*, -(x_\nu^1)^*, (x_\nu^4)^*, -(x_\nu^3)^*)$$

$$V_3^c(\mathbf{x}_\nu) = (x_\nu^3, -x_\nu^4, -x_\nu^1, x_\nu^2, (x_\nu^3)^*, -(x_\nu^4)^*, -(x_\nu^1)^*, (x_\nu^2)^*)$$

$$V_4^c(\mathbf{x}_\nu) = (x_\nu^4, x_\nu^3, -x_\nu^2, -x_\nu^1, (x_\nu^4)^*, (x_\nu^3)^*, -(x_\nu^2)^*, -(x_\nu^1)^*) \tag{7.59}$$

These vectors are orthogonal to each other and they form an orthogonal basis for the four dimensional subspace of any arbitrary four dimensional constellation symbols and their conjugates in an eight dimensional space [8]. Note that the vectors $V_i^c(\mathbf{x}_\nu)$ can be expressed as

$$V_i^c(\mathbf{x}_\nu) = \big(V_i(\mathbf{x}_\nu), V_i^*(\mathbf{x}_\nu)\big) \quad i = 1, 2, 3, 4 \tag{7.60}$$

where $V_i^*(\mathbf{x}_\nu)$ denotes the conjugates of $V_i(\mathbf{x}_\nu)$. The differential encoding rule (7.58) can be rewritten as

$$V_1^c(\mathbf{x}_{\nu+1}) = \sum_{i=1}^{4} R_i \cdot V_i^c(\mathbf{x}_\nu) \tag{7.61}$$

If the both sides of (7.61) are multiplied by $\left(V_i^c(\mathbf{x}_v)\right)^H$, $i = 1, 2, 3$, and 4, due to orthogonality of the vectors of $V_i^c(\mathbf{x}_v)$, we can compute the differential coefficients as

$$
R_i = \tfrac{1}{2}(V_1(\mathbf{x}_{v+1})V_i^H(\mathbf{x}_v) + V_1^*(\mathbf{x}_{v+1})V_i^T(\mathbf{x}_v))
$$

$$
= Re\{V_1(\mathbf{x}_{v+1})V_i^H(\mathbf{x}_v)\} \tag{7.62}
$$

7.5.2 Differential Decoding

Let us consider a system with a single receive antenna. We denote the received signals for the vth block message by $r_1^v, r_2^v, \ldots, r_8^v$, and the received signals for the $(v + 1)$th block message by $r_1^{v+1}, r_2^{v+1}, \ldots, r_8^{v+1}$. The received signals can be written in the following vector form

$$
\mathbf{r}_v^1 = (r_1^v, r_2^v, r_3^v, r_4^v, (r_5^v)^*, (r_6^v)^*, (r_7^v)^*, (r_8^v)^*)
$$

$$
= V_1(\mathbf{x}_v)\Lambda^c + (n_1^v, n_2^v, n_3^v, n_4^v, (n_5^v)^*, (n_6^v)^*, (n_7^v)^*, (n_8^v)^*) \tag{7.63}
$$

$$
\mathbf{r}_v^2 = (-r_2^v, r_1^v, r_4^v, -r_3^v, -(r_6^v)^*, (r_5^v)^*, (r_8^v)^*, -(r_7^v)^*)
$$

$$
= V_2(\mathbf{x}_v)\Lambda^c + (-n_2^v, n_1^v, n_4^v, -n_3^v, -(n_6^v)^*, (n_5^v)^*, (n_8^v)^*, -(n_7^v)^*) \tag{7.64}
$$

$$
\mathbf{r}_v^3 = (-r_3^v, -r_4^v, r_1^v, r_2^v, -(r_7^v)^*, -(r_8^v)^*, (r_5^v)^*, (r_6^v)^*)
$$

$$
= V_3(\mathbf{x}_v)\Lambda^c + (-n_3^v, -n_4^v, n_1^v, n_2^v, -(n_7^v)^*, -(n_8^v)^*, (n_5^v)^*, (n_6^v)^*) \tag{7.65}
$$

$$
\mathbf{r}_v^4 = (-r_4^v, r_3^v, -r_2^v, r_1^v, -(r_8^v)^*, (r_7^v)^*, -(r_6^v)^*, (r_5^v)^*)
$$

$$
= V_4(\mathbf{x}_v)\Lambda^c + (-n_4^v, n_3^v, -n_2^v, n_1^v, -(n_8^v)^*, (n_7^v)^*, -(n_6^v)^*, (n_5^v)^*) \tag{7.66}
$$

$$
\mathbf{r}_{v+1} = (r_1^{v+1}, r_2^{v+1}, r_3^{v+1}, r_4^{v+1}, (r_5^{v+1})^*, (r_6^{v+1})^*, (r_7^{v+1})^*, (r_8^{v+1})^*)
$$

$$
= V_1(\mathbf{x}_{v+1})\Lambda^c + (n_1^{v+1}, n_2^{v+1}, n_3^{v+1}, n_4^{v+1}, (n_5^{v+1})^*, (n_6^{v+1})^*,
$$

$$
(n_7^{v+1})^*, (n_8^{v+1})^*) \tag{7.67}
$$

where Λ^c is a 8×4 matrix defined by

$$
\Lambda^c = \begin{bmatrix} h_1 & h_2 & h_3 & h_4 & h_1^* & h_2^* & h_3^* & h_4^* \\ h_2 & -h_1 & -h_4 & h_3 & h_2^* & -h_1^* & -h_4^* & h_3^* \\ h_3 & h_4 & -h_1 & -h_2 & h_3^* & h_4^* & -h_1^* & -h_2^* \\ h_4 & -h_3 & h_2 & -h_1 & h_4^* & -h_3^* & h_2^* & -h_1^* \end{bmatrix} \tag{7.68}
$$

Let us define the noise vectors in (7.63)–(7.67) as $N_v^1, N_v^2, N_v^3, N_v^4$, and N_{v+1}, respectively. Now, let us construct the decision statistics signals $\tilde{R}_i$, $i = 1, 2, 3, 4$, as the real part of the inner product of vectors $\mathbf{r}_{v+1}$ and $\mathbf{r}_v^i$. The decision statistics signal can be computed as [8]

$$
\tilde{R}_i = Re\left\{\mathbf{r}_{v+1} \cdot \mathbf{r}_v^i\right\}
$$

$$
= 2\sum_{j=1}^{4} |h_j|^2 \cdot Re\left\{V_1(\mathbf{x}_{v+1})V_i^H(\mathbf{x}_v)\right\} + \tilde{N}_i
$$

$$
= 2\sum_{j=1}^{4} |h_j|^2 \cdot R_i + \tilde{N}_i \tag{7.69}
$$

where $\tilde{N}_i$ is a noise term given by

$$\tilde{N}_i = Re \left\{ V_1(\mathbf{x}_{\nu+1}) \Lambda^c (N_\nu^i)^H + N_{\nu+1}(\Lambda^c)^H V_i^H(\mathbf{x}_\nu) + N_{\nu+1}(N_\nu^i)^H \right\}. \tag{7.70}$$

The decision statistics signals in a vector form are given by

$$(\tilde{R}_1, \tilde{R}_2, \tilde{R}_3, \tilde{R}_4) = 2 \left(\sum_{j=1}^{4} |h_j|^2 \right) \cdot (R_1, R_2, R_3, R_4) + (\tilde{N}_1, \tilde{N}_2, \tilde{N}_3, \tilde{N}_4) \tag{7.71}$$

The decision statistics signal vector is only a function of the differential coefficient vector. Because all the elements in the coefficient vector set V have equal length, the receiver can choose the closest coefficient vector from V to the decision statistics signal vector $\hat{\mathcal{R}}$ as the differential detection output. The inverse mapping from the coefficient vector to the block of information bits is then applied to decode the transmitted signal [8]. From (7.71), we can observe that the scheme provides four-level transmit diversity with four transmit antennas.

The same differential encoding and decoding algorithms for $\mathbf{X}_4^c$ can be used directly for the code $\mathbf{X}_3^c$. In this case, the received signal can also be represented by (7.63)–(7.67), where h_4 in matrix Λ^c is set to zero.

7.5.3 Performance Simulation

Performance comparison between coherent and differential QPSK STBC with three and four transmit antennas on slow Rayleigh fading channels is illustrated in Figs. 7.8 and 7.9, respectively [13]. The frame length is 130 symbols and one receive antenna is used in the simulations. We can see that for both cases, the differential scheme is worse by about 3 dB relative to the corresponding coherent scheme.

7.6   Unitary Space-Time Modulation

Unitary space-time modulation was introduced in [10] [11] for MIMO systems with no CSI at either the transmitter or the receiver. A unitary space-time modulation for systems with n_T transmit antennas and spectral efficiency of η bits/s/Hz is defined by a set of M distinct $n_T \times n_T$ unitary signal matrices

$$\mathbf{V}_0, \mathbf{V}_1, \mathbf{V}_2, \ldots, \mathbf{V}_{M-1} \tag{7.72}$$

where

$$\mathbf{V}_i \cdot \mathbf{V}_i^H = \mathbf{V}_i^H \cdot \mathbf{V}_i = \mathbf{I}_{n_T} \tag{7.73}$$

and

$$M = 2^{\eta n_T} \tag{7.74}$$

The unitary signal matrices satisfy the orthogonal design criteria for space-time block codes. Therefore, they can be used as space-time transmission matrices to achieve a full transmit diversity. The i-th row of each matrix represents the signal sequence transmitted from i-th transmit antenna over n_T symbol periods.

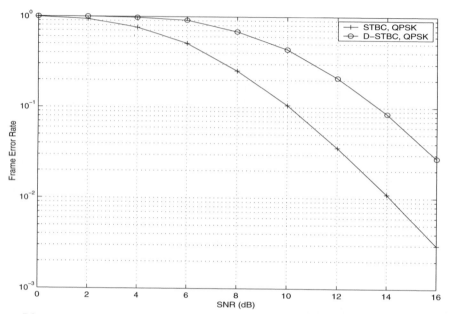

Figure 7.8 Performance comparison of the coherent and differential QPSK STBC with three transmit antennas on slow Rayleigh fading channels

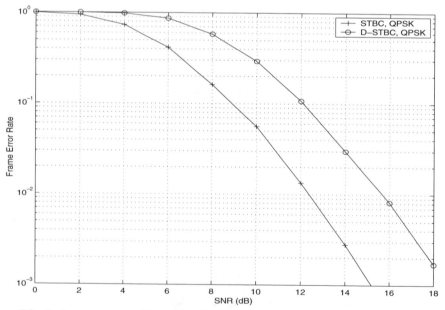

Figure 7.9 Performance comparison of the coherent and differential QPSK STBC with four transmit antennas on slow Rayleigh fading channels

information	$\mathbf{c}_1$	$\mathbf{c}_2$	$\mathbf{c}_3$	$\cdots$
space-time modulation	$\mathbf{V}_{z_1}$	$\mathbf{V}_{z_2}$	$\mathbf{V}_{z_3}$	$\cdots$
differential modulation	$\mathbf{X}_0 = \mathbf{I}_{n_T}$ $\quad$ $\mathbf{X}_1 = \mathbf{V}_{z_1}$	$\mathbf{X}_2 = \mathbf{V}_{z_1}\mathbf{V}_{z_2}$	$\mathbf{X}_3 = \mathbf{V}_{z_1}\mathbf{V}_{z_2}\mathbf{V}_{z_3}$	$\cdots$

Figure 7.10 A differential space-time modulation scheme

A differential transmission scheme for unitary space-time modulation is shown in Fig. 7.10. Let us denote by t the index of the modulation block where the block length is n_T. At t-th modulation block, a group of ηn_T bits $\mathbf{c}_t$ is mapped into the modulation signal set and chooses a unitary matrix $\mathbf{V}_{z_t}$, where $z_t \in \{0, 1, 2, \cdots, M-1\}$. We denote by $\mathbf{X}_t$ the transmission matrix at block t. The initial transmission matrix is determined by an identity matrix $\mathbf{X}_0 = \mathbf{I}_{n_T}$, which is sent across n_T transmit antennas over n_T symbol periods. The initial transmission matrix does not carry any data information and it is regarded as a reference. After the initial transmission, data are sent in a differential manner. The differential operation can be represented as

$$\mathbf{X}_t = \mathbf{X}_{t-1} \cdot \mathbf{V}_{z_t} \tag{7.75}$$

We assume that the fading coefficients are constant across every two transmission blocks. In a receiver with n_R receive antennas, the received signals for the t-th transmission block are represented by an $n_R \times n_T$ matrix, denoted by $\mathbf{R}_t$ and given by

$$\mathbf{R}_t = \mathbf{H}\mathbf{X}_t + \mathbf{N}_t \tag{7.76}$$

where $\mathbf{H}$ is an $n_R \times n_T$ channel matrix and $\mathbf{N}_t$ is an $n_R \times n_T$ noise matrix with independent complex noise sample entries. The noise samples have a zero mean and variance σ^2. Substituting (7.75) into (7.76), we get [11]

$$\mathbf{R}_t = \mathbf{H}\mathbf{X}_t + \mathbf{N}_t$$
$$= \mathbf{H}\mathbf{X}_{t-1}\mathbf{V}_{z_t} + \mathbf{N}_t$$
$$= \mathbf{R}_{t-1}\mathbf{V}_{z_t} + \mathbf{N}'_t \tag{7.77}$$

where

$$\mathbf{N}'_t = \mathbf{N}_t - \mathbf{N}_{t-1}\mathbf{V}_{z_t} \tag{7.78}$$

As $\mathbf{V}_{z_t}$ is a unitary matrix, $\mathbf{N}'_t$ is an $n_R \times n_T$ noise matrix with independent complex noise sample entries. The variance of the noise samples in $\mathbf{N}'_t$ is $2\sigma^2$.

Since the channel fading coefficients matrix $\mathbf{H}$ does not appear in (7.77), we can perform space-time differential demodulation based on the current and previous received signal matrices without channel state information. The maximum-likelihood differential demodulation rule is given by [11]

$$\hat{z}_t = \arg\max_{l \in Q} \|\mathbf{R}_{t-1} + \mathbf{R}_t \mathbf{V}_l^H\| \tag{7.79}$$

where $Q = \{0, 1, 2, \ldots, M-1\}$, and $\|\cdot\|$ denotes the Frobenius norm of a matrix, which is the sum of the norms of all the matrix elements.

Because the noise variances in (7.77) increase by a factor of 2 relative to the noise variance in (7.76), the differential scheme is 3 dB worse compared to the case when the channel is known at the receiver [11].

7.7 Unitary Group Codes

Hughes proposed unitary space-time codes with a group structure [9]. Let $\mathcal{G}$ be a group of $L \times L$ unitary matrices, where $L \geq n_T$.

$$\mathbf{G}^H \cdot \mathbf{G} = \mathbf{G} \cdot \mathbf{G}^H = \mathbf{I}_L, \quad \text{for } \mathbf{G} \in \mathcal{G} \tag{7.80}$$

Let us consider a system with n_T transmit antennas and modulation constellation $\mathcal{A}$. We assume that there is an $n_T \times L$ matrix $\mathbf{D}$, such that for any unitary matrix $\mathbf{G}$ in the group, $\mathbf{DG}$ generates an $n_T \times L$ matrix, whose entries are the elements of the signal constellation set $\mathcal{A}$. That is

$$\mathbf{DG} \in \mathcal{A}^{n_T \times L}, \quad \text{for all } \mathbf{G} \in \mathcal{G} \tag{7.81}$$

The set of matrices

$$\mathbf{D}\mathcal{G} = \{\mathbf{DG} | \mathbf{G} \in \mathcal{G}\} \tag{7.82}$$

forms a space-time group code. Each matrix in the set specifies a space-time transmission codeword of length L for n_T transmission antennas over the signal constellation $\mathcal{A}$. Assume that the number of codewords is denoted by $|\mathcal{G}|$. The spectral efficiency of the code is given by

$$\eta = \frac{1}{L} \log_2 |\mathcal{G}| \tag{7.83}$$

If matrix $\mathbf{D}$ satisfies

$$\mathbf{D} \cdot \mathbf{D}^H = L\mathbf{I}_{n_T} \tag{7.84}$$

We have

$$(\mathbf{DG}) \cdot (\mathbf{DG})^H = L\mathbf{I}_{n_T} \tag{7.85}$$

In this case, the group code in (7.82) is a unitary space-time group code.

Example 2.1

For $n_T = L = 2$, let $\mathbf{D} = \begin{bmatrix} 1 & -1 \\ 1 & 1 \end{bmatrix}$.

$$\mathcal{G} = \left\{ \begin{bmatrix} 1 & 0 \\ 0 & 1 \end{bmatrix}, \begin{bmatrix} 0 & 1 \\ -1 & 0 \end{bmatrix} \right\} \tag{7.86}$$

$\mathbf{D}\mathcal{G}$ is a group code over the BPSK modulation constellation $\mathcal{A} = \{1, -1\}$.

Example 2.2

For $n_T = L = 2$, let $\mathbf{D} = \begin{bmatrix} 1 & -1 \\ 1 & 1 \end{bmatrix}$.

$$\mathcal{G} = \left\{ \begin{bmatrix} 1 & 0 \\ 0 & 1 \end{bmatrix}, \begin{bmatrix} 0 & 1 \\ -1 & 0 \end{bmatrix}, \begin{bmatrix} j & 0 \\ 0 & -j \end{bmatrix}, \begin{bmatrix} 0 & j \\ j & 0 \end{bmatrix} \right\} \tag{7.87}$$

Then, $\mathbf{D}\mathcal{G}$ forms a group code over the QPSK modulation constellation $\mathcal{A} = \{1, j, -1, -j\}$.

In the above examples, it is assumed that $L = n_T$. In general, the space-time codeword length L can be greater than or equal to n_T.

The differential encoding/decoding principles for unitary space-time modulation schemes discussed in the previous section can be applied to the space-time unitary group codes. The differential transmission scheme for a space-time unitary group code is illustrated in Fig. 7.11.

At the t-th encoding block, $\log_2 |\mathcal{G}|$ bits are mapped into the group code $\mathcal{G}$ and they select a unitary matrix $\mathbf{G}_{z_t}$, where $z_t \in \{0, 1, 2, \ldots, |\mathcal{G}| - 1\}$. To initialize the differential transmission, $\mathbf{X}_0 = \mathbf{D}$ is sent from n_T transmit antennas over L symbol periods. The differential encoding rule is given by [9]

$$\mathbf{X}_t = \mathbf{X}_{t-1} \cdot \mathbf{G}_{z_t} \tag{7.88}$$

The group structure ensures that $\mathbf{X}_t \in \mathcal{A}^{n_T \times L}$ if $\mathbf{X}_{t-1} \in \mathcal{A}^{n_T \times L}$.

The received signals for the t-th transmission block are represented by an $n_R \times L$ matrix $\mathbf{R}_t$. The differential space-time decoding based on the current and previous received signal

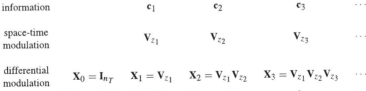

information	$\mathbf{c}_1$	$\mathbf{c}_2$	$\mathbf{c}_3$	$\cdots$
space-time modulation	$\mathbf{V}_{z_1}$	$\mathbf{V}_{z_2}$	$\mathbf{V}_{z_3}$	$\cdots$
differential modulation	$\mathbf{X}_0 = \mathbf{I}_{n_T}$ $\quad \mathbf{X}_1 = \mathbf{V}_{z_1}$	$\mathbf{X}_2 = \mathbf{V}_{z_1}\mathbf{V}_{z_2}$	$\mathbf{X}_3 = \mathbf{V}_{z_1}\mathbf{V}_{z_2}\mathbf{V}_{z_3}$	$\cdots$

Figure 7.11   A differential space-time group code

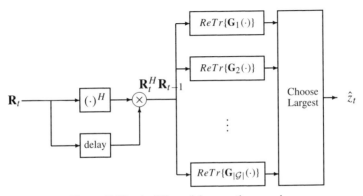

Figure 7.12   A differential space-time receiver

matrices is given by [9]

$$\hat{z}_t = \arg\max_{l \in Q} ReTr\{\mathbf{R}_{t-1}\mathbf{G}_l\mathbf{R}_t^H\}$$

$$= \arg\max_{l \in Q} ReTr\{\mathbf{G}_l\mathbf{R}_t^H\mathbf{R}_{t-1}\} \qquad (7.89)$$

where $ReTr$ denotes the real part of the trace. The receiver with maximum-likelihood differential decoding for a space-time unitary group code is shown in Fig. 7.12 [9].

Bibliography

[1] S. M. Alamouti, "A simple transmit diversity technique for wireless communications", *IEEE Journal Select. Areas Commun.*, vol. 16, no. 8, pp. 1451–1458, Oct. 1998.

[2] A. Wittneben, "A new bandwidth efficient transmit antenna modulation diversity scheme for linear digital modulation", in *Proc. IEEE ICC93*, pp. 1630–1634, 1993.

[3] V. Tarokh, H. Jafarkhani and A. R. Calderbank, "Space-time block codes from orthogonal designs", *IEEE Trans. Inform. Theory*, vol. 45, no. 5, pp. 1456–1467, July 1999.

[4] V. Tarokh, H. Jafarkhani and A. R. Calderbank, "Space-time block coding for wireless communications: performance results", *IEEE J. Select. Areas Commun.*, vol. 17, no. 3, pp. 451–460, Mar. 1999.

[5] V. Tarokh, A. Naguib, N. Seshadri, and A. R. Calderbank, "Combined array processing and space-time coding", *IEEE Trans. Inform. Theory*, vol. 45, no. 4, pp. 1121–1128, May 1999.

[6] V. Tarokh, A. Naguib, N. Seshadri and A. R. Calderbank, "Space-time codes for high data rate wireless communication: Performance criteria in the presence of channel estimation errors, mobility, and multiple paths", *IEEE Trans. Commun.*, vol. 47, no. 2, pp. 199–207, Feb. 1999.

[7] V. Tarokh and H. Jafarkhani, "A differential detection scheme for transmit diversity", *IEEE J. Select. Areas Commun.*, vol. 18, pp. 1169–1174, July 2000.

[8] H. Jafarkhani and V. Tarokh, "Multiple transmit antenna differential detection from generalized orthogonal designs", *IEEE Trans. Inform. Theory*, vol. 47, no. 6, pp. 2626–2631, Sep. 2001.

[9] B. L. Hughes, "Differential space-time modulation", *IEEE Trans. Inform. Theory*, vol. 46, no. 7, pp. 2567–2578, Nov. 2000.

[10] B. M. Hochwald and T. L. Marzetta, "Unitary space-time modulation for multiple-antenna communications in Rayleigh flat fading", *IEEE Trans. Inform. Theory*, vol. 46, no. 2, pp. 543–564, Mar. 2000.

[11] B. M. Hochwald and W. Sweldens, "Differential unitary space-time modulation", *IEEE Trans. Communi.*, vol. 48, no. 12, Dec. 2000.

[12] B. Hochwald, T. L. Marzetta and C. B. Papadias, "A transmitter diversity scheme for wideband CDMA systems based on space-time spreading", *IEEE Journal on Selected Areas in Commun.*, vol. 19, no. 1, Jan. 2001, pp. 48–60.

[13] J. Yuan and X. Shao, "New differential space-time coding schemes with two, three and four transmit antennas", in *Proc. ICCS 2002*, Singapore, Nov. 25–28, 2002.

[14] T. S. Rappaport, *Wireless Communications: Principles and Practice*, Prentice Hall, 1996.

8

Space-Time Coding for Wideband Systems

8.1 Introduction

In the previous chapters, the design and performance analysis of various space-time coding schemes have been discussed for narrow band wireless systems, which are characterized by frequency-nonselective flat fading channels. Recently, there has been an increasing interest in providing high data rate services such as video conference, multimedia, and mobile computing over wideband wireless channels. In wideband wireless communications, the symbol period becomes smaller relative to the channel delay spread, and consequently, the transmitted signals experience frequency-selective fading. Space-time coding techniques could be used to achieve very high data rates in wideband systems. Therefore, it is desirable to investigate the effect of frequency-selective fading on space-time code performance.

In this chapter, we present the performance of space-time codes on wideband wireless channels with frequency-selective fading. Various space-time coding schemes are investigated in wideband OFDM and CDMA systems.

8.2 Performance of Space-Time Coding on Frequency-Selective Fading Channels

8.2.1 Frequency-Selective Fading Channels

Frequency-selective fading channels can be modeled by a tapped-delay line. For a multipath fading channel with L_p different paths, the time-variant impulse response at time t to an impulse applied at time $t - \tau$ is expressed as [1]

$$h(t; \tau) = \sum_{\ell=1}^{L_p} h^{t,\ell} \delta(\tau - \tau_\ell) \tag{8.1}$$

Space-Time Coding Branka Vucetic and Jinhong Yuan
© 2003 John Wiley & Sons, Ltd ISBN: 0-470-84757-3

where τ_ℓ represents the time delay of the ℓ-th path and $h^{t,\ell}$ represents the complex amplitude of the ℓ-th path.

Without loss of generality, we assume that $h(t; \tau)$ is wide-sense stationary, which means that the mean value of the channel random process is independent of time and the autocorrelation of the random process depends only on the time difference [1]. Then, $h^{t,\ell}$ can be modeled by narrowband complex Gaussian processes, which are independent for different paths. The autocorrelation function of $h(t; \tau)$ is given by [1]

$$\phi_h(\Delta t; \tau_i, \tau_j) = \tfrac{1}{2} E[h^*(t, \tau_i) h(t + \Delta t, \tau_j)] \tag{8.2}$$

where Δt denotes the observation time difference. If we let $\Delta t = 0$, the resulting autocorrelation function, denoted by $\phi_h(\tau_i, \tau_j)$, is a function of the time delays τ_i and τ_j. Due to the fact that scattering at two different paths is uncorrelated in most radio transmissions, we have

$$\phi_h(\tau_i, \tau_j) = \phi_h(\tau_i)\delta(\tau_i - \tau_j) \tag{8.3}$$

where $\phi_h(\tau_i)$ represents the average channel output power as a function of the time delay τ_i. We can further assume that the L_p different paths have the same normalized autocorrelation function, but different average powers. Let us denote the average power for the ℓ-th path by $P(\tau_\ell)$. Then we have

$$P(\tau_\ell) = \phi_h(\tau_\ell) = \tfrac{1}{2} E[h^*(t, \tau_\ell) h(t, \tau_\ell)] \tag{8.4}$$

Here, $P(\tau_\ell)$, $\ell = 1, 2, \ldots, L_p$, represent the *power delay profile* of the channel.

The root mean square (rms) *delay spread* of the channel is defined as [2]

$$\tau_d = \sqrt{\frac{\sum\limits_{\ell=1}^{L_p} P(\tau_\ell)\tau_\ell^2}{\sum\limits_{\ell=1}^{L_p} P(\tau_\ell)} - \left[\frac{\sum\limits_{\ell=1}^{L_p} P(\tau_\ell)\tau_\ell}{\sum\limits_{\ell=1}^{L_p} P(\tau_\ell)}\right]^2} \tag{8.5}$$

In wireless communication environments, the channel power delay profile can be Gaussian, exponential or two-ray equal-gain [8]. For example, the two-ray equal-gain profile can be represented by

$$P(\tau) = \tfrac{1}{2}\delta\tau + \delta(\tau - 2\tau_d) \tag{8.6}$$

where $2\tau_d$ is the delay difference between the two paths and τ_d is the rms delay spread. We can further denote the delay spread normalized by the symbol duration T_s by $\overline{\tau}_d = \frac{\tau_d}{T_s}$.

8.2.2 Performance Analysis

In this section, we consider the performance analysis of space-time coding in multipath and frequency-selective fading channels. In the analysis, we assume that the delay spread τ_d is relatively small compared with the symbol duration. In order to investigate the effect of frequency-selective fading on the code performance, we assume that no equalization is used at the receiver.

Consider a system with n_T transmit and n_R receiver antennas. Let $h_{j,i}(t, \tau)$ denote the channel impulse response between the i-th transmit antenna and j-th receive antenna. At time t, the received signal at antenna j after matched filtering is given by [8]

$$r_t^j = \frac{1}{T_s} \int_{tT_s}^{(t+1)T_s} \left[\sum_{i=1}^{n_T} \int_0^\infty u^i(t' - \tau_i) h_{j,i}(t', \tau_i) d\tau_i \right] dt' + n_t^j \tag{8.7}$$

where T_s is the symbol period, n_t^j is an independent sample of a zero-mean complex Gaussian random process with the single-sided power spectrum density N_0 and $u^i(t)$ represents the transmitted signal from antenna i, given by

$$u^i(t) = \sum_{k=-\infty}^{\infty} x_k^i g(t - kT_s) \tag{8.8}$$

where x_k^i is the message for the i-th antenna at the k-th symbol period and $g(t)$ is the pulse shaping function. The received signal can be decomposed into the following three terms [7][8]

$$r_t^j = \alpha \sum_{i=1}^{n_T} \sum_{\ell=1}^{L_p} h_{j,i}^{t,\ell} x_t^i + I_t^j + n_t^j \tag{8.9}$$

where I_t^j is a term representing the intersymbol interference (ISI), and α is a constant dependent on the channel power delay profile, which can be computed as

$$\alpha = \frac{1}{T_s} \int_{-T_s}^{T_s} P(\tau)(T_s - |\tau|) d\tau \tag{8.10}$$

For different power delay profiles, the values of α are given by [8]

$$\alpha = \begin{cases} 1 - \bar{\tau}_d & \text{Exponential or two-ray equal-gain profile} \\ 1 - \sqrt{2/\pi}\,\bar{\tau}_d & \text{Gaussian profile} \end{cases} \tag{8.11}$$

The mean value of the ISI term I_t^j is zero and the variance is given by [8]

$$\sigma_I^2 = \begin{cases} 3n_T \bar{\tau}_d E_s & \text{Exponential or two-ray equal-gain profile} \\ 2n_T(1 - 1/\pi)\bar{\tau}_d^2 E_s & \text{Gaussian profile} \end{cases} \tag{8.12}$$

where E_s is the energy per symbol. For simplicity, the ISI term is approximated by a Gaussian random variable with a zero-mean and single-sided power spectral density $N_I = \sigma_I^2 T_s$. Let us denote the sum of the additive noise and the ISI by $\bar{n}_t^j$.

$$\bar{n}_t^j = I_t^j + n_t^j \tag{8.13}$$

The received signal can be rewritten as

$$r_t^j = \alpha \sum_{i=1}^{n_T} \sum_{\ell=1}^{L_p} h_{j,i}^{t,\ell} x_t^i + \bar{n}_t^j \tag{8.14}$$

where $\bar{n}_t^j$ is a complex Gaussian random variable with a zero mean and the single-sided power spectral density $N_I + N_0$. Note that the additive noise and the ISI are uncorrelated with the signal term. The pairwise error probability under this approximation is given by [8]

$$P(\mathbf{X}, \hat{\mathbf{X}}) \leq \left[\prod_{i=1}^{n_T} \left(1 + \lambda_i \frac{\alpha^2 E_s}{4(N_0 + N_I)} \right) \right]^{-n_R}$$

$$\leq \left[\prod_{i=1}^{r} \left(\lambda_i \frac{\alpha^2}{N_I/N_0 + 1} i \right) \right]^{-n_R} \left(\frac{E_s}{4N_0} \right)^{-rn_R} \tag{8.15}$$

where r is the rank of the codeword distance matrix, and $\lambda_i, i = 1, 2, \ldots, r$, are the nonzero eigenvalues of the matrix. From the above upper bound, we can observe that the diversity gain achieved by the space-time code on multipath and frequency-selective fading channels is rn_R, which is the same as that on frequency-nonselective fading channels. The coding gain is

$$G_{\text{coding}} = \frac{\left(\prod_{i=1}^{r} \lambda_i \right)^{1/r} \dfrac{\alpha^2}{N_I/N_0 + 1}}{d_u^2} \tag{8.16}$$

The coding gain is reduced by a factor of $\left(\frac{\alpha^2}{N_I/N_0+1} \right)$ compared to the one on frequency flat fading channels. Furthermore, it is reported that at high SNRs, there exists an irreducible error rate floor [7] [8].

Note that the above performance analysis is performed under the assumptions that the time delay spread is small and no equalizer is used at the receiver. When the delay spread becomes relatively high, the coding gain will decrease considerably due to ISI, and cause a high performance degradation. In order to improve the code performance over frequency-selective fading channels, additional processing is required to remove or prevent ISI.

It is shown in [4] that a space-time code on frequency-selective fading channels can achieve at least the same diversity gain as that on frequency-nonselective fading channels provided that maximum likelihood decoding is performed at the receiver. In other words, an optimal space-time code on frequency-selective fading channels may achieve a higher diversity gain than on frequency-nonselective fading channels. As the maximum likelihood decoding on frequency-selective channels is prohibitively complex, a reasonable solution to improve the performance of space-time codes on frequency-selective fading channels is to mitigate ISI. By mitigating ISI, one can convert frequency-selective channels into frequency-nonselective channels. Then, good space-time codes for frequency-nonselective fading channels can be applied [9].

A conventional approach to mitigate ISI is to use an adaptive equalizer at the receiver. An optimum space-time equalizer can suppress ISI, and therefore, the frequency-selective fading channels become intersymbol interference free. The main drawback of this approach is a high receiver complexity because a multiple-input/multiple-output equalizer (MIMO-EQ) has to be used at the receiver [17] [18] [19].

An alternative approach is to use orthogonal frequency division multiplexing (OFDM) techniques [5] [6]. In OFDM, the entire channel is divided into many narrow parallel sub-channels, thereby increasing the symbol duration and reducing or eliminating the ISI caused by the multipath environments [15]. Since MIMO-EQ is not required in OFDM systems, this approach is less complex.

An OFDM technique transforms a frequency-selective fading channel into parallel correlated frequency-nonselective fading channels. OFDM has been chosen as a standard for various wireless communication systems, including European digital audio broadcasting (DAB) and digital video broadcasting (DVB), IEEE broadband wireless local area networks (WLAN) IEEE802.11 and European HIPERLAN [26] [27]. In OFDM systems, there is a high error probability for those sub-channels in deep fades and therefore, error control coding is combined with OFDM to mitigate the deep fading effects. For a MIMO frequency-selective fading channel, the combination of space-time coding with wideband OFDM has the potential to exploit multipath fading and to achieve very high data rate robust transmissions [5][10][11][14][15][16]. In the next section, we will discuss space-time coding in wideband OFDM systems, which is called STC-OFDM.

8.3   STC in Wideband OFDM Systems

8.3.1   OFDM Technique

In a conventional serial data system such as microwave digital radio data transmission and telephone lines, in which the symbols are transmitted sequentially, adaptive equalization techniques have been introduced to combat ISI. However, the system complexity precludes the equalization implementation if the data rate is as high as a few megabits per second.

A parallel data system can alleviate ISI even without equalization. In such a system the high-rate data stream is demultiplexed into a large number of sub-channels with the spectrum of an individual data element occupying only a small part of the total available bandwidth. A parallel system employing conventional frequency division multiplexing (FDM) without sub-channel overlapping is bandwidth inefficient. A much more efficient use of bandwidth can be obtained with an OFDM system in which the spectra of the individual sub-channels are permitted to overlap and the carriers are orthogonal. A basic OFDM system is shown in Fig. 8.1 [32].

Let us assume that the serial data symbols after the encoder have a duration of $T_s = \frac{1}{f_s}$ seconds each, where f_s is the input symbol rate. Each OFDM frame consists of K coded symbols, denoted by $d[0], d[1], \ldots, d[K-1]$, where $d[n] = a[n] + jb[n]$ and $a[n]$ and $b[n]$ denote the real and imaginary parts of the sampling values at discrete time n, respectively. After the serial-to-parallel converter, the K parallel data modulate K sub-carrier frequencies, $f_0, f_1, \ldots, f_{K-1}$, which are then frequency division multiplexed. The sub-carrier frequencies are separated by multiples of $\Delta f = \frac{1}{KT_s}$, making any two carrier frequencies orthogonal. Because the carriers are orthogonal, data can be detected on each of these closely spaced carriers without interference from the other carriers. In addition, after the serial-to-parallel converter, the signaling interval is increased from T_s to KT_s, which makes the system less susceptible to delay spread impairments.

The OFDM transmitted signal $D(t)$ can be expressed as

$$D(t) = \sum_{n=0}^{K-1} \{a[n]\cos(\omega_n t) - bn\sin(\omega_n t)\} \tag{8.17}$$

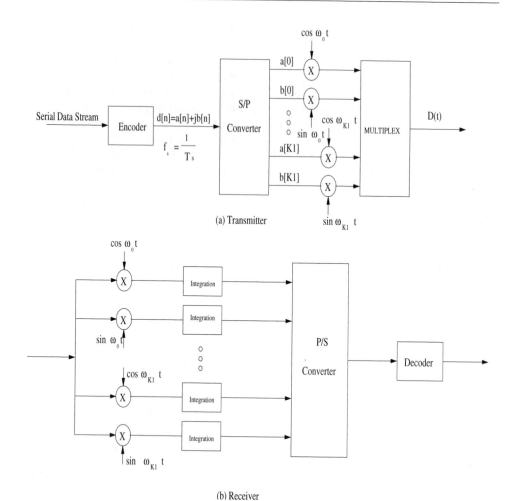

(a) Transmitter

(b) Receiver

Figure 8.1   A basic OFDM system

where

$$\omega_n = 2\pi f_n$$
$$f_n = f_0 + n\Delta f \tag{8.18}$$

Substituting (8.18) into (8.17), the transmitted signal can be rewritten as

$$D(t) = Re\left\{e\sum_{n=0}^{K-1}\{d[n]e^{j\omega_n t}\}\right\}$$

$$= Re\left\{\sum_{n=0}^{K-1}\{d[n]e^{j2\pi n\Delta f t}e^{j2\pi f_0 t}\}\right\}$$

$$= Re\{\tilde{D}(t)e^{j2\pi f_0 t}\} \tag{8.19}$$

where

$$\tilde{D}(t) = \sum_{n=0}^{K-1} \{d[n]e^{j2\pi n\Delta f t}\} \tag{8.20}$$

represents the complex envelope of the transmitted signal $D(t)$.

At the receiver, correlation demodulators (or matched filters) are employed to recover the symbol for each sub-channel. However, the complexity of the equipment, such as filters and modulators, makes the direct implementation of the OFDM system in Fig. 8.1 impractical, when N is large.

Now consider that the complex envelope signal $\tilde{D}(t)$ in (8.19) is sampled at a sampling rate of f_s. Let $t = mT_s$, where m is the sampling instant. The samples of $\tilde{D}(t)$ in an OFDM frame, $\tilde{D}[0], \tilde{D}[1], \ldots, \tilde{D}[K-1]$, are given by

$$\tilde{D}[m] = \sum_{n=0}^{K-1} \{d[n]e^{j2\pi n\Delta f m T_s}\}$$

$$= \sum_{n=0}^{K-1} d[n]e^{j(2\pi/K)\,\mathrm{nm}}$$

$$= \mathrm{IDFT}\{d[n]\}, \tag{8.21}$$

Equation (8.21) indicates that the OFDM modulated signal is effectively the inverse discrete Fourier transform (IDFT) of the original data stream and, similarly, we may prove that a bank of coherent demodulators in Fig. 8.1 is equivalent to a discrete Fourier transform (DFT). This makes the implementation of OFDM system completely digital and the equipment complexity is decreased to a large extent [30]. If the number of sub-channels K is large, fast Fourier transform (FFT) can be employed to bring in further reductions in complexity [31]. An OFDM system employing FFT algorithm is shown in Fig. 8.2. Note that FFT and IFFT can be exchanged between the transmitter and receiver, depending on the initial phase of the carriers.

8.3.2 STC-OFDM Systems

We consider a baseband STC-OFDM communication system with K OFDM sub-carriers, n_T transmit and n_R receive antennas. The total available bandwidth of the system is W Hz. It is divided into K overlapping sub-bands. The system block diagram is shown in Fig. 8.3.

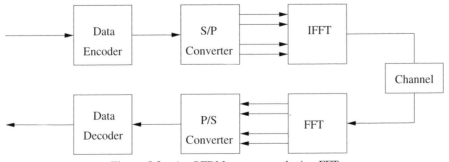

Figure 8.2 An OFDM system employing FFT

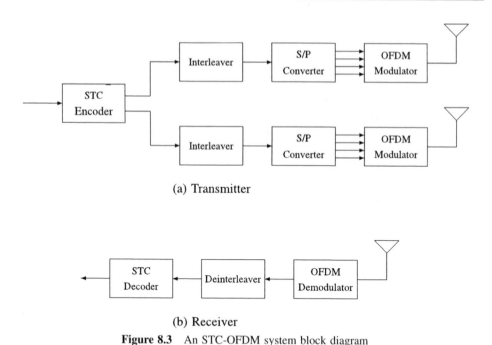

(a) Transmitter

(b) Receiver

Figure 8.3   An STC-OFDM system block diagram

At each time t, a block of information bits is encoded to generate a space-time codeword which consists of $n_T L$ modulated symbols. The space-time codeword is given by

$$
\mathbf{X}_t = \begin{bmatrix}
x_{t,1}^1 & x_{t,2}^1 & \cdots & x_{t,L}^1 \\
x_{t,1}^2 & x_{t,2}^2 & \cdots & x_{t,L}^2 \\
\vdots & \vdots & \ddots & \vdots \\
x_{t,1}^{n_T} & x_{t,2}^{n_T} & \cdots & x_{t,L}^{n_T}
\end{bmatrix}
\tag{8.22}
$$

where the i-th row $\mathbf{x}_t^i = x_{t,1}^i, x_{t,2}^i, \ldots, x_{t,L}^i$, $i = 1, 2, \ldots, n_T$, is the data sequence for the i-th transmit antenna. For the sake of simplicity, we assume that the codeword length is equal to the number of OFDM sub-carriers, $L = K$. Signals $x_{t,1}^i, x_{t,2}^i, \ldots, x_{t,L}^i$ are OFDM modulated on K different OFDM sub-carriers and transmitted from the i-th antenna simultaneously during one OFDM frame, where $x_{t,k}^i$ is sent on the k-th OFDM sub-carrier.

In OFDM systems, in order to avoid ISI due to the delay spread of the channel, a cyclic prefix is appended to each OFDM frame during the guard time interval. The cyclic prefix is a copy of the last L_p samples of the OFDM frame, so that the overall OFDM frame length is $L + L_p$, where L_p is the number of multipaths in fading channels.

In the performance analysis, we assume ideal frame and symbol synchronization between the transmitter and the receiver. A sub-channel is modeled by quasi-static Rayleigh fading. The fading process remains constant during each OFDM frame. It is also assumed that channels between different antennas are uncorrelated.

At the receiver, after matched filtering, the signal from each receive antenna is sampled at a rate of W Hz and the cyclic prefix is discarded from each frame. Then these samples are applied to an OFDM demodulator. The output of the OFDM demodulator for the k-th

OFDM sub-carrier, $k = 1, 2, \ldots, K$, at receive antenna j, $j = 1, 2, \ldots, n_R$, is given by [5]

$$R_{t,k}^j = \sum_{i=1}^{n_T} H_{j,i}^{t,k} x_{t,k}^i + N_{t,k}^j \tag{8.23}$$

where $H_{j,i}^{t,k}$ is the channel frequency response for the path from the i-th transmit antenna to the j-th receive antenna on the k-th OFDM sub-channel, and $N_{t,k}^j$ is the OFDM demodulation output for the noise sample at the j-th receive antenna and the k-th sub-channel with power spectral density N_0. Assuming that perfect channel state information is available at the receiver, the maximum likelihood decoding rule is given by

$$\hat{\mathbf{X}}_t = \arg \min_{\hat{\mathbf{X}}} \sum_{j=1}^{n_R} \sum_{k=1}^{K} \left| R_{t,k}^j - \sum_{i=1}^{n_T} H_{j,i}^{t,k} x_{t,k}^i \right|^2 \tag{8.24}$$

where the minimization is performed over all possible space-time codewords.

Recall that the channel impulse response in the time domain is modeled as a tapped-delay line. The channel impulse response between the i-th transmit antenna to the j-th receive antenna is given by

$$h_{j,i}(t; \tau) = \sum_{\ell=1}^{L_p} h_{j,i}^{t,\ell} \delta(\tau - \tau_\ell) \tag{8.25}$$

where L_p is the number of multipaths, τ_ℓ is the time delay of the ℓ-th path and $h_{j,i}^{t,\ell}$ is the complex amplitude of the ℓ-th path. Let us denote by T_f the time duration of each OFDM frame and by Δf the separation between the OFDM sub-carriers. We have

$$T_f = KT_s$$

$$T_s = \frac{1}{W} = \frac{1}{K\Delta f} \tag{8.26}$$

Now the delay of the ℓ-th path can be represented as

$$\tau_\ell = n_\ell T_s = \frac{n_\ell}{K\Delta f} \tag{8.27}$$

where n_ℓ is an integer. Performing the Fourier transform of the channel impulse response, we can get the channel frequency response at time t as

$$H_{j,i}^{t,k} \triangleq H_{j,i}(tT_f, k\Delta f)$$

$$= \int_{-\infty}^{+\infty} h_{j,i}(tT_f, \tau) e^{-j2\pi k\Delta f \tau} d\tau$$

$$= \sum_{\ell=1}^{L_p} h_{j,i}(tT_f, n_\ell T_s) e^{-j2\pi k n_\ell / K}$$

$$= \sum_{\ell=1}^{L_p} h_{j,i}(t, n_\ell) e^{-j2\pi k n_\ell / K} \tag{8.28}$$

Let

$$h^t_{j,i} = [h^{t,1}_{j,i}, h^{t,2}_{j,i}, \ldots, h^{t,L_p}_{j,i}]^H$$

$$w_k = [e^{-j2\pi k n_1/K}, e^{-j2\pi k n_2/K}, \ldots, e^{-j2\pi k n_{L_p}/K}]^T \qquad (8.29)$$

The equation (8.28) can be rewritten as

$$H^{t,k}_{j,i} = (h^t_{j,i})^H \cdot w_k \qquad (8.30)$$

From (8.28), we can see that the channel frequency response $H^{t,k}_{j,i}$ is the digital Fourier transform of the channel impulse response $h^t_{j,i}$. The transform is specified by the vector w_k for the k-th OFDM sub-carrier, $k = 1, 2, \ldots, K$.

8.4   Capacity of STC-OFDM Systems

In this section, we consider the capacity of an OFDM-based MIMO channels. We assume that the fading is quasi-static and the channel is unknown at the transmitter but perfectly known at the receiver. Since the channel is described by a non-ergodic random process, we define the instantaneous channel capacity as the mutual information conditioned on the channel responses [10]. The instantaneous channel capacity is a random variable. For each realization of the random channel frequency response $H^{t,k}_{j,i}$, the instantaneous channel capacity of an OFDM based MIMO system is given by [13]

$$C = \frac{1}{K} \sum_{k=1}^{K} \log_2 [\det(\mathbf{I}_{n_R} + \text{SNR} \mathbf{H}^k \cdot (\mathbf{H}^k)^H)] \qquad (8.31)$$

where $\mathbf{I}_{n_R}$ is the identity matrix of size n_R, $\mathbf{H}^k$ is an $n_R \times n_T$ channel matrix with its (j, i)-th entry $H^{t,k}_{j,i}$, and SNR is the signal-to-noise ratio per receive antenna. The instantaneous channel capacity in (8.31) can be estimated by simulation. If the channel is ergodic, the channel capacity can be calculated as the average of the instantaneous capacity over the random channel values. For quasi-static fading channels, the random process of the channel is non-ergodic. In this case, we calculate the outage capacity, from the instantaneous channel capacity in (8.31).

Now we consider the following three different OFDM system settings.

- OFDM-1: The total available bandwidth is 1 MHz and 256 sub-carriers are used. The corresponding sub-channel separation is 3.9 KHz and OFDM frame duration is 256 μ s. For each frame, a guard interval of 40 μ s is added to mitigate the effect of ISI.
- OFDM-2: The total available bandwidth is 20 MHz with 64 sub-carriers. This corresponds to the sub-channel separation of 312.5 KHz and the OFDM frame length of 3.2 μ s. For each frame, a guard period of 0.8 μ s is added and a total of 48 sub-carriers are used for data transmission. Additional 4 sub-carriers are assigned for transmission of pilot tones. Note that OFDM-2 represents the standard specifications for IEEE802.11a and HIPERLAN/2 systems.
- OFDM-3: The total available bandwidth 4.2224 MHz is divided into 528 sub-channels, each of which has the bandwidth of 8 KHz. The OFDM frame length is 125 μ s, and a guard time of 31.25 μ s is introduced for each OFDM frame.

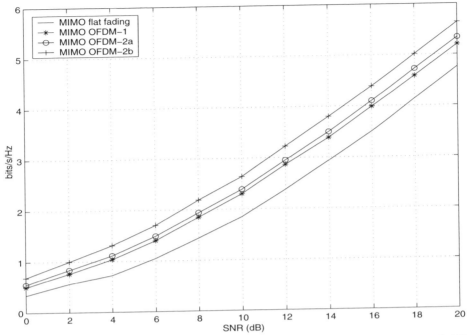

Figure 8.4 Outage capacity for MIMO channels with OFDM modulation and the outage probability of 0.1

The outage capacity for these OFDM systems over various MIMO channels with two transmit antennas and one receive antenna was evaluated. Figure 8.4 shows the outage capacity for OFDM-1 and OFDM-2 systems with the outage probability of 10% [14]. For OFDM-1, a two-ray equal gain delay profile is used. For OFDM-2, two different channel models were employed. In OFDM-2a systems, a 6-path ITU-B indoor office channel model was chosen, while a 18-path ETSI BRAN-B channel model for a large open space office environment was used in OFDM-2b systems [14].

From Fig. 8.4, we can observe that frequency-selective MIMO channels have higher capacity than frequency flat fading channels and that increasing the delay spread in MIMO systems increases the channel capacity. To achieve the channel capacity, space-time codes should be carefully designed to exploit MIMO multipath fading channel properties.

8.5 Performance Analysis of STC-OFDM Systems

Let us consider the maximum likelihood decoding of the STC-OFDM systems as shown in (8.24). Assuming that ideal CSI is available at the receiver, for a given realization of the fading channel $\mathbf{H}_t$, the pairwise error probability of transmitting $\mathbf{X}_t$ and deciding in favor of another codeword $\hat{\mathbf{X}}_t$ at the decoder conditioned on $\mathbf{H}_t$ is given by

$$P(\mathbf{X}_t, \hat{\mathbf{X}}_t | \mathbf{H}_t) \leq \exp\left(-d_H^2(\mathbf{X}_t, \hat{\mathbf{X}}_t)\frac{E_s}{4N_0}\right) \qquad (8.32)$$

where E_s is the average symbol energy, N_0 is the noise power spectral density, and $d_H^2(\mathbf{X}_t, \hat{\mathbf{X}}_t)$ is given by

$$d_H^2(\mathbf{X}_t, \hat{\mathbf{X}}_t) = \sum_{j=1}^{n_R} \sum_{k=1}^{K} \left| \sum_{i=1}^{n_T} H_{j,i}^{t,k} (x_{t,k}^i - \hat{x}_{t,k}^i) \right|^2$$

$$= \sum_{j=1}^{n_R} \sum_{k=1}^{K} \left| \sum_{i=1}^{n_T} (h_{j,i}^t)^H \cdot w_k \cdot (x_{t,k}^i - \hat{x}_{t,k}^i) \right|^2$$

$$= \sum_{j=1}^{n_R} \sum_{k=1}^{K} |\mathbf{h}_j \mathbf{W}_k \mathbf{e}_k|^2 \tag{8.33}$$

where

$$\mathbf{h}_j = (h_{j,1}^t)^H, (h_{j,2}^t)^H, \dots, (h_{j,n_T}^t)^H{}_{1 \times L_p n_T}$$

$$\mathbf{W}_k = \begin{bmatrix} w_k & 0 & \cdots & 0 \\ 0 & w_k & \cdots & 0 \\ \vdots & \vdots & \ddots & \vdots \\ 0 & 0 & \cdots & w_k \end{bmatrix}_{L_p n_T \times n_T}$$

$$\mathbf{e}_k = \begin{bmatrix} x_{t,k}^1 - \hat{x}_{t,k}^1 \\ x_{t,k}^2 - \hat{x}_{t,k}^2 \\ \vdots \\ x_{t,k}^{n_T} - \hat{x}_{t,k}^{n_T} \end{bmatrix}_{n_T \times 1} \tag{8.34}$$

Equation (8.33) can be rewritten as

$$d_H^2(\mathbf{X}_t, \hat{\mathbf{X}}_t) = \sum_{j=1}^{n_R} \sum_{k=1}^{K} \mathbf{h}_j \mathbf{W}_k \mathbf{e}_k \mathbf{e}_k^H \mathbf{W}_k^H \mathbf{h}_j^H$$

$$= \sum_{j=1}^{n_R} \mathbf{h}_j \left[\sum_{k=1}^{K} \mathbf{W}_k \mathbf{e}_k \mathbf{e}_k^H \mathbf{W}_k^H \right] \mathbf{h}_j^H$$

$$= \sum_{j=1}^{n_R} \mathbf{h}_j \mathbf{D}_H(\mathbf{X}_t, \hat{\mathbf{X}}_t) \mathbf{h}_j^H \tag{8.35}$$

where $\mathbf{D}_H(\mathbf{X}_t, \hat{\mathbf{X}}_t)$ is an $L_p n_T \times L_p n_T$ matrix given by

$$\mathbf{D}_H(\mathbf{X}_t, \hat{\mathbf{X}}_t) = \sum_{k=1}^{K} \mathbf{W}_k \mathbf{e}_k \mathbf{e}_k^H \mathbf{W}_k^H \tag{8.36}$$

It is clear that matrix $\mathbf{D}_H(\mathbf{X}_t, \hat{\mathbf{X}}_t)$ is a variable depending on the codeword difference and the channel delay profile. Let us denote the rank of $\mathbf{D}_H(\mathbf{X}_t, \hat{\mathbf{X}}_t)$ by r_h. Since $\mathbf{D}_H(\mathbf{X}_t, \hat{\mathbf{X}}_t)$

is nonnegative definite Hermitian, the eigenvalues of the matrix can be ordered as

$$\lambda_1 \geq \lambda_2 \geq \cdots \geq \lambda_{r_h} > 0 \tag{8.37}$$

Now we consider matrix $\mathbf{e}_k \mathbf{e}_k^H$ in (8.36). In the case that the symbols of codewords $\mathbf{X}_t$ and $\hat{\mathbf{X}}_t$ for the k-th sub-carrier and n_T transmit antennas are the same, $x_{t,k}^1 x_{t,k}^2 \cdots x_{t,k}^{n_T} = \hat{x}_{t,k}^1 \hat{x}_{t,k}^2 \cdots \hat{x}_{t,k}^{n_T}$, $\mathbf{e}_k \mathbf{e}_k^H$ is an all zero matrix. On the other hand, if $x_{t,k}^1 x_{t,k}^2 \cdots x_{t,k}^{n_T} \neq \hat{x}_{t,k}^1 \hat{x}_{t,k}^2 \cdots \hat{x}_{t,k}^{n_T}$, $\mathbf{e}_k \mathbf{e}_k^H$ is a rank-one matrix. Let δ_H denote the number of instances k, $k = 1, 2, \ldots, K$, such that $x_{t,k}^1 x_{t,k}^2 \cdots x_{t,k}^{n_T} \neq \hat{x}_{t,k}^1 \hat{x}_{t,k}^2 \cdots \hat{x}_{t,k}^{n_T}$. Obviously, the rank of $\mathbf{D}_H(\mathbf{X}_t, \hat{\mathbf{X}}_t)$ is determined by

$$r_h \leq \min(\delta_H, L_p n_T) \tag{8.38}$$

δ_H is called the symbol-wise Hamming distance. Using a similar analytical method as in Chapter 2, we can obtain the pairwise error probability of an STC-OFDM system over a frequency-selective fading channel by averaging (8.32) with respect to the channel coefficients $h_{t,\ell}^{i,j}$. It is upper bounded by [10]

$$P(\mathbf{X}_t, \hat{\mathbf{X}}_t) \leq \left(\frac{1}{\prod\limits_{j=1}^{r_h} \left(1 + \lambda_j \dfrac{E_s}{4N_0}\right)} \right)^{n_R}$$

$$\leq \left(\prod_{j=1}^{r_h} \lambda_j \right)^{-n_R} \left(\frac{E_s}{4N_0} \right)^{-r_h n_R} \tag{8.39}$$

Note that this performance upper-bound is similar to the upper-bound on slow Rayleigh fading channels. The STC-OFDM on frequency-selective fading channels can achieve a diversity gain of $r_h n_R$ and a coding gain of $(\prod_{j=1}^{r_h} \lambda_j)^{1/r_h}/d_u^2$. To minimize the code error probability, one need to choose a code with the maximum diversity gain and coding gain.

Consider the rank of $\mathbf{D}_H(\mathbf{X}_t, \hat{\mathbf{X}}_t)$ in (8.38). The maximum possible diversity gain for a space-time code on frequency-selective fading channels is $L_p n_T n_R$, which is the product of the transmit diversity n_T, receive diversity n_R and the time diversity L_p. To achieve this maximum possible diversity, the code symbol-wise Hamming distance δ_H must be equal to or greater than $L_p n_T$. In this case, the space-time code is able to exploit both the transmit diversity and the multipath channel delay spread. When δ_H is less than $L_p n_T$, the achieved diversity gain is $\delta_H n_R$. In this situation, the multipath channel delay spread effectively enables a slow fading channel to approach a fast fading channel. Therefore, the diversity gain is equal to the one for fast fading channels.

In communication systems, the number of multipath delays is usually unknown at the transmitter. In code design it is desirable to construct space-time codes with the largest minimum symbol-wise Hamming distance δ_H [10].

It is worth noting that since the matrix $\mathbf{D}_H(\mathbf{X}_t, \hat{\mathbf{X}}_t)$ depends on both the code structure and the channel delay profile, it is not possible to design a good code for various channels with different delay profiles. Usually using an interleaver between a space-time encoder and an OFDM modulator may help to achieve reasonable robust code performance on various channels [10].

8.6   Performance Evaluation of STC-OFDM Systems

In this section, we evaluate the performance of STC-OFDM systems by simulation. In the simulations, we choose a 16-state space-time trellis coded QPSK with two transmit antennas. The OFDM-1 modulation format is employed. The OFDM has 256 sub-carriers. During each OFDM frame, a block of 512 information bits is encoded to generate two coded QPSK sequences of length 256, each of which is interleaved and OFDM modulated on 256 sub-carriers. The two modulated sequences are transmitted from two transmit antennas simultaneously. In the trellis encoder, we require that the initial and the final states of each frame are all-zero states. This can be done by setting the last four bits of the input block to be zero. Considering the tail bits of the trellis encoder and the guard interval of the OFDM modulation, the bandwidth efficiency of the STC-OFDM system is

$$\eta = 2 \times \frac{256}{296} \times \frac{508}{512} = 1.72 \text{ bits/s/Hz} \tag{8.40}$$

8.6.1   Performance on A Single-Path Fading Channel

A single-path fading channel is conceptually equivalent to a quasi-static frequency-nonselective fading channel [5]. In Fig. 8.5, the performance of the STC-OFDM on a single-path fading channel is shown. In the simulation, one receive antenna is employed. Since $n_T = 2, n_R = 1$, and $L_p = 1$, the scheme achieves a diversity gain of $L_p n_T n_R = 2$. The figure shows that no benefit can be obtained with OFDM on a quasi-static frequency-nonselective fading channel. Also, interleavers cannot improve the code performance, since the channel is quasi-static.

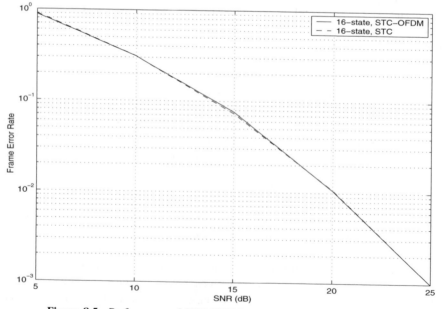

Figure 8.5   Performance of STC-OFDM on a single-path fading channel

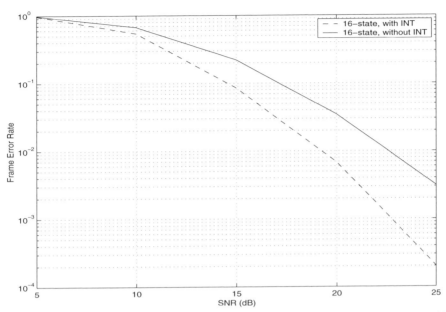

Figure 8.6 Performance of STC-OFDM on a two-path equal-gain fading channel with and without interleavers

8.6.2   The Effect of The Interleavers on Performance

Figure 8.6 shows the performance comparison for the 16-state STC-OFDM scheme on a two-path equal-gain fading channel with and without interleavers in the transmitter [10]. The delay between the two paths is 5 μ s. It is obvious that the random interleavers help to improve the code performance significantly. At the FER of 10^{-2}, the STC-OFDM with interleavers is 3.8 dB better than the scheme without interleavers.

8.6.3   The Effect of Symbol-Wise Hamming Distance on Performance

Figure 8.8 shows the performance of two STC-OFDM schemes on a two-path equal-gain fading channel [10]. The delay between the two paths is 5 μ s. The first scheme is a 16-state space-time trellis coded QPSK, whose symbol-wise Hamming distance is 3. The other scheme is a 256-state space-time trellis coded QPSK, which is modified based on the conventional optimum rate 2/3, 256-state trellis coded 8-PSK scheme on flat fading channels with single transmit antenna [12]. In this modification, the original 8-PSK mapper is split into two QPSK mappers and the original rate 2/3 8-PSK scheme for single transmit antenna is transformed into a rate 2/4 $2\times$ QPSK code for two transmit antennas as shown in Fig. 8.7 [10]. After the modification, the space-time code has the same symbol-wise Hamming distance as the original code. For the 256-state code, the symbol-wise Hamming distance is 6. Comparing the performance in Fig. 8.8, we can see that the 256-state code performs much better than the 16-state code due to a larger symbol-wise Hamming distance. At the FER of 10^{-2}, the performance gain is about 4 dB. In this system, as $n_T = 2, n_R = 1$, and $L_p = 2$, the maximum possible diversity is $L_p n_T n_R = 4$. For the 256-state code, $\delta_H = 6$, which is larger than $L_p n_T = 4$, so that the diversity gain is $L_p n_T$. It can achieve the maximum

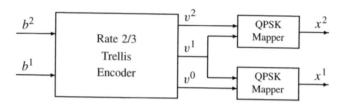

Figure 8.7 An STTC encoder structure

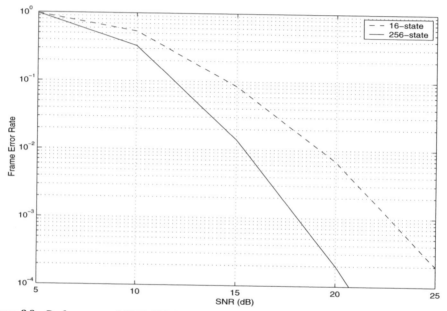

Figure 8.8 Performance of STC-OFDM with various number of states on a two-path equal-gain fading channel

diversity of 4. However, for the 16-state code, $\delta_H = 3$, which is less than $L_p n_T = 4$, so the diversity gain is $\delta_H n_R = 3$. This code cannot achieve the maximum diversity. Therefore, we can conclude that the symbol-wise Hamming distance of the code plays an important role in the STC-OFDM performance on frequency-selective fading channels.

8.6.4 The Effect of The Number of Paths on Performance

In this section, we briefly discuss the impact of the number of paths L_p on the system performance. Figure 8.9 depicts the performance of the 16-state STC-OFDM scheme on a two-path and six-path equal-gain fading channel. For the two-path channel, the delay between the two paths is 40 μ s. For the six-path channel, six paths are equally spread with the delay of 6.5 μ s between adjacent paths. The figure indicates that the code performance slightly improves when the number of paths increases.

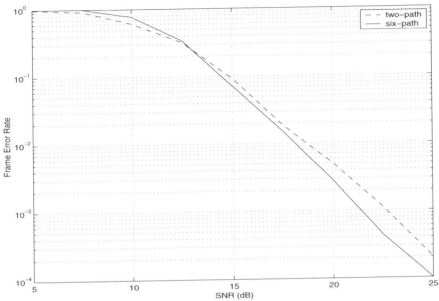

Figure 8.9   Performance of STC-OFDM on various MIMO fading channels

8.7   Performance of Concatenated Space-Time Codes Over OFDM Systems

In order to further improve the code performance, we can use concatenated space-time codes. In this section, three different concatenation schemes are considered. They are serial concatenated RS codes with space-time codes (RS-STC), serial concatenated convolutional codes with space-time codes (CONV-STC) and space-time turbo trellis codes.

8.7.1   Concatenated RS-STC over OFDM Systems

An outer (72, 64, 9) RS code over $GF(2^7)$ is serially concatenated with the 16-state space-time trellis coded QPSK scheme. In this simulation, OFDM-1 is used as the modulation format. The codeword length for the outer code is $72 \times 7 = 504$ bits. The bandwidth efficiency of the concatenated scheme is

$$\eta = 2 \times \frac{256}{296} \times \frac{504}{512} \times \frac{64}{72} = 1.514 \text{ bits/s/Hz} \tag{8.41}$$

The performance of the scheme on various two-path equal-gain fading channels is shown in Fig. 8.10 [5]. The delay is chosen to be 5 μ s and 40 μ s. We can observe that the code performance improves when the delay spread increases. In this figure, we also plot the performance of the STC scheme. It can be observed that the concatenated scheme achieves a better performance than the STC schemes. Note that these results were obtained without interleavers.

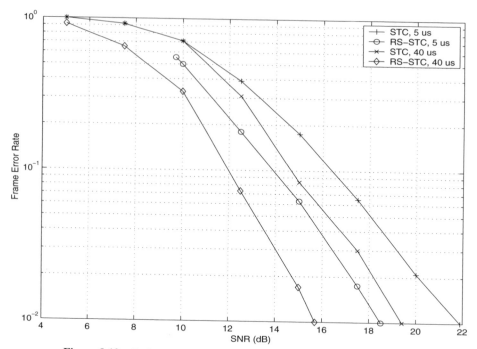

Figure 8.10   Performance of concatenated RS-STC over OFDM systems

8.7.2   Concatenated CONV-STC over OFDM Systems

Figure 8.11 depicts the performance of a serially concatenated convolutional and space-time trellis codes with OFDM-1 modulation on a two-path equal-gain fading channel. The outer code was a rate half convolutional code. Its generator polynomials in octal form are (37, 21). A 16-state QPSK space-time trellis code with two transmit antennas was chosen as an inner code. The bandwidth efficiency of the scheme is 0.851 bits/s/Hz. The figure shows that the code performance improves with the increasing delay spread for systems without interleaving. However, if an interleaver is used between the outer and the inner encoders, the relative delay between the multipaths does not affect the code performance.

8.7.3   ST Turbo TC over OFDM Systems

The serial concatenated schemes can achieve better performance relative to the STC schemes. However, there is a loss in bandwidth efficiency. This loss can be avoided with space-time turbo trellis codes. Here, we consider a turbo scheme based on parallel concatenation of two 8-state recursive space-time QPSK codes linked by a bit interleaver. Figure 8.12 illustrates the performance of the turbo scheme with OFDM-1 modulation on a two-path equal-gain fading channel. We also show the performance curves of some STC-OFDM schemes with the same bandwidth efficiency 1.72 bits/s/Hz, for comparison. It is clear that the turbo scheme outperforms all these codes. At a FER of 10^{-2}, the turbo scheme is superior to the 256, 32, 16-state STC schemes by about 2 dB, 3.2 dB and 5.0 dB, respectively. The turbo scheme performs within 2.5 dB of the 10% outage capacity.

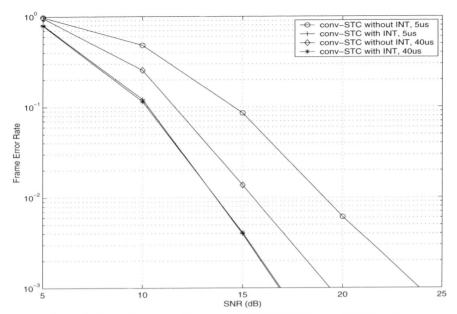

Figure 8.11   Performance of concatenated CONV-STC over OFDM systems

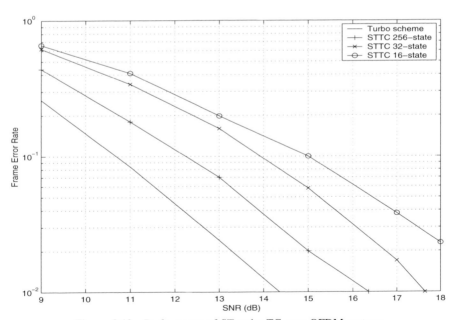

Figure 8.12   Performance of ST turbo TC over OFDM systems

8.8 Transmit Diversity Schemes in CDMA Systems

Direct-sequence code division multiple access (DS-CDMA) has been chosen in most proposals for the next generation cellular mobile communication standards. To provide high data rate services to a large number of users in wideband CDMA systems, it is essential to employ multiple antennas at the base stations. In this section, we consider various transmit diversity techniques for the down-link of wideband CDMA systems.

8.8.1 System Model

Let us first describe a baseband CDMA system without transmit diversity. Assume that the base station is communicating with K active users in the down-link. The k-th user transmits a binary sequence, denoted by $\mathbf{d}^k$, at a symbol rate $1/T_s$, where T_s is the symbol interval. The data stream for the k-th user is BPSK modulated, generating a real modulated sequence $\mathbf{b}^k$. Then the modulated sequence is spread by the user specific spreading sequence. The spreading sequence can be represented by an $N_c \times 1$ vector

$$\mathbf{s}_k = [s_{k,1}, s_{k,2}, \dots, s_{k,N_c}]^T \tag{8.42}$$

where $s_{k,j} \in \{-1/\sqrt{N_c}, +1/\sqrt{N_c}\}$ is the j-th chip of the spreading sequence for user k, N_c is the spreading gain, defined as the ratio of the symbol interval to the chip interval. If T_c is the chip interval, then $T_s = N_c T_c$. After spreading, the signal sequence is transmitted at the chip rate $1/T_c$. For simplicity, we assume that the spreading sequences of different users are orthogonal. That is $\mathbf{s}_k^T \mathbf{s}_j = \delta_{kj}$, where δ_{kj} is the Kronecker delta.

The transmitted chip signals for the symbol at time t, denoted by an $N_c \times 1$ vector $\mathbf{x}_t$, can be expressed as

$$\mathbf{x}_t = \sum_{k'=1}^{K} b_{k',t} \mathbf{s}_{k'} \tag{8.43}$$

where $b_{k,t}$ is the coded symbol for user k at time t. The received signal for the k-th user, sampled at the chip rate $1/T_c$, represented by an $N_c \times 1$ vector $\mathbf{r}_{k,t}$, is given by

$$\mathbf{r}_{k,t} = h^{k,t} \mathbf{x}_t + \mathbf{n}_{k,t}$$

$$= h^{k,t} \sum_{k'=1}^{K} b_{k',t} \mathbf{s}_{k'} + \mathbf{n}_{k,t} \tag{8.44}$$

where $h^{k,t}$ is the complex-valued channel fading coefficient between the transmitter and the k-th user at time t (assuming that the channel is constant during each symbol interval T_s) and $\mathbf{n}_{k,t}$ is an $N_c \times 1$ vector representing samples of the additive white Gaussian noises with zero mean and variance σ_n^2. The received signal is despread by multiplying with $(\mathbf{s}_k)^H$ and can be expressed as

$$\tilde{b}_{k,t} = (\mathbf{s}_k)^H \mathbf{r}_{k,t}$$

$$= h^{k,t} b_{k,t} + w_{k,t} \tag{8.45}$$

where

$$w_{k,t} = (\mathbf{s}_k)^H \mathbf{n}_{k,t} \tag{8.46}$$

The instantaneous signal-to-noise ratio of the received signal is

$$\text{SNR} = \frac{|h^{k,t}|^2}{\sigma_n^2} \tag{8.47}$$

where σ_n^2 is the noise variance.

Transmit diversity techniques can be employed to increase the received SNR. A simple form of transmit diversity involves sending the same message over two transmit antennas. This can be implemented as open-loop or closed-loop schemes. In the following analysis, we assume that two transmit and one receive antennas are employed, and channels are modeled by flat fading.

8.8.2 Open-Loop Transmit Diversity for CDMA

The block diagram of an open-loop transmit diversity scheme is shown in Fig. 8.13. In this scheme, two different spreading sequences are assigned to each user. The same BPSK modulated symbols are transmitted from two transmit antennas. Each of them has a half of the total transmitted power and a different spreading sequence. Assume that the spreading sequences of user k are $\mathbf{s}_k^1$ and $\mathbf{s}_k^2$ for antennas 1 and 2, respectively. The transmitted signal at antenna i, $i = 1, 2$, can be expressed as

$$\mathbf{x}_t^i = \frac{1}{\sqrt{2}} \sum_{k'=1}^{K} b_{k',t} \mathbf{s}_{k'}^i \tag{8.48}$$

Obviously, the transmitted signals from the two antennas are orthogonal, since the spreading sequences are orthogonal. Thus, this scheme is also known as *orthogonal transmit diversity* (OTD).

The received signal at the k-th user is given by

$$\mathbf{r}_{k,t} = \frac{1}{\sqrt{2}} \sum_{k'=1}^{K} (h_1^{k,t} \mathbf{s}_{k'}^1 + h_2^{k,t} \mathbf{s}_{k'}^2) b_{k',t} + \mathbf{n}_{k,t} \tag{8.49}$$

where $h_i^{k,t}$, $i = 1, 2$, represents the fading coefficients for the path from transmit antenna i to user k at time t. The received signal is despread by the user's two spreading sequences,

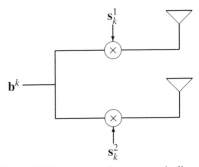

Figure 8.13   An open-loop transmit diversity

generating the signals

$$\tilde{b}_{k,t}^1 = (\mathbf{s}_k^1)^H \mathbf{r}_{k,t} = \frac{1}{\sqrt{2}} h_1^{k,t} b_{k,t} + (\mathbf{s}_k^1)^H \mathbf{n}_{k,t}$$

$$\tilde{b}_{k,t}^2 = (\mathbf{s}_k^2)^H \mathbf{r}_{k,t} = \frac{1}{\sqrt{2}} h_2^{k,t} b_{k,t} + (\mathbf{s}_k^2)^H \mathbf{n}_{k,t} \qquad (8.50)$$

Assuming that the channel state information is perfectly known at the receiver, we can combine the two despread signals as

$$\tilde{b}_{k,t} = (h_1^{k,t})^* \tilde{b}_{k,t}^1 + (h_2^{k,t})^* \tilde{b}_{k,t}^2$$

$$= \frac{1}{\sqrt{2}} (|h_{k,t}^1|^2 + |h_{k,t}^1|^2) b_{k,t} + w_{k,t} \qquad (8.51)$$

where

$$w_{k,t} = ((h_1^{k,t})^* (\mathbf{s}_k^1)^H + (h_2^{k,t})^* (\mathbf{s}_k^2)^H) \mathbf{n}_{k,t} \qquad (8.52)$$

The instantaneous SNR of the receiver signal is

$$\mathrm{SNR}_o = \frac{|h_1^{k,t}|^2 + |h_2^{k,t}|^2}{2\sigma_n^2} \qquad (8.53)$$

This scheme can achieve a two-fold diversity gain with a simple receiver for two transmit antennas and one receive antenna. It can be extended to n_T transmit antennas to achieve an n_T-fold diversity gain. Its major drawback is that each user requires n_T spreading sequences for n_T transmit antennas. Since the spreading sequences are the resources in CDMA systems and the number of orthogonal codes is limited for a given spreading gain, this open-loop diversity scheme significantly reduces the number of users that can be simultaneously supported by the system [21].

8.8.3   Closed-Loop Transmit Diversity for CDMA

The block diagram of a closed-loop transmit diversity is shown in Fig. 8.14. Assume that the ideal channel information is fed back to the transmitter by the receiver through the feedback channel. In the closed-loop scheme, the same symbol is transmitted from two transmit antennas with the same spreading sequence but different weighting factors. The weighting factors depend on the feedback channel information and are chosen in such a way that the received signal has the maximum SNR. It is shown in [21] that the weighting factor for the i-th transmit antenna is given by

$$w_{k,t}^i = \frac{(h_i^{k,t})^*}{\sqrt{|h_1^{k,t}|^2 + |h_2^{k,t}|^2}} \qquad (8.54)$$

The transmitted signal from the i-th transmit antenna is

$$\mathbf{x}_t^i = \sum_{k'=1}^K b_{k',t} \mathbf{s}_{k'} \frac{(h_i^{k',t})^*}{\sqrt{|h_1^{k',t}|^2 + |h_2^{k',t}|^2}} \qquad (8.55)$$

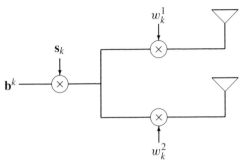

Figure 8.14   A closed-loop transmit diversity

The received signal for the k-th user is given by

$$\mathbf{r}_{k,t} = \sum_{i=1}^{2} h_i^{k,t} \mathbf{x}_t^i + \mathbf{n}_{k,t} \tag{8.56}$$

After despreading with $\mathbf{s}_k$, the received signal can be expressed as

$$\tilde{b}_{k,t} = \mathbf{s}_k^H \mathbf{r}_{k,t}$$

$$= \sqrt{|h_1^{k,t}|^2 + |h_2^{k,t}|^2} b_{k,t} + \mathbf{s}_k^H \mathbf{n}_{k,t} \tag{8.57}$$

The instantaneous SNR at the receiver output is

$$\mathrm{SNR}_c = \frac{|h_1^{k,t}|^2 + |h_2^{k,t}|^2}{\sigma_n^2} \tag{8.58}$$

This scheme can provide a diversity gain of two. Compared to the open-loop scheme, the SNR for the closed-loop scheme is improved by 3 dB. This gain is obtained by utilizing channel state information at the transmitter and transmitting signals coherently [21]. However, this scheme can only be used when reliable channel estimation is available.

8.8.4   Time-Switched Orthogonal Transmit Diversity (TS-OTD)

The open-loop transmit diversity in the previous section provides orthogonal transmissions by using two spreading sequences for two transmit antennas, which limits the number of users.

The number of users can be increased by an alternative scheme, called time-switched orthogonal transmit diversity. In this scheme, the encoded symbols are transmitted from antennas one and two alternately. Since only one antenna is active at each time slot and only one spreading sequence is required for each user, there is no waste of transmission resources.

The block diagram of the scheme is shown in Fig. 8.15. Let us assume that the spreading sequence for user k is $\mathbf{s}_k$ with length N_c chips. The coded symbol sequence $\{b_{k,t}\}$ is split into odd and even sub-streams, $\{b_{k,2t+1}\}$ and $\{b_{k,2t+2}\}$, respectively. Each of them is transmitted from one transmit antenna. At time $2t + 1$, $b_{k,2t+1}$ is spread by $\mathbf{s}_k$ and then transmitted from antenna one. At time $2t + 2$, $b_{k,2t+2}$ is spread by $\mathbf{s}_k$ and then transmitted from antenna two.

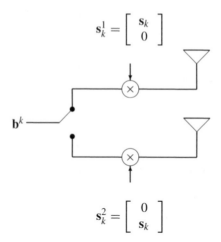

Figure 8.15   A time-switched orthogonal transmit diversity

Let us denote the transmitted chip signals at times $2t + 1$ and $2t + 2$ by an $2N_c \times 1$ vector $\mathbf{x}_t^i$ for antenna i, $i = 1, 2$. We have

$$\mathbf{x}_t^1 = \sum_{k'}^{K} b_{k',2t+1} \mathbf{s}_{k'}^1$$

$$\mathbf{x}_t^2 = \sum_{k'}^{K} b_{k',2t+2} \mathbf{s}_{k'}^2 \tag{8.59}$$

where

$$\mathbf{s}_k^1 = \begin{bmatrix} s_k \\ 0 \end{bmatrix} \quad \text{and} \quad \mathbf{s}_k^2 = \begin{bmatrix} 0 \\ s_k \end{bmatrix} \tag{8.60}$$

and $\mathbf{0}$ is an $N_c \times 1$ all-zero vector. Since $\mathbf{s}_k^1$ and $\mathbf{s}_k^2$ are orthogonal and each has a length of $2N_c$ chips, orthogonal transmit diversity is achieved. Note that $\mathbf{s}_k^1$ and $\mathbf{s}_k^2$ are extended from s_k. It is obvious that only one spreading sequence of length N_c is required for each user.

The received signal for the k-th user is given by

$$\mathbf{r}_{k,t} = h_1^{k,t} \mathbf{x}_t^1 + h_2^{k,t} \mathbf{x}_t^2 + \mathbf{n}_{k,t} \tag{8.61}$$

This signal is despread with $\mathbf{s}_k^1$ and $\mathbf{s}_k^2$, and then multiplied by $(h_1^{k,t})^*$ and $(h_2^{k,t})^*$, respectively, to generate two decision statistics

$$\tilde{b}_{k,2t+1} = (h_1^{k,t})^* (\mathbf{s}_k^1)^H \mathbf{r}_{k,t} = |h_1^{k,t}|^2 b_{k,2t+1} + (h_1^{k,t})^* (\mathbf{s}_k^1)^H \mathbf{n}_{k,t}$$

$$\tilde{b}_{k,2t+2} = (h_2^{k,t})^* (\mathbf{s}_k^2)^H \mathbf{r}_{k,t} = |h_2^{k,t}|^2 b_{k,2t+2} + (h_2^{k,t})^* (\mathbf{s}_k^2)^H \mathbf{n}_{k,t} \tag{8.62}$$

The instantaneous SNR for the two sub-streams denoted by SNR_1 and SNR_2 can be computed as

$$\text{SNR}_i = \frac{|h_i^{k,t}|^2}{\sigma_n^2}, \quad i = 1, 2 \tag{8.63}$$

One can see that the SNR of the scheme is different from other schemes. Although it uses spreading sequences efficiently, there is a performance loss due to data splitting. In fact, this scheme can be viewed as a form of interleaving performed to the coded sequence over the space domain, whereby different coded bits are transmitted over different antennas. The diversity gain is obtained from combining the two sub-streams, which are faded independently. When the channel fading is fast, the performance of the TS-OTD scheme is similar to that of the open-loop OTD scheme since

$$\text{SNR}_o = \tfrac{1}{2}(\text{SNR}_1 + \text{SNR}_2) \qquad (8.64)$$

However, when fading is slow, the data sub-stream on the channel will be lost during a deep fade [21]. Therefore, the scheme is not reliable for systems with slow mobility users.

8.8.5 Space-Time Spreading (STS)

In space-time spreading scheme as shown in Fig. 8.16, we split the coded symbol sequence for user k, $\{b_{k,t}\}$, into two sub-streams $\{b_{k,2t+1}\}$ and $\{b_{k,2t+2}\}$. Assume $\mathbf{s}_k^1$ and $\mathbf{s}_k^2$ are two orthogonal spreading sequences with length of $2N_c$ chips. The transmitted signals are constructed as a linear combination of the two sub-streams spread by the two spreading sequences.

At times $2t + 1$ and $2t + 2$, the transmitted chip signals for user k from antenna i, $\mathbf{x}_{k,t}^i$, $i = 1, 2$, are given by

$$\mathbf{x}_{k,t}^1 = \frac{1}{\sqrt{2}}(b_{k,2t+1}\mathbf{s}_k^1 - b_{k,2t+2}\mathbf{s}_k^2)$$

$$\mathbf{x}_{k,t}^2 = \frac{1}{\sqrt{2}}(b_{k,2t+1}\mathbf{s}_k^2 + b_{k,2t+2}\mathbf{s}_k^1) \qquad (8.65)$$

The received signal at the k-th user is given by

$$\mathbf{r}_{k,t} = h_1^{k,t}\mathbf{x}_t^1 + h_2^{k,t}\mathbf{x}_t^2 + \mathbf{n}_{k,t} \qquad (8.66)$$

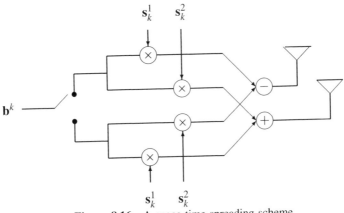

Figure 8.16 A space-time spreading scheme

The received signals after despreading with $\mathbf{s}_k^1$ and $\mathbf{s}_k^2$ are

$$d_k^1 = \frac{1}{\sqrt{2}}(h_1^{k,t}b_{k,2t+1} + h_2^{k,t}b_{k,2t+2}) + (\mathbf{s}_k^1)^*\mathbf{n}_{k,t}$$

$$d_k^2 = \frac{1}{\sqrt{2}}(-h_1^{k,t}b_{k,2t+2} + h_2^{k,t}b_{k,2t+1}) + (\mathbf{s}_k^2)^*\mathbf{n}_{k,t} \qquad (8.67)$$

In order to recover the transmitted symbols, we assume that channel state information is perfectly known at the receiver. Then, we construct the decision statistics as

$$\tilde{d}_k^1 = (h_1^{k,t})^*d_k^1 + (h_2^{k,t})^*d_k^2$$

$$\tilde{d}_k^2 = (h_2^{k,t})^*d_k^1 - (h_1^{k,t})^*d_k^2 \qquad (8.68)$$

The real parts of the decision statistics can be computed as

$$\tilde{b}_{k,2t+1} = Re\{\tilde{d}_k^1\}$$

$$= \frac{1}{\sqrt{2}}(|h_1^{k,t}|^2 + |h_2^{k,t}|^2)b_{k,2t+1} + Re\{((h_1^{k,t})^*\mathbf{s}_k^1 + (h_2^{k,t})^*\mathbf{s}_k^2)\mathbf{n}_{k,t}\}$$

$$\tilde{b}_{k,2t+2} = Re\{\tilde{d}_k^2\}$$

$$= \frac{1}{\sqrt{2}}(|h_1^{k,t}|^2 + |h_2^{k,t}|^2)b_{k,2t+2} + Re\{((h_2^{k,t})^*\mathbf{s}_k^1 - (h_1^{k,t})^*\mathbf{s}_k^2)\mathbf{n}_{k,t}\} \qquad (8.69)$$

The decoder chooses a pair of signals $b_{k,2t+1}$ and $b_{k,2t+2}$ from the modulation constellation as the decoder output, which is closest to the decision statistics in (8.69).

In this scheme, two spreading sequences are used to spread the odd and even coded symbols. With the linear combining operation performed at the transmitter and the receiver, this scheme achieves the full transmit diversity. Note that the spreading is performed over every two symbol periods. The two spreading sequences of length $2N_c$ can be constructed as an extension of one spreading sequence of length N_c, similar to the spreading sequences in (8.60). Therefore, only one spreading sequence of length N_c is required for each user and there is no waste of transmission resources.

The scheme is called *space-time spreading* (STS) since the coded data are spread in a different way for different transmit antennas [21].

8.8.6 STS for Three and Four Antennas

In this section, we extend the space-time spreading schemes to a larger number of transmit antennas. The transmitted signals are arranged in a matrix form. Since we only consider a single user in this analysis, we may drop the index k henceforth. Let

$$\mathbf{X} = [\mathbf{x}_t^1, \mathbf{x}_t^2]$$

$$\mathbf{S} = [\mathbf{s}^1, \mathbf{s}^2]$$

$$\mathbf{B} = \begin{bmatrix} b_{k,2t+1} & b_{k,2t+2} \\ -b_{k,2t+2} & b_{k,2t+1} \end{bmatrix} \qquad (8.70)$$

Then, (8.65) can be rewritten as

$$\mathbf{X} = \frac{1}{\sqrt{2}}\mathbf{SB} \qquad (8.71)$$

where matrix $\mathbf{X}$ with the size of $2N_c \times 2$ represents the transmitted chip signals over two symbol periods from two transmit antennas. Note that $\mathbf{B}^T$ is the transmission matrix for the space-time block code with two transmit antennas over real signal constellations. The STS scheme then can be viewed as the application of the Alamouti transmit diversity scheme for CDMA systems by using two orthogonal spreading sequences [21]. For example, if we choose the spreading gain $N_c = 1$ (no spreading), then

$$\mathbf{S} = [\mathbf{s}^1, \mathbf{s}^2] = \begin{bmatrix} 1 & | & 0 \\ 0 & | & 1 \end{bmatrix} \tag{8.72}$$

and

$$\mathbf{X} = \frac{1}{\sqrt{2}} \begin{bmatrix} b_{k,2t+1} & b_{k,2t+2} \\ -b_{k,2t+2} & b_{k,2t+1} \end{bmatrix} \tag{8.73}$$

The transmitted spread symbol matrix $\mathbf{X}^T$ is equivalent to the Alamouti scheme with real signals. Furthermore, it is worth noting that since the Alamouti scheme is a rate one orthogonal code, the generalization of this scheme for CDMA systems results in no bandwidth expansion or waste of transmission resources. If we denote by $\mathbf{s}$ a spreading sequence of length N_c, we can construct orthogonal matrix $\mathbf{S}$ as [21]

$$\mathbf{S} = \begin{bmatrix} \mathbf{s} & | & \mathbf{s} \\ \mathbf{s} & | & -\mathbf{s} \end{bmatrix}_{2N_c \times 2} \tag{8.74}$$

Another possible construction of $\mathbf{S}$ is

$$\mathbf{S} = \begin{bmatrix} \mathbf{s} & | & \mathbf{0} \\ \mathbf{0} & | & \mathbf{s} \end{bmatrix}_{2N_c \times 2} \tag{8.75}$$

The choice of $\mathbf{S}$ in (8.74) was recommended for the IS-2000 standard while the choice of $\mathbf{S}$ in (8.75) was proposed for the UMTS W-CDMA standard [21]. Both of them have the same diversity gain. As in space-time block codes, the schemes can be applied to systems with two transmit and n_R receive antennas to achieve a diversity gain of $2n_R$.

Let us define matrices

$$\mathbf{d} = \begin{bmatrix} d_{k,t}^1 \\ d_{k,t}^2 \end{bmatrix} \quad \mathbf{b} = \begin{bmatrix} b_{k,2t+1} \\ b_{k,2t+2} \end{bmatrix} \quad \mathbf{H} = \begin{bmatrix} h_1^{k,t} & h_2^{k,t} \\ h_2^{k,t} & -h_1^{k,t} \end{bmatrix} \tag{8.76}$$

$$\tilde{\mathbf{d}} = \begin{bmatrix} \tilde{d}_{k,t}^1 \\ \tilde{d}_{k,t}^2 \end{bmatrix} \quad \tilde{\mathbf{b}} = \begin{bmatrix} \tilde{b}_{k,2t+1} \\ \tilde{b}_{k,2t+2} \end{bmatrix} \quad v = \begin{bmatrix} (\mathbf{s}_k^1)^* \mathbf{n}_{k,t} \\ (\mathbf{s}_k^2)^* \mathbf{n}_{k,t} \end{bmatrix} \tag{8.77}$$

The received signals in (8.67) can be rewritten as

$$\mathbf{d} = \mathbf{H}\mathbf{b} + v \tag{8.78}$$

Then, the decision statistics can be represented as

$$\tilde{\mathbf{d}} = \mathbf{H}^H \mathbf{d}$$

$$= \mathbf{H}^H \mathbf{H} \cdot \mathbf{b} + \mathbf{H}^H v \tag{8.79}$$

Obviously, in order to decouple the odd and even symbols from the decision statistics and perform the maximum-likelihood decoding for each of them separately, the real part of $\mathbf{H}^H\mathbf{H}$ should be a diagonal matrix, for real signals $\mathbf{b}$. In other words, the columns of matrix $\mathbf{H}$ should be orthogonal to each other if $\mathbf{H}$ is real [21].

Now let us define

$$\mathbf{h} = \begin{bmatrix} h_1^{k,t} \\ h_2^{k,t} \end{bmatrix} \tag{8.80}$$

Note that

$$\mathbf{Bh} = \mathbf{Hb} \tag{8.81}$$

One can see that the columns of matrix $\mathbf{B}$ are orthogonal since matrix $\mathbf{B}$ has a similar property as matrix $\mathbf{H}$ [21].

We now consider a general space-time spreading scheme for CDMA systems with n_T transmit antennas and a single receive antenna. Assume the spreading gain is N_c. In a space-time spreading scheme, the coded sequence is split into Q sub-streams, $b_1, b_2, \ldots, b_Q$. At each STS operation, Q coded symbols are spread by L_s spreading sequences, $\mathbf{s}^1, \mathbf{s}^2, \ldots, \mathbf{s}^{L_s}$, each of them has length of $L_s N_c$ chips, where $L_s \geq Q$. The transmitted signals for the n_T transmit antennas are linear combinations of the coded signals after spreading, which can be represented by

$$\mathbf{X} = \mathbf{SB} \tag{8.82}$$

where $\mathbf{S} = [\mathbf{s}^1, \mathbf{s}^2, \ldots, \mathbf{s}^{L_s}]$ and $\mathbf{B}$ is an $L_s \times n_T$ orthogonal transmission matrix, whose elements are $b_1, b_2, \ldots, b_Q$ and their linear combinations. Since L_s spreading sequences with length $L_s N_c$ are employed for transmitting Q coded symbols, the coded symbols after spreading are transmitted over L_s symbol periods. Therefore, the transmission rate is Q/L_s [21].

At the receiver, the received signal $\mathbf{r}$ is despread with L_s spreading sequences

$$\tilde{\mathbf{d}} = \mathbf{S}^H\mathbf{r} \tag{8.83}$$

Then, an $L_s \times Q$ matrix $\mathbf{H}$, whose elements are channel coefficients $h_1, h_2, \ldots, h_{n_T}$ and their linear combinations, is used to decouple the transmitted coded symbols as (8.79). This means that each coded symbol can be detected independently.

In general, to design STS for CDMA systems achieving n_T transmit diversity with a simple decoding algorithm, matrices $\mathbf{B}$ and $\mathbf{H}$ should satisfy [21]

$$Re\{\mathbf{H}^H\mathbf{H}\} = \sum_{i=1}^{n_T} |h_i|^2 \mathbf{I}_Q$$

$$\mathbf{B}^T\mathbf{B} = \sum_{i=1}^{Q} |b_i|^2 \mathbf{I}_{n_T} \tag{8.84}$$

for real signals, where $\mathbf{I}_Q$ and $\mathbf{I}_{n_T}$ are identity matrices of size Q and n_T, respectively. Based on orthogonal designs, the STS schemes with three and four transmit antennas are given by [21]

$$
\mathbf{H}_3 = \begin{bmatrix} h_1 & h_2 & h_3 & 0 \\ -h_2 & h_1 & 0 & h_3 \\ -h_3 & 0 & h_1 & -h_2 \\ 0 & -h_3 & h_2 & h_1 \end{bmatrix} \qquad \mathbf{B}_3 = \begin{bmatrix} b_1 & b_2 & b_3 \\ b_2 & -b_1 & b_4 \\ b_3 & -b_4 & -b_1 \\ b_4 & b_3 & -b_2 \end{bmatrix}
$$

$$
\mathbf{H}_4 = \begin{bmatrix} h_1 & h_2 & h_3 & h_4 \\ -h_2 & h_1 & -h_4 & h_3 \\ -h_3 & h_4 & h_1 & -h_2 \\ -h_4 & -h_3 & h_2 & h_1 \end{bmatrix} \qquad \mathbf{B}_4 = \begin{bmatrix} b_1 & b_2 & b_3 & b_4 \\ b_2 & -b_1 & b_4 & -b_3 \\ b_3 & -b_4 & -b_1 & b_2 \\ b_4 & b_3 & -b_2 & -b_1 \end{bmatrix} \qquad (8.85)
$$

For these two schemes, $L_s = Q = 4$, and the transmission rate is one. Therefore, there is no waste of transmission resources.

Similarly, the STS schemes for CDMA with complex signals require that

$$
\mathbf{H}^H \mathbf{H} = \sum_{i=1}^{n_T} |h_i|^2 \mathbf{I}_Q
$$

$$
\mathbf{B}^H \mathbf{B} = \sum_{i=1}^{Q} |b_i|^2 \mathbf{I}_{n_T} \qquad (8.86)
$$

For two transmit antennas, the scheme is given by [21]

$$
\mathbf{H}_2^c = \begin{bmatrix} h_1 & h_2 \\ -h_2^* & h_1^* \end{bmatrix} \qquad \mathbf{B}_2^c = \begin{bmatrix} b_1 & b_2 \\ b_2^* & -b_1^* \end{bmatrix} \qquad (8.87)
$$

For $n_T > 2$, the STS scheme with $L_s = Q = n_T$ does not exist. For three transmit antennas, the STS scheme with $Q = 3$ and $L_s = 4$ is given by [21]

$$
\mathbf{H}_3^c = \begin{bmatrix} h_1 & -h_2 & -h_3 \\ h_2^* & h_1^* & 0 \\ h_3^* & 0 & h_1^* \\ 0 & h_3^* & -h_2^* \end{bmatrix} \qquad \mathbf{B}_3^c = \begin{bmatrix} b_1 & -b_2 & -b_3 \\ b_2^* & b_1^* & 0 \\ b_3^* & 0 & b_1^* \\ 0 & -b_3^* & b_2^* \end{bmatrix} \qquad (8.88)
$$

It is worth noting that among all these STS schemes, $\mathbf{B}_3$, $\mathbf{B}_4$ and $\mathbf{B}_2^c$ are the space-time block codes directly applied to CDMA systems.

8.9 Space-Time Coding for CDMA Systems

In the previous section, various transmit diversity schemes for wideband CDMA systems were discussed. The feature of the schemes is that they can achieve a full diversity but no coding gain. In this section, we consider space-time coding for CDMA systems, which can provide both diversity and coding gain.

In a wideband system, due to the multipath delay spread, the maximum possible diversity order is $L_p n_T n_R$, where L_p is the number of multipaths. For CDMA systems, the presence of multiple access and spreading makes the full diversity easily achievable. Therefore, the code design for these systems should focus on coding gain [24]. On the other hand, since the total diversity $L_p n_T n_R$ is usually large for wideband systems, the coding gain in this scenario is determined by the code minimum Euclidean distance. Therefore, good codes with a high minimum Euclidean distance for narrowband systems tend to perform well in wideband CDMA systems.

A number of space-time coding schemes have been designed for narrowband systems. Now we evaluate their performance in multipath fading CDMA systems. In particular, we consider the STTC and layered STC in the following sections.

8.10 Performance of STTC in CDMA Systems

We consider a QPSK STTC coded WCDMA system with K users in a cell [36]. The transmitter block diagram is shown in Fig. 8.17. The binary information data $\{b_k\}$ for user k are STTC encoded and the coded QPSK symbols are transmitted from n_T transmit antennas. The encoded symbol sequence transmitted from the ith ($i = 1, 2, \ldots, n_T$) antenna is denoted by $\mathbf{x}_k^i$, and it is represented as

$$\mathbf{x}_k^i = [x_k^i(1), x_k^i(2), \ldots, x_k^i(L)]. \tag{8.89}$$

where L is the sequence length. Each coded symbol $x_k^i(n)$ of duration T is then modulated by a spreading waveform $s_k(t)$ and transmitted from antenna i. The transmitted signal for user k and antenna i is given by

$$d_k^i(t) = A_k^i \sum_n x_k^i(n) s_k(t - nT) \tag{8.90}$$

where A_k^i denotes the amplitude of the signal from antenna i for the kth user and $s_k(t)$ is the normalized spreading waveform given by

$$s_k(t) = \sum_{q=1}^{N_c} s_{k,q} p(t - (q - 1)T_c), \tag{8.91}$$

which is identical for all antennas. Here, $s_{k,q}$ is the qth chip of the spreading sequence for user k, N_c is the spreading gain, $p(t)$ is the chip waveform, and T_c is the chip duration. The spreading sequence can be represented by an $N_c \times 1$ vector

$$\mathbf{s}_k = [s_{k,1}, s_{k,2}, \ldots, s_{k,N_c}]^T \tag{8.92}$$

Considering multiple receive antennas, the kth user's signal from the ith transmit antenna to the jth ($j = 1, 2, \ldots, n_R$) receive antenna propagates through a multipath channel with the impulse response

$$h_{j,i}^k(t) = \sum_{l=1}^{L_{j,i}^k} h_{j,i}^{k,l} \delta(t - \tau_{j,i}^{k,l}) \tag{8.93}$$

where $L_{j,i}^k$ is the number of paths in the (j, i)th channel for user k and $h_{j,i}^{k,l}$ and $\tau_{j,i}^{k,l}$ are the complex gain and delay of the lth path from the ith transmit to the jth receive antenna

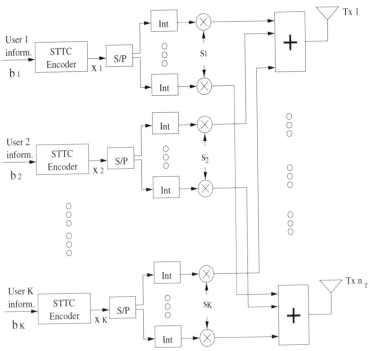

Figure 8.17 Block diagram of a space-time trellis coded CDMA transmitter

for user k, respectively. The received signal at the jth receive antenna can be expressed as

$$r^j(t) = \sum_{k=1}^{K} \sum_{i=1}^{n_T} d_k^i(t) \star h_{j,i}^k(t)$$

$$= \sum_{k=1}^{K} \sum_{i=1}^{n_T} \sum_{l=1}^{L_{j,i}^k} h_{j,i}^{k,l} d_k^i(t - \tau_{j,i}^{k,l}) + n^j(t) \tag{8.94}$$

where $\star$ denotes the convolution and $n^j(t)$ is an additive white Gaussian noise process with a zero mean and the variance σ_n^2.

In this analysis, we focus on a synchronous down-link WCDMA system in a frequency-selective fading channel. We assume that the number of resolvable multipaths for all the channels is the same, and it is denoted by L_p. For simplicity, we also assume that the multipath delay $\tau_{j,i}^{k,l}$ is independent on the transmit and receive antennas. That is, $\tau_{j,i}^{k,l} = \tau^{k,l}$, and $\tau^{k,1} = 0$. Let

$$s = [s_{1,1}, \ldots, s_{1,L_p}, \ldots \ldots, s_{K,1}, \ldots, s_{K,L_p}], \tag{8.95}$$

where $s_{k,l}$ is the spreading sequence that corresponds to the lth resolvable multipath component of the kth user's signal. It is obtained as a delayed version of the spreading sequence s_k by $\lfloor \tau^{k,l}/T_c \rfloor$, given by

$$s_{k,l} = [0_b^{k,l}, s_{k,1}, s_{k,2}, \ldots, s_{k,N_c}, 0_e^{k,l}]^T \tag{8.96}$$

where $\mathbf{0}_b^{k,l}$ is a row vector with $b = \lfloor \tau^{k,l}/T_c \rfloor$ zeros as elements, $\mathbf{0}_e^{k,l}$ is a row vector with $e = \lfloor \tau^{\max}/T_c \rfloor - \lfloor \tau^{k,l}/T_c \rfloor$ zeros as elements, and $\tau^{\max} = \max\{\tau^{k,L_p} | k = 1, 2, \ldots, K\}$. Considering n_T transmit antennas, we can arrange n_T replicas of $\mathbf{s}$ into an $N_c' \times n_T K L_p$ spreading sequence matrix $\mathbf{S}' = (\mathbf{s}, \mathbf{s}, \ldots, \mathbf{s})$, where $N_c' = N_c + \lfloor \tau^{\max}/T_c \rfloor$. Following a chip matched filter, the discrete-time complex baseband received signal in (8.94) at the jth antenna during a given symbol period can be written as a complex $N_c' \times 1$ column vector

$$\mathbf{r}^j = \mathbf{S}' \mathbf{H}'^j \mathbf{A} \mathbf{x} + \mathbf{n}^j \tag{8.97}$$

where the qth component of the vector represents the qth chip of the received signal, $\mathbf{H}'^j$ is a block diagonal $n_T K L_p \times n_T K$ channel matrix defined by

$$\mathbf{H}'^j = \text{diag}\left(\begin{bmatrix} h_{j,1}^{1,1} \\ \vdots \\ h_{j,1}^{1,L_p} \end{bmatrix}, \ldots, \begin{bmatrix} h_{j,1}^{K,1} \\ \vdots \\ h_{j,1}^{K,L_p} \end{bmatrix}, \ldots, \begin{bmatrix} h_{j,n_T}^{1,1} \\ \vdots \\ h_{j,n_T}^{1,L_p} \end{bmatrix}, \ldots, \begin{bmatrix} h_{j,n_T}^{K,1} \\ \vdots \\ h_{j,n_T}^{K,L_p} \end{bmatrix}\right). \tag{8.98}$$

$\mathbf{A}$ is a real $n_T K \times n_T K$ diagonal matrix of signal amplitudes, given by

$$\mathbf{A} = \text{diag}(A_1^1, A_2^1, \ldots, A_K^1, \ldots\ldots, A_1^{n_T}, A_2^{n_T}, \ldots, A_K^{n_T}), \tag{8.99}$$

$\mathbf{x}$ is a complex $n_T K \times 1$ column vector of the transmitted QPSK space-time coded symbols given by

$$\mathbf{x} = [x_1^1, x_2^1, \ldots, x_K^1, \ldots\ldots, x_1^{n_T}, x_2^{n_T}, \ldots, x_K^{n_T}]^T, \tag{8.100}$$

and $\mathbf{n}^j$ is an $N_c' \times 1$ complex Gaussian noise vector with independent identically distributed (i.i.d.) components whose real and imaginary components each have a zero mean and variance σ^2.

If we define the stacked received signal as $\mathbf{r} = [(\mathbf{r}^1)^T, (\mathbf{r}^2)^T, \ldots, (\mathbf{r}^{n_R})^T]^T$, we can write

$$\mathbf{r} = \mathbf{S}\mathbf{H}\mathbf{A}\mathbf{x} + \mathbf{n} \tag{8.101}$$

where $\mathbf{S}$ is an $n_R N_c' \times n_R n_T K L_p$ block diagonal spreading sequence matrix with duplicate diagonal blocks $\mathbf{S}'$, $\mathbf{H} = [(\mathbf{H}'^1)^T, (\mathbf{H}'^2)^T, \ldots, (\mathbf{H}'^{n_R})^T]^T$ is a complex $n_R n_T K L_p \times n_T K$ stacked channel matrix, and $\mathbf{n} = [(\mathbf{n}^1)^T, (\mathbf{n}^2)^T, \ldots, (\mathbf{n}^{n_R})^T]^T$ is a stacked received noise vector.

8.10.1   Space-Time Matched Filter Detector

Receiver Structure

In a maximum likelihood space-time multiuser receiver [37], the detector selects an estimate of the transmitted symbols for all users and all transmit antennas, represented by an $n_T K$ vector

$$\hat{\mathbf{x}} = [\hat{x}_1^1, \hat{x}_2^1, \ldots, \hat{x}_K^1, \ldots\ldots, \hat{x}_1^{n_T}, \hat{x}_2^{n_T}, \ldots, \hat{x}_K^{n_T}]^T, \tag{8.102}$$

which maximizes the likelihood function [29]

$$\exp[-(1/2\sigma^2)\|\mathbf{r} - \mathbf{S}\mathbf{H}\mathbf{A}\hat{\mathbf{x}}\|^2]. \tag{8.103}$$

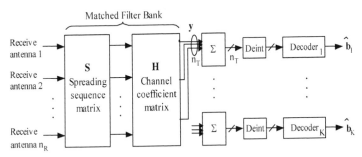

Figure 8.18 Block diagram of the space-time matched filter receiver

This is equivalent to maximizing

$$2Re(\hat{\mathbf{x}}^H \mathbf{A} \mathbf{H}^H \mathbf{S}^H \mathbf{r}) - \hat{\mathbf{x}}^H \mathbf{A} \mathbf{H}^H \mathbf{S}^H \mathbf{S} \mathbf{H} \mathbf{A} \hat{\mathbf{x}}. \tag{8.104}$$

As the maximum-likelihood detector is too complex, we consider a simple receiver structure as shown in Fig. 8.18 [36]. The receiver consists of a space-time matched filter detector and a bank of STTC decoders, one for each user. Assuming the knowledge of the channel matrices, the matched filter detector generates decision statistics of the transmitted space-time symbols for all users and all transmit antennas at a given symbol period. The matched filter is represented by an $n_T K \times n_R N_c$ matrix $\mathbf{H}^H \mathbf{S}^H$. The decision statistics at the output of the detector can be represented by a complex $n_T K \times 1$ column vector, given by

$$\mathbf{y} = (\mathbf{H}^H \mathbf{S}^H) \mathbf{r}$$

$$= (\mathbf{H}^H \mathbf{S}^H)(\mathbf{S} \mathbf{H} \mathbf{A} \mathbf{x} + \mathbf{n}) \tag{8.105}$$

$$= \mathbf{M} \mathbf{A} \mathbf{x} + \mathbf{n}_y$$

where $\mathbf{M} = \mathbf{H}^H \mathbf{S}^H \mathbf{S} \mathbf{H}$ is the space-time correlation matrix and $\mathbf{n}_y = \mathbf{H}^H \mathbf{S}^H \mathbf{n}$ is the resulting noise vector. The $(K(i-1)+k)$th element of the decision statistics vector $\mathbf{y}$, denoted by y_k^i, is simply the space-time matched filter output for the signal of the ith antenna and user k, obtained by correlating each of the n_R received signals with its L_p multipath spreading sequences, $(\mathbf{s}_{k,1}, \mathbf{s}_{k,2}, \dots, \mathbf{s}_{k,L_p})$, weighting them by the complex conjugate of the corresponding channel coefficients $(h_{j,i}^{k,1}, h_{j,i}^{k,2}, \dots, h_{j,i}^{k,L_p}, j = 1, 2, \dots, n_R)$, and summing over the multipath indices l and receive antenna j [33].

The decision statistics for user k, $y_k^1, y_k^2, \dots, y_k^{n_T}$, are then passed to the user's STTC decoder, which estimates the transmitted binary information data $\hat{b}_k$.

Error Probability for The Space-Time Matched Filter Detector

The space-time matched filter detector in (8.105) demodulates the received signal using the knowledge of the kth user's spreading sequence, timing, and channel information for each transmit antenna. It does not take into account the structure of the multiple access interference (MAI). The error probability for the signals of the ith antenna of the kth user conditioned on the other users' data and on the channel coefficients is

$$P_k^i = Q \left(\frac{(\mathbf{M} \mathbf{A} \mathbf{x})_{k'}}{\sigma \sqrt{(\mathbf{M})_{k',k'}}} \right), k' = K(i-1)+k \tag{8.106}$$

where the subscript $(\cdot)_{k'}$ denotes the k'th element of the vector, and $(\cdot)_{k',k'}$ denotes the k'th diagonal element of the matrix.

STTC Decoder

Now we consider the decoding problem in MIMO channels. For STTC, the decoder employs the Viterbi algorithm to perform maximum likelihood decoding for each user. Assuming that perfect CSI is available at the receiver, for a branch labelled by $x_k^1(t), x_k^2(t), \ldots, x_k^{n_T}(t)$, the branch metric is computed as the squared Euclidean distance between the hypothesized received symbols and the actual received signals as

$$\sum_{j=1}^{n_R} \left| r_k^j(t) - \sum_{i=1}^{n_T} \sum_{l=1}^{L_{j,i}^k} h_{j,i}^{k,l} x_k^i(t) \right|^2 \tag{8.107}$$

where $r_k^j(t)$ is the received signal at receive antenna j at time t after chip synchronization and despreading with the kth user's spreading sequence. The Viterbi algorithm selects the path with the minimum path metric as the decoded sequence.

When the matched filter detection is considered as a multipath diversity reception technique for frequency-selective fading in a MIMO system, it introduces interference from multiple antennas and multipaths. The output of the matched filter detector, $\mathbf{y}$, does not only have a diversity gain which is obtained from the diagonal element of the correlation matrix $(\mathbf{M})_{k',k'}, k' = 1, 2, \ldots, n_T K$, but also has the multiple antenna and multipath interference from the off-diagonal elements of $(\mathbf{M})_{k',u'}, u' = 1, 2, \ldots, n_T K (u' \neq k')$. Therefore, to reduce the effect of the multiple antennas interference of the user, we reconstruct the trellis branch labels as $\tilde{x}_k^1(t), \tilde{x}_k^2(t), \ldots, \tilde{x}_k^{n_T}(t)$, where [36]

$$\tilde{x}_k^i(t) = \sum_{j=1}^{n_T} (\mathbf{MA})_{K(i-1)+k, K(j-1)+k} \cdot x_k^j(t). \tag{8.108}$$

After matched filtering, the branch metric of (8.107) in the Viterbi decoder is replaced by

$$\sum_{i=1}^{n_T} |y_k^i(t) - \tilde{x}_k^i(t)|^2. \tag{8.109}$$

where $y_k^i(t)$ is the matched filter output for the ith antenna of user k.

8.10.2 Space-Time MMSE Multiuser Detector

Space-Time MMSE Multiuser Detector

In order to reduce the effects of multipath, multiuser, and multiple antennas interference, we consider a space-time MMSE detector [29] [33] as shown in Fig. 8.19. Given the decision statistics vector $\mathbf{y}$ in (8.105), the space-time MMSE detector applies a linear transformation $\mathbf{W}$ to $\mathbf{y}$ so that the mean-squared error between the resulting vector and the data vector $\mathbf{x}$ is minimized. The space-time MMSE detection matrix $\mathbf{W}$ of size $n_T K \times n_T K$ should satisfy

$$\mathbf{W} = \arg \min_{\mathbf{W}} \{ E \| \mathbf{W}^H \mathbf{y} - \mathbf{A}\mathbf{x} \|^2 \} \tag{8.110}$$

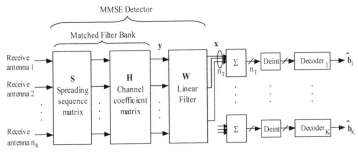

Figure 8.19 Block diagram of the STTC MMSE receiver

which results in the standard Wiener solution

$$W^H = E[\mathbf{A}xy^H](E[yy^H])^{-1}$$

$$= \mathbf{A}^2 \mathbf{M}^H (\mathbf{M}\mathbf{A}^2\mathbf{M}^H + \sigma^2\mathbf{M})^{-1},$$

(8.111)

where

$$\mathbf{A}^2 = \text{diag}((A_1^1)^2, (A_2^1)^2, \dots, (A_K^1)^2, \dots, (A_1^{n_T})^2, (A_2^{n_T})^2, \dots, (A_K^{n_T})^2). \quad (8.112)$$

If all of the K users' $n_T L_p$ spreading sequences are linearly independent, then $\mathbf{S}'^H\mathbf{S}'$ has a full rank. Under this assumption, it can be shown that with probability one, $\mathbf{H}'^{jH}\mathbf{S}'^H\mathbf{S}'\mathbf{H}'^j$ has a full rank for any j. It follows that matrix $\mathbf{M} = \mathbf{H}^H\mathbf{S}^H\mathbf{S}\mathbf{H}$ is of a full rank and invertible. Then the space-time MMSE matrix $\mathbf{W}$ in (8.111) is simplified to

$$\mathbf{W}^H = [\mathbf{M} + \sigma^2\mathbf{A}^{-2}]^{-1}, \quad (8.113)$$

where

$$\mathbf{A}^{-2} = \text{diag}\left(\frac{1}{(A_1^1)^2}, \frac{1}{(A_2^1)^2}, \dots, \frac{1}{(A_K^1)^2}, \dots, \frac{1}{(A_1^{n_T})^2}, \frac{1}{(A_2^{n_T})^2}, \dots, \frac{1}{(A_K^{n_T})^2}\right).$$

(8.114)

Error Probability of the Space-Time MMSE with STTC

We now consider the error probability conditioned on the interfering users' data and on the channel realization for the space-time MMSE receiver. The space-time linear MMSE detector takes into account both the interference and the background noise. However, it does not completely eliminate MAI. The space-time MMSE detector output for antenna i of user k in the synchronous system can be written as

$$(\mathbf{W}^H y)_{k'} = ([\mathbf{M} + \sigma^2\mathbf{A}^{-2}]^{-1}y)_{k'}$$

$$= B_k^i \left(x_k^i + \sum_{\substack{p \ u \\ (p,u)\neq(i,k)}}^{n_T} \sum^{K} \beta_u^p x_u^p \right) + n_{k'}, \quad (8.115)$$

with

$$\beta_u^p = \frac{B_u^p}{B_k^i}$$

$$B_u^p = A_u^p (\mathbf{W}^H \mathbf{M})_{k',u'} \tag{8.116}$$

$$\text{Var}\{n_{k'}\} = (\mathbf{W}^H \mathbf{M} \mathbf{W})_{k',k'} \sigma^2$$

where $k' = K(i - 1) + k$ and $u' = K(p - 1) + u$. The leakage coefficient β_u^p quantifies the contribution of the pth antenna component of the uth interferer to the decision statistics relative to the contribution of the ith antenna of the desired user k. The average error probability at the output of the space-time MMSE detector for antenna i of user k is then given by

$$P_k^i = \frac{1}{2^{n_T K - 1}} \sum_{\forall x_{u'(u' \neq k')} \in \{-1,1\}^{n_T K - 1}}$$

$$Q \left(\frac{A_k^i}{\sigma} \frac{(\mathbf{W}^H \mathbf{M})_{k',k'}}{\sqrt{(\mathbf{W}^H \mathbf{M} \mathbf{W})_{k',k'}}} \left[1 + \sum_{\substack{p \ u \\ (p,u) \neq (i,k)}}^{n_T \ K} \beta_u^p x_u^p \right] \right). \tag{8.117}$$

The complexity of calculating the error probability from the above expression is exponential in the number of users and the number of transmit antennas. This computational burden is mainly due to the leakage coefficients calculation. The error probability can be approximated by replacing the multiple access interference with a Gaussian random variable with the same variance [29]. Thus, the error probability in (8.117) for the space-time MMSE detection can be represented by

$$P_k^i \approx Q \left(\frac{\mu}{\sqrt{1 + \chi^2}} \right), \tag{8.118}$$

where

$$\mu = \frac{A_k^i}{\sigma} \frac{(\mathbf{W}^H \mathbf{M})_{k',k'}}{\sqrt{(\mathbf{W}^H \mathbf{M} \mathbf{W})_{k',k'}}}$$

$$\chi^2 = \mu^2 \sum_{\substack{p \ u \\ (p,u) \neq (i,k)}}^{n_T \ K} \beta_u^{p2}. \tag{8.119}$$

After the trellis decoding, the average pairwise error probability of the STTC on a slow Rayleigh fading channel can be written as [3]

$$P(\mathbf{x}, \hat{\mathbf{x}}) \leq \left(\prod_{i=1}^{r} \lambda_i \right)^{-n_R} (E_s / 4N_0)^{-r n_R}$$

$$\leq \left(\prod_{i=1}^{r} \lambda_i \right)^{-n_R} \left(\frac{\mu}{8\sqrt{1 + \chi^2}} \right)^{-r n_R} \tag{8.120}$$

where r denotes the rank of codeword distance matrix $\mathbf{A}(\mathbf{x}, \hat{\mathbf{x}})$ and λ_i is the nonzero eigenvalue of the codeword distance matrix.

8.10.3 Space-Time Iterative MMSE Detector

An iterative MMSE receiver [34] is also considered in a multipath MIMO system. The interference estimate for the ith antenna of the kth user is formed by adding the regenerated signals of all users and all transmit antennas, except the one for the desired user k and antenna i. After each decoding iteration, the soft decoder outputs are used to update the a priori probabilities of the transmitted symbols. These updated probabilities are applied in the calculation of the MMSE filter feedforward and feedback coefficients. Assuming that $z_k^i(t)$ is the input to the kth user decoder corresponding to the ith transmit antenna at time t, it is given by

$$z_k^i(t) = (\mathbf{w}_{f,k}^i(t))^H \mathbf{r}(t) + (\mathbf{w}_{b,k}^i(t))^H \hat{\underline{\mathbf{x}}}_k^i \tag{8.121}$$

where $\mathbf{w}_{f,k}^i(t)$ is an $n_R N_c' \times 1$ optimized feedforward coefficients matrix, $\mathbf{w}_{b,k}^i(t)$ is an $(n_T K - 1) \times 1$ feedback coefficients matrix, and $\hat{\underline{\mathbf{x}}}_k^i$ is an $(n_T K - 1) \times 1$ vector representing the feedback soft decisions for all users and all transmit antennas except the one for the ith transmit antenna of user k. Note that the feedback coefficients appear only through their sum in (8.121). We can assume, without loss of generality, that

$$w_{b,k}^i(t) = (\mathbf{w}_{b,k}^i(t))^H \hat{\underline{\mathbf{x}}}_k^i \tag{8.122}$$

where $w_{b,k}^i(t)$ is a single coefficient that represents the sum of the feedback terms.

The coefficients $\mathbf{w}_{f,k}^i(t)$ and $w_{b,k}^i(t)$ are obtained by minimizing the mean square value of the error ϵ between the data symbols and its estimates, given by

$$
\begin{aligned}
\epsilon &= E[|z_k^i(t) - x_k^i(t)|^2] \\
&= E[|(\mathbf{w}_{f,k}^i(t))^H \mathbf{r}(t) + w_{b,k}^i(t) - x_k^i(t)|^2] \\
&= E[|(\mathbf{w}_{f,k}^i(t))^H \{\mathbf{h}_k^i x_k^i(t) + \mathbf{H}_k^i \underline{\mathbf{x}}_k^i(t) + \mathbf{n}(t)\} \\
&\quad + w_{b,k}^i(t) - x_k^i(t)|^2]
\end{aligned} \tag{8.123}
$$

where

$$\mathbf{h}_k^i = (\mathbf{SHA})_{K(i-1)+k} \tag{8.124}$$

is an $n_R N_c' \times 1$ signature matrix for the ith antenna of the kth user, $\mathbf{H}_k^i = (\mathbf{SHA})_k^i$ is an $n_R N_c' \times (n_T K - 1)$ matrix composed of the signature vectors of all users and antennas except the ith antenna of the kth user, and $\underline{\mathbf{x}}_k^i(t)$ is the $(n_T K - 1) \times 1$ transmitted data vector from all users and antennas except the ith antenna of the kth user. The optimum feedforward and feedback coefficients $\mathbf{w}_{f,k}^i(t)$ and $w_{b,k}^i(t)$ can be represented by

$$\mathbf{w}_{f,k}^i(t) = (A + B + R_n - FF^H)^{-1} \mathbf{h}_k^i \tag{8.125}$$

$$w_{b,k}^i(t) = -(\mathbf{w}_{f,k}^i(t))^H F \tag{8.126}$$

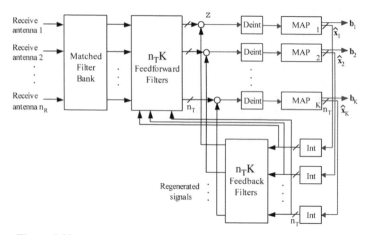

Figure 8.20 Block diagram of the space-time iterative MMSE receiver

where

$$A = \mathbf{h}_k^i (\mathbf{h}_k^i)^H$$

$$B = \underline{\mathbf{H}}_k^i \mathbf{I}_{n_T K-1} - \text{Diag}\,(\underline{\mathbf{x}}_k^{E_i}(\underline{\mathbf{x}}_k^{E_i})^H)$$

$$+\, \underline{\mathbf{x}}_k^{E_i}(\underline{\mathbf{x}}_k^{E_i})^H\,(\underline{\mathbf{H}}_k^i)^H \qquad\qquad (8.127)$$

$$F = \underline{\mathbf{H}}_k^i \underline{\mathbf{x}}_k^{E_i}$$

$$R_n = \sigma_n^2 \mathbf{I}_{n_R N_c}$$

where $\mathbf{I}_N$ denotes the identity matrix of size N, $\underline{\mathbf{x}}_k^{E_i}$ is the $(n_T K - 1) \times 1$ vector of the expected values of the transmitted symbols from the other $n_T K - 1$ users and their antennas.

Figure 8.20 shows the space-time iterative MMSE receiver structure [36]. In the first decoding iteration, we assume that the a priori probabilities for transmitting all symbols are equal, and hence, $\underline{\mathbf{x}}_k^{E_i} = \mathbf{0}$. The feedforward filter coefficients vector $\mathbf{w}_{f,k}^i(t)$ in this iteration is given by the MMSE equations and the feedback coefficient $w_{b,k}^i(t) = 0$. After each iteration, $\underline{\mathbf{x}}_k^{E_i}$ is recalculated from the decoders' soft outputs and then used to generate the new set of filter coefficients.

8.10.4 Performance Simulations

In this section, we illustrate the performance of the STTC WCDMA system in frequency-selective MIMO fading channels [36]. The performance is measured in terms of the BER and FER as a function of E_b/N_0 per receive antenna. Table 8.1 lists the simulation environment parameters. The generator polynomials for the 16-state QPSK STTC with two transmit antennas are obtained from Chapter 4. Figure 8.21 illustrates the error performance of a single-user Rake matched filter, a single-user MMSE receiver, and a multiuser MMSE receiver for 1 to 32 simultaneous users over a flat fading channel. The single-user MMSE receiver is the MMSE detector that considers only the desired signal's spreading sequence. It is shown that the three different detectors provide similar performance regardless of the

Table 8.1 Parameters for system environments

Multiple Access	WCDMA / Forward link
Chip rate	3.84 Mcps
Spreading/Scrambling	OVSF codes/PN sequence
Spreading gain	32 chip
Frame interleaving	10 ms (2400 bits/frame)
Fading rate	1.5×10^{-4}
MIMO channels	2 Tx. 2 Rx. antennas
STTC encoder	16-state QPSK
Generator polynomial	(1,2), (1,3), (3,2) / (2,0), (2,2), (2,0)

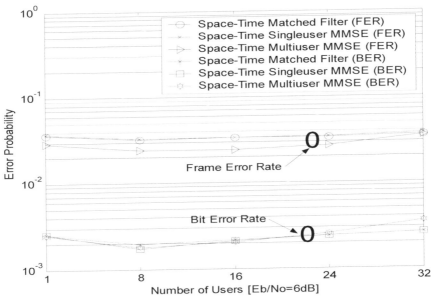

Figure 8.21 Error performance of an STTC WCDMA system on a flat fading channel

number of users, since there is no MAI due to synchronous transmission and the orthogonal spreading sequences.

Figure 8.22 depicts the FER performance of various receivers versus the number of users on a two-multipath fading channel. The performance curves show that the space-time MMSE multiuser receiver improves the error performance significantly compared to the matched filter or the single-user receiver. Figure 8.23 represents the BER performance versus the number of users on a two-multipath fading channel. From the results, we can see that the space-time MMSE multiuser receiver increases the number of users of the system about 3 times than that of the single-user or the matched filter receiver at a BER of 10^{-3}.

The performance of STTC WCDMA systems with iterative MMSE receivers is also evaluated by simulations. We assume that the number of users is $K = 4$ and the spreading factor is $N_c = 7$. The spreading sequences assigned to different users were chosen randomly. The WCDMA chip rate is set to be 3.84 Mcps and each frame is composed of 130 symbols. The fading coefficients are constant within each frame and a 2×2 MIMO channel is considered. Figures 8.24 and 8.25 depict the FER performance of various receivers on flat fading and

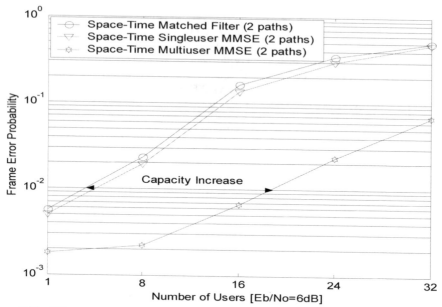

Figure 8.22   FER performance of an STTC WCDMA system on frequency-selective fading channels

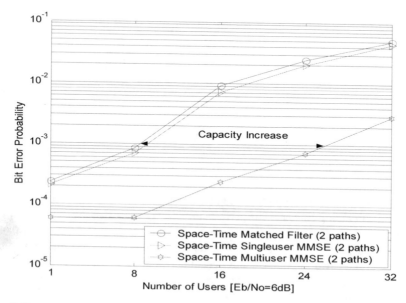

Figure 8.23   BER performance of an STTC WCDMA system on frequency-selective fading channels

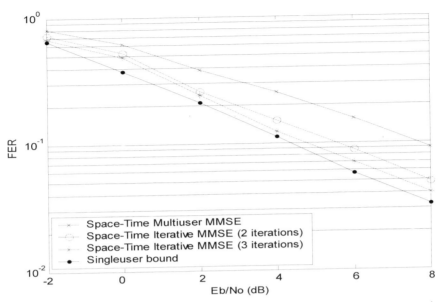

Figure 8.24 FER performance of an STTC WCDMA system with the iterative MMSE receiver on a flat fading channel

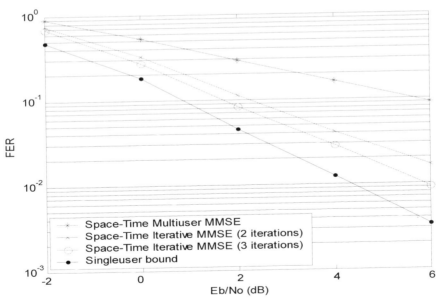

Figure 8.25 FER performance of an STTC WCDMA system with the iterative MMSE receiver on a two-path Rayleigh fading channel

two-path Rayleigh fading channels, respectively. The figures show that the iterative MMSE receiver achieves a remarkable gain compared to the LMMSE receiver.

8.11   Performance of Layered STC in CDMA Systems

In this section we consider a synchronous DS-CDMA LST encoded system with both random and orthogonal sequences over a multipath Rayleigh fading channel. The transmitter block diagram is shown in Fig. 8.26.

There are K active users in the system. The signal transmitted from each of the active users is encoded, interleaved and multiplexed into n_T parallel streams. All layers of the same user are spread by the same random or orthogonal Walsh spreading sequence assigned to that user. Various layers of each user are transmitted simultaneously from n_T antennas.

The delay spread of the multipath Rayleigh fading channel is assumed to be uniformly distributed between $[0, \lceil N_c T_c/2 \rceil]$ for random and $[0, \lceil N_c T_c/4 \rceil]$ for orthogonal sequences, where T_c is the chip duration, N_c is a spreading gain defined as a ratio of the symbol and the chip durations and $\lceil x \rceil$ denotes integer part of x. The delay of the lth multipath, denoted by $\tau^{k,l}$ for user k, is an integer multiple of the chip interval.

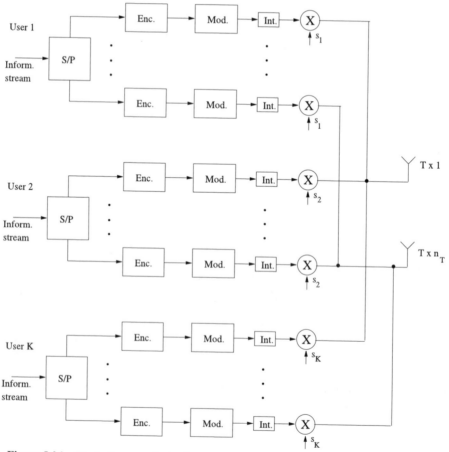

Figure 8.26   Block diagram of a horizontal layered CDMA space-time coded transmitter

The receiver has n_R antennas and employs an IPIC-DSC or an IPIC-STD multiuser detector/decoder, described in Chapter 6. We assume that the receiver knows all user spreading sequences and perfectly recovers the channel coefficients.

In the discrete time model, the spreading sequence and a vector with channel gains on the paths from transmit antenna m to all receive antennas for the lth multipath are combined into a composite spreading sequence of length $n_R N_c$, where N_c is the spreading gain. This composite spreading sequence for user k and transmit antenna m, shifted for the delay corresponding to the jth symbol and multipath l, can be expressed by a column vector as

$$\mathbf{h}_m^{k,l}(j) = \left[\begin{array}{cccccc} \mathbf{0}_b, & s_{k,1}(j)h_{1,m}^{k,l}, & s_{k,2}(j)h_{1,m}^{k,l}, & \cdots, & s_{k,N_c}(j)h_{1,m}^{k,l}, & \mathbf{0}_e, \\ \mathbf{0}_b, & s_{k,1}(j)h_{2,m}^{k,l}, & s_{k,2}(j)h_{2,m}^{k,l}, & \cdots, & s_{k,N_c}(j)h_{2,m}^{k,l}, & \mathbf{0}_e, \\ & & \cdots & & \cdots & \\ \mathbf{0}_b, & s_{k,1}(j)h_{n_R,m}^{k,l}, & s_{k,2}(j)h_{n_R,m}^{k,l}, & \cdots, & s_{k,N_c}(j)h_{n_R,m}^{k,l}, & \mathbf{0}_e \end{array} \right]^T$$

$$(8.128)$$

where $\mathbf{0}_b$ is a row vector with $b = (j-1)N_c + \lceil \tau^{k,l}/T_c \rceil$ zeros as elements, $s_{k,q}(j)$ is the qth chip of the kth user spreading sequence for symbol at discrete time j, $h_{n,m}^{k,l}$ is the channel gain on the path from transmit antenna m to receive antenna n for user k and multipath l and $\mathbf{0}_e$ is a row vector with $e = (L-j)N_c + \lceil \tau^{max}/T_c \rceil - \lceil \tau^{k,l}/T_c \rceil$ zeros as elements, where $\tau^{max} = \max\{\tau^{k,L_p} | k = 1, 2, \dots, K\}$. The composite spreading sequences $\mathbf{h}_m^{k,l}(j)$ for the kth user and the lth multipath at discrete time j are given by

$$\mathbf{h}^{k,l}(j) = [\mathbf{h}_1^{k,l}, \mathbf{h}_2^{k,l}, \dots, \mathbf{h}_{n_T}^{k,l}] \qquad (8.129)$$

All the composite spreading sequences are arranged into the combined channel and spreading matrix $\mathbf{H}$ as

$$\mathbf{H} = \left[\begin{array}{ccccccc} \mathbf{h}^{1,1}(1), & \cdots, & \mathbf{h}^{1,L_p}(1), & \cdots\cdots & \mathbf{h}^{K,1}(1), & \cdots, & \mathbf{h}^{K,L_p}(1), \\ \mathbf{h}^{1,1}(2), & \cdots, & \mathbf{h}^{1,L_p}(2), & \cdots\cdots & \mathbf{h}^{K,1}(2), & \cdots, & \mathbf{h}^{K,L_p}(2), \\ & & \cdots & & \cdots & & \\ \mathbf{h}^{1,1}(L), & \cdots, & \mathbf{h}^{1,L_p}(L), & \cdots\cdots & \mathbf{h}^{K,1}(L), & \cdots, & \mathbf{h}^{K,L_p}(L) \end{array} \right]$$

$$(8.130)$$

A block diagram of the CDMA iterative receiver is shown in Fig. 8.27.

The received chip matched signal sequence at antenna n is denoted by $\mathbf{r}^n$ and can be expressed as

$$\mathbf{r}^n = \left[\begin{array}{cccc} r_{1,1}^n, & r_{1,2}^n, & \cdots & r_{1,N_c}^n, \\ r_{2,1}^n, & r_{2,2}^n, & \cdots & r_{2,N_c}^n, \\ & \cdots & \cdots & \\ r_{L,1}^n, & r_{L,2}^n, & \cdots, & r_{L,N_c}^n, \\ r_{L+1,1}^n, & r_{L+1,2}^n, & \cdots & r_{L+1,\lceil \tau_{max}/T_c \rceil}^n \end{array} \right]^T$$

$$(8.131)$$

where $r_{j,q}^n$ denotes the received qth chip at discrete time j at receive antenna n.

The received signal sequences for n_R receive antennas are arranged into a vector $\mathbf{r}$ as

$$\mathbf{r} = \left[\begin{array}{cccc} (\mathbf{r}^1)^T, & (\mathbf{r}^2)^T, \dots, & (\mathbf{r}^{n_R})^T \end{array} \right]^T \qquad (8.132)$$

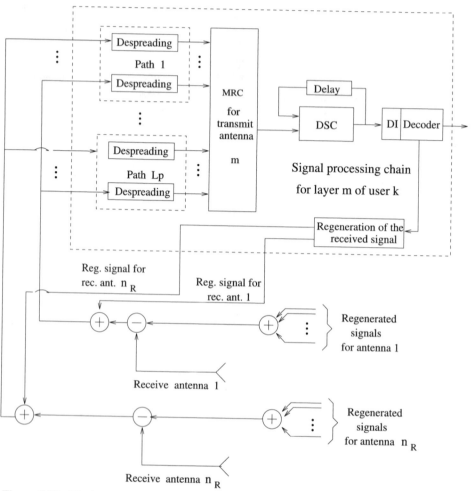

Figure 8.27 Block diagram of a horizontal layered CDMA space-time coded iterative receiver

The transmitted symbols for user k at time j are arranged into a vector $\mathbf{x}^k(j)$ as

$$\mathbf{x}^k(j) = \begin{bmatrix} x_j^{k,1}, & x_j^{k,2}, & \ldots & x_j^{k,n_T} \end{bmatrix}^T \tag{8.133}$$

where $x_j^{k,m}$ is the symbol transmitted at discrete time j by user k and antenna m.

In order to incorporate the multipath effects in a system model with L_p multipaths, we introduce a vector $\mathbf{x}_P^k(j)$ which is a column vector with L_p replicas of vector $\mathbf{x}^k(j)$, given by

$$\mathbf{x}_P^k(j) = \left[(\mathbf{x}^k(j))^T, \ldots, (\mathbf{x}^k(j))^T \right]^T \tag{8.134}$$

The transmitted symbols with L_p replicas for all users and all antennas at time j are arranged into a vector

$$\mathbf{x}_P(j) = \left[(\mathbf{x}_P^1(j))^T, (\mathbf{x}_P^2(j))^T, \ldots, (\mathbf{x}_P^K(j))^T \right]^T \tag{8.135}$$

Let us denote the transmitted signals for a frame of L time intervals by $\mathbf{x}'$.

$$\mathbf{x}' = \begin{bmatrix} (\mathbf{x}_P(1))^T, & (\mathbf{x}_P(2))^T, & \dots, & (\mathbf{x}_P(L))^T \end{bmatrix}^T \tag{8.136}$$

The chip sampled received signal for a frame of L symbols can now be expressed as

$$\mathbf{r} = \mathbf{H}\mathbf{x}' + \mathbf{n} \tag{8.137}$$

where $\mathbf{r}$ and $\mathbf{H}$ are given by Eqs. (8.132) and (8.130), respectively, and $\mathbf{n}$ is a column vector with AWGN samples.

The output of the IPIC for user k, transmit antenna m and iteration i can be expressed as

$$y^{k,m,i}(j) = \sum_{l=1}^{L_p} (\mathbf{h}_m^{k,l}(j))^H (\mathbf{r} - \mathbf{H}\hat{\underline{x}}^{k,m,i-1}) \tag{8.138}$$

where $\hat{\underline{x}}^{k,m,i-1}$ is a vector with transmitted symbol estimates in iteration $i-1$ as elements, except for the elements corresponding to the estimates of the kth user's mth transmit antenna symbols. The latter are set to zero.

In the IPIC-STD receiver the output of the PIC is approximated by a Gaussian random variable with the mean $\mu^{k,m,i}$ and the variance $(\sigma^{k,m,i})^2$ and fed into the decoder for a particular user and a transmit antenna.

The mean of the decoder input is calculated as

$$\mu^{k,m,i} = \sum_{l=1}^{L_p} (\mathbf{h}_m^{k,l}(j))^H \mathbf{h}_m^{k,l}(j) \tag{8.139}$$

Its variance is estimated as

$$(\sigma^{k,m,i})^2 = E[(y^{k,m,i}(j) - \sum_{l=1}^{L_p} (\mathbf{h}_m^{k,l}(j))^H \mathbf{h}_m^{k,l}(j) \hat{x}^{k,m,i-1}(j))^2] \tag{8.140}$$

where $\hat{x}^{k,m,i-1}(j)$ is an LLR estimate in iteration $i-1$ for the kth user's symbol transmitted at time j by the mth transmit antenna.

The IPIC-DSC receiver performs soft parallel interference cancellation and decision statistics combining for each user and each transmit antenna.

In the IPIC-DSC receiver the input to the decoder is formed as

$$y_c^{k,m,i}(j) = p_1^{k,m,i} y^{k,m,i}(j) + p_2^{k,m,i} y_c^{k,m,i-1}(j) \tag{8.141}$$

where $y^{k,m,i}(j)$ and $y_c^{k,m,i-1}(j)$ are outputs of the PIC in the iteration i and the DSC in iteration $i-1$, respectively.

The DSC coefficients $p_1^{k,m,i}$ and $p_2^{k,m,i}$ are given by

$$p_1^{k,m,i} = \frac{\mu^{k,m,i}(\sigma_c^{k,m,i-1})^2}{(\mu^{k,m,i})^2(\sigma_c^{k,m,i-1})^2 + (\mu_c^{k,m,i-1})^2(\sigma^{k,m,i})^2} \tag{8.142}$$

$$p_2^{k,m,i} = \frac{1 - p_1^{k,m,i}}{\mu_c^{k,m,i-1}} \tag{8.143}$$

where $\mu^{k,m,i}$ and $(\sigma^{k,m,i})^2$ are the mean and the variance of $y^{k,m,i}(j)$, and $\mu_c^{k,m,i}$ and $(\sigma_c^{k,m,i-1})^2$ are the mean and the variance of $y_c^{k,m,i-1}(j)$.

The performance of an HLSTC encoded down-link DS-CDMA system with PIC-DSC and PIC-STD detectors is evaluated by simulation. The HLST code employs $n_T = 4$ transmit and $n_R = 4$ receive antennas and each layer's signal is encoded by an $R = 1/2$ rate, 4-state convolutional code. The convolutional code is terminated to the all-zero state. A frame for each layer consists of $L = 206$ coded symbols. Assuming that BPSK modulation is used, the spectral efficiency of the system is $\eta = 2$ bits/s/Hz. The spreading sequences are either random with the spreading gain of $N_c = 7$ or Walsh orthogonal sequences with the spreading gain of 16 and a long scrambling code. The number of users in a system with random codes was variable and adjusted to achieve the FER close to the interference free performance while in the system with orthogonal sequences the multiple access interference is low and the maximum number of users equal to the spreading gain 16 was adopted. The channel is represented by a frequency-selective multipath Rayleigh fading model with $L_p = 2$ equal power paths. The signal transmissions are synchronous. The channel is quasi-static, i.e. the delay and the path attenuations are constant for a frame duration and change independently from frame to frame. It is assumed that $E[\sum_{l=1}^{L_p}(\mathbf{h}_m^{k,l}(j))^H\mathbf{h}_m^{k,l}(j)] = 1$.

Figures 8.28 and 8.29 show the bit error rate and frame error rate curves versus the number of the users for the multiuser system with PIC-STD and PIC-DSC receiver in a

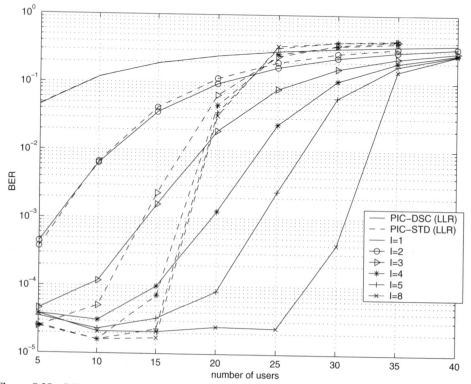

Figure 8.28 BER performance of a DS-CDMA system with (4,4) HLSTC in a two-path Rayleigh fading channel, $E_b/N_0 = 9$ dB

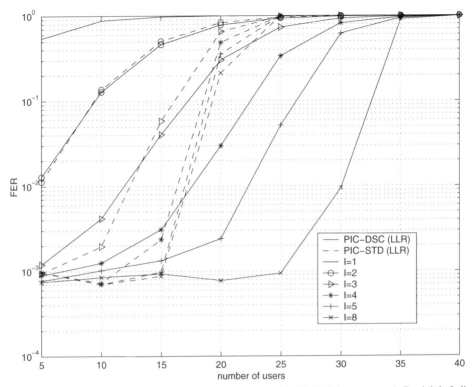

Figure 8.29 FER performance of a DS-CDMA system with HLSTC in a two-path Rayleigh fading channel, $E_b/N_0 = 9$ dB

Table 8.2 Spectral efficiency of CDMA HLST systems with random sequences and interference free performance

HLSTC	η_{hlstc} (bits/sec/Hz)	K	$\eta_{tot} = \frac{K}{N_c}\eta_{hlstc}$ (bits/sec/Hz)	E_b/N_0	FER
(2,2)	1	25	3.571	12 dB	0.0013
(4,2)	2	10	2.8571	12 dB	0.0012
(6,2)	3	5	2.14	12 dB	0.0011
(8,2)	4	2	1.14	12 dB	0.0011
(1,2)	0.5	1	0.071	12 dB	0.0009
(4,4)	2	25	7.14	9 dB	0.0009
(1,4)	0.5	1	0.071	9 dB	0.0007

(4,4) HLSTC CDMA system. The performance is examined for $E_b/N_0 = 9$ dB. The results show that PIC-DSC successfully removes both the multiuser and multilayer interference. The capacity improvement of PIC-DSC, expressed by the maximum number of users supported by the system for a target BER, is 67% for BER $= 10^{-5}$ and 78% for BER$=10^{-3}$ for 8 iterations, relative to the PIC-STD.

Table 8.2 shows the number of users and the achievable spectral efficiency in CDMA HLST systems with random codes for a specified FER equal to the single user MIMO interference free performance and a variable number of antennas. In systems with random sequences, a more efficient interference cancellation provides a larger number of users and thus an improved spectral efficiency as shown in Figs. 8.28 and 8.29. On the other hand, increasing the number of transmit antennas, while the number of receive antennas remains constant, introduces interference from transmit antennas, which limits the achievable throughput. The spectral efficiency of a single user HLST system is denoted by η_{hlst}, the

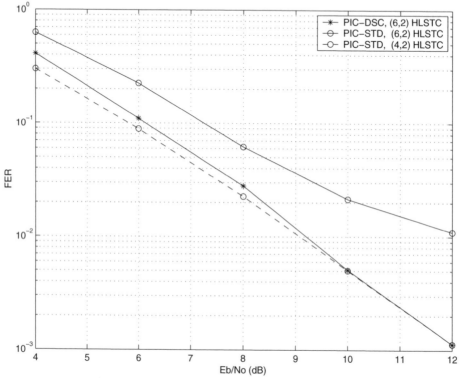

Figure 8.30 FER performance of IPIC-STD and IPIC-DSC in a synchronous CDMA with orthogonal Walsh codes of length 16, with K = 16 users and (6,2) and (4,2) HLSTC on a two-path Rayleigh fading channel

Table 8.3 Spectral efficiency of HLST CDMA systems with orthogonal sequences and interference free performance

HLSTC	η_{hlstc} (bits/sec/Hz)	K	$\eta_{tot} = \frac{K}{N_c}\eta_{hlstc}$ (bits/sec/Hz)	E_b/N_0	FER
(2,2)	1	16	1	12 dB	0.0001
(4,2)	2	16	2	12 dB	0.001
(6,2)	3	16	3	12 dB	0.001
(8,2)	4	16	4	12 dB	0.001
(1,2)	0.5	16	0.5	12 dB	0.001
(4,4)	2	16	2	9 dB	0.001
(1,4)	0.5	16	0.5	9 dB	0.001

spectral efficiency of the CDMA HLST system by η_{tot} and the number of users by K. The highest CDMA HLST throughput of 7.14 bits/sec/Hz is obtained for the (4,4) system at E_b/N_0 of 9 dB, while the (2,2) system has a spectral efficiency of 3.57 bits/sec/Hz for E_b/N_0 of 12 dB. The (1,2) and (1,4) systems, shown in the table, are interference free and provide the reference for the systems with two and four transmit antennas, respectively, and a variable number of receive antennas. In these systems it is desirable to keep the number of receive antennas as large as the number of transmit antennas to limit the combined MIMO and MA interference.

As the third generation cellular standards use orthogonal spreading sequences, we also show the performance results in a system with orthogonal sequences. Due to multipath fading, the orthogonality is violated and there is considerable multiple access interference. The performance of the CDMA HLST schemes with PIC-STD and PIC-DSC, for various numbers of antennas in a two-path Rayleigh fading channel is depicted in Fig. 8.30. It is clear that the PIC-DSC is about 3 dB better than the PIC-STD at the FER of 10^{-2} in a (6,2) CDMA HLST system. The relative performance of TLST versus HLST is similar to the one in a single user system with no spreading.

In systems with orthogonal sequences, the number of users is fixed and equal to the spreading gain. In such systems the spectral efficiency can be increased by increasing the number of transmit antennas as shown in the examples in Table 8.3. The number of receive antennas in this example is kept fixed, as it is limited for mobile receivers. In (2,2) and (4,2) CDMA HLST coded systems, the spectral efficiency is 1 and 2 bits/sec/Hz, respectively, and the multiuser interference can be effectively removed by using only a standard PIC receiver. However, starting with a (6,2) system, which has the spectral of 3 bits/sec/Hz, the PIC-DSC is needed to obtain the interference free performance. Clearly, the proposed PIC-DSC enables maximization of the spectral efficiency in systems with orthogonal codes, by eliminating interference coming from an increased number of antennas.

Bibliography

[1] J. G. Proakis, *Digital Communications*, 4th Edition, McGRAW-HILL, 2001.

[2] T. S. Rappaport, *Wireless Communications: Principles and Practice*, Prentice Hall, 1996.

[3] V. Tarokh, N. Seshadri and A. Calderbank, "Space-time codes for high data rate wireless communication: performance criterion and code construction", *IEEE Trans. Inform. Theory*, vol. 44, no. 2, pp. 744–765, 1998.

[4] V. Tarokh, A. Naguib, N. Seshadri and A. R. Calderbank, "Space-time codes for high data rate wireless communication: Performance criteria in the presence of channel estimation errors, mobility, and multiple paths", *IEEE Trans. Commun.*, vol. 47, no. 2, Feb. 1999, pp. 199–207.

[5] D. Agrawal, V. Tarokh, A. Naguib and N. Seshadri, "Space-time coded OFDM for high data rate wireless communication over wide-band channels", in *Proc. IEEE VTC'98*, Ottawa, Canada, May 1998, pp. 2232–2236.

[6] L. J. Cimini and N. R. Sollenberger, "OFDM with diversity and coding for high bit-rate mobile data applications", *AT & T Technical Memorandum*, HA6131000-961015-07TM.

[7] Y. Gong and K. B. Letaief, "Performance evaluation and analysis of space-time coding for high data rate wireless personal communication", in *Proc. 1999 IEEE VTC*, 1999, pp. 1331–1335.

[8] Y. Gong and K. B. Letaief, "Performance evaluation and analysis of space-time coding in unequalized multipath fading links", *IEEE Trans. Commun.*, vol. 48, no. 11, Nov. 2000, pp. 1778–1782.

[9] Z. Liu, G. B. Giannakis, S. Zhuo and B. Muquet, "Space-time coding for broadband wireless communications", *Wireless Communications and Mobile Computing*, vol. 1, no. 1, pp. 35–53, Jan. 2001.

[10] B. Lu and X. Wang, "Space-time code design in OFDM systems", in *Proc. IEEE GLOBECOM'00*, San Francisco, Nov. 2000, pp. 1000–1004.

[11] B. Lu, X. Wang and K. R. Narayanan, "LDPC-based space-time coded OFDM systems over correlated fading channels: performance analysis and receiver design", to appear in *IEEE Trans. Commun.*, 2001.

[12] C. Schlegel and D. J. Costello, "Bandwidth efficient coding for fading channels: code construction and performance analysis", *IEEE J. Selected Areas Commun.*, vol. 7, no. 12, pp. 1356–1368, Dec. 1989.

[13] H. Boleskei, D. Gesbert and A. J. Paulraj, "On the capacity of OFDM-based multi-antenna systems", in *Proc. ICASSP'00*, 2000, pp. 2569–2572.

[14] D. Tujkovic, M. Juntti and M. Latva-aho, "Space-time turbo coded modulation: design and applications", *EURASIP Journal on Applied Signal Processing*, vol. 2002, no. 3, pp. 236–248, Mar. 2002.

[15] Y. Li, N. Seshadri and S. Ariyavisitakul, "Transmit diversity for OFDM systems with mobile wireless channels", in *Proc. IEEE GLOBECOM'98*, Sydney, Nov. 1998, pp. 968–973.

[16] Y. Li, J. C. Chuang and N. R. Sollenberger, "Transmit Diversity for OFDM systems and its impact on high-rate data wireless networks", *IEEE Journal on Selected Areas in Communications,* vol. 17, no. 7, July 1999, pp. 1233–1243.

[17] W. Choi and J. M. Cioffi, "Multiple input/multiple output (MIMO) equalization for space-time block coding", in *Proc. IEEE Pacific Rim Conf. on Commun., Comput., and Signal Processing*, Victoria, Canada, Aug. 1999, pp. 341–344.

[18] W. Choi and J. M. Cioffi, "Space-time block codes over frequency-selective Rayleigh fading channels", in *Proc. IEEE VTC'99*, Amsterdam, Netherlands, Sep. 1999, pp. 2541–2545.

[19] S. L. Ariyavisitakul, J. H. Winters and I. Lee, "Optimum space-time processors with dispersive interference: unified analysis and required filter span", *IEEE Trans. Commun.*, vol. 47, July 1999, pp. 1073–1083.

[20] R. S. Blum, Y. Li, J. H. Winters and Q. Yan, "Improved space-time coding for MIMO-OFDM Wireless Communications", *IEEE Trans. Commun.*, vol. 49, no. 11, Nov. 2001, pp. 1873–1878.

[21] B. Hochwald, T. L. Marzetta and C. B. Papadias, "A transmitter diversity scheme for wideband CDMA systems based on space-time spreading", *IEEE Journal on Selected Areas in Commun.*, vol. 19, no. 1, Jan. 2001, pp. 48–60.

[22] K. Rohani, M. Harrison and K. Kuchi, "A comparison of base station transmit diversity methods for third generation cellular standards", in *Proc. IEEE VTC'99*, pp. 351–355.

[23] L. M. A. Jalloul, K. Rohani, K. Kuchi and J. Chen, "Performance analysis of CDMA transmit diversity methods", in *Proc. IEEE VTC'99 Fall,* pp. 1326–1330.

[24] J. Geng, U. Mitra and M. P. Fitz, "Space-time block codes in multipath CDMA systems", submitted to *IEEE Trans. Inform. Theory*.

[25] J. Yuan, Z. Chen, B. Vucetic and W. Firmanto, "Performance of analysis and design of space-time coding on fading channels", submitted to *IEEE Trans. Commun.*, 2001.

[26] IEEE 802.11, "Wireless LAN medium access control (MAC) and physical layer (PHY) specification", 1997.

[27] ETSI TS 101 475 V1.1.1, "Broadband radio access networks (BRAN); Hiperlan type 2; physical (PHY) layer", Nov. 1998.

[28] 3rd Generation Partnership Project (3GPP) Technical Specification *TS 25series*, v3.

[29] S. Verdu, *Multiuser Detection*, Cambridge, MA: Cambridge University Press, Sept. 1998.

[30] S. B. Weinstein and P. M. Ebert, "Data transmission by frequency-division multiplexing using the discrete Fourier transform", *IEEE Transactions on Communications Technologies*, vol. COM-19, no. 5, Oct. 1971.

[31] E. C. Ifeachor and B. W. Jervis, *Digital signal processing: a practical approach*, Addison-Wesley Publishing Company, 1993.

[32] L. J. Cimini, Jr, "Analysis and simulation of a digital mobile channel using orthogonal frequency division multiplexing", *IEEE Transactions on Communications*, vol. COM-33, no. 7, Jul. 1985.

[33] C. B. Papadias and H. Huang, "Linear space-time multiuser detection for multipath CDMA channels", *IEEE Journal on Selected Areas in Communications*, vol. 19, no. 2, Feb. 2001.

[34] H. E. Gamal and E. Geraniotis, "Iterative multiuser detection for coded CDMA signals in AWGN and fading channels", *IEEE Journal on Selected Areas in Communications*, vol. 18, no. 1, Jan. 2000.

[35] S. Marinkovic, B. Vucetic and A. Ushirokawa, "Space-time iterative and multistage receiver structures for CDMA mobile communication systems", *IEEE Journal on Selected Areas in Communications*, vol. 19, no. 8, Aug. 2001.

[36] H. Yang, J. Yuan and B. Vucetic, "Performance of space-time trellis codes in frequency selective WCDMA systems", in *Proc. IEEE VTC 2002-Fall*, Vancouver, Canada, Sept. 24–28, 2002.

[37] M. Nagatsuka and R. Kohno, "A spatially and temporally optimal multiuser receiver using an antenna array for DS/CDMA", *IEICE Trans. Commun.*, vol. E78-B, pp. 1489–1497, Nov. 1995.

Index

Space-Time Coding Branka Vucetic and Jinhong Yuan
© 2003 John Wiley & Sons, Ltd ISBN: 0-470-84757-3